T0310445

Ceramics for Environmental and Energy Applications II

Ceramics for Environmental and Energy Applications II

Ceramic Transactions, Volume 246

A Collection of Papers Presented at the
10th Pacific Rim Conference on
Ceramic and Glass Technology
June 2–6, 2013
Coronado, California

Edited by
Fatih Dogan
Terry M. Tritt
Tohru Sekino
Yutai Katoh
Aleksander J. Pyzik
Ilias Belharouak
Aldo R. Boccaccini
James Marra

Volume Editor
Hua-Tay Lin

WILEY

Published by John Wiley & Sons, Inc., Hoboken, New Jersey.
Published simultaneously in Canada.

Library of Congress Cataloging-in-Publication Data is available.

ISBN: 978-1-118-77124-2
ISSN: 1042-1122

Printed in the United States of America.

10 9 8 7 6 5 4 3 2 1

Contents

CERAMICS FOR NEXT GENERATION NUCLEAR ENERGY

ADVANCES IN PHOTOCATALYTIC MATERIALS FOR ENERGY AND ENVIRONMENTAL APPLICATIONS

CERAMICS ENABLING ENVIRONMENTAL PROTECTION: CLEAN AIR AND WATER

ADVANCED MATERIALS AND TECHNOLOGIES FOR ELECTROCHEMICAL ENERGY STORAGE SYSTEMS

GLASSES AND CERAMICS FOR NUCLEAR AND HAZARDOUS WASTE TREATMENT

Preface

This Ceramic Transactions volume represents 25 selected papers based on presentations in eight symposia during the 10th Pacific Rim Conference on Ceramic and Glass Technology, June 2–6, 2013 in Coronado, California. The symposia include:

- Solid Oxide Fuel Cells and Hydrogen Technology
- Direct Thermal to Electrical Energy Conversion Materials and Applications
- Photovoltaic Materials and Technologies
- Ceramics for Next Generation Nuclear Energy
- Advances in Photocatalytic Materials for Energy and Environmental Applications
- Ceramics Enabling Environmental Protection: Clean Air and Water
- Advanced Materials and Technologies for Electrochemical Energy Storage Systems
- Glasses and Ceramics for Nuclear and Hazardous Waste Treatment

The editors wish to extend their gratitude and appreciation to all the co-organizers for their help and support, to all the authors for their cooperation and contributions, to all the participants and session chairs for their time and efforts, and to all the reviewers for their valuable comments and suggestions. Thanks are due to the staff of the meetings and publication departments of The American Ceramic Society for their invaluable assistance. We also acknowledge the skillful organization and leadership of Dr. Hua-Tay Lin, PACRIM 10 Program Chair.

Fatih Dogan, Missouri University of Science and Technology, USA
Terry M. Tritt, Clemson University, USA
Tohru Sekino, Tohoku University, Japan
Yutai Katoh, Oak Ridge National Laboratory, USA
Aleksander J. Pyzik, The Dow Chemical Company, USA
Ilias Belharouak, Argonne National Laboratory, USA
Aldo R. Boccaccini, University of Erlangen-Nuremberg, Germany
James Marra, Savannah River National Laboratory, USA

RECENT RESEARCH ACTIVITIES FOR FUTURE CHALLENGES IN GLOBAL ENERGY AND ENVIRONMENT IN TOYOTA CENTRAL R&D LABS., INC. (TCRDL)

Tomoyoshi Motohiro[1,2,3]
[1]TOYOTA Central R&D Labs.,Inc.,
41-1, Yokomichi, Nagakute, Aichi, 480-1192, Japan
[2]Graduate School of Engineering, TOYOTA Technological Institute,
2-12-1, Hisakata, Tenpaku-ku, Nagoya, Aichi, 468-8511,Japan
[3]Green Mobility Collaborative Research Center & Graduate School of Engineering, Nagoya University,
Furo-cho, Chikusa-ku, Nagoya, 464-8603, Japan

ABSTRACT

Possible decrease in global supply of conventional oil and greenhouse warming make demands for reduction of oil consumption and diversification of fuels. In association with this, automotive powertrains are also diversifying from conventional internal combustion engine (ICE) into HV, plug-in HV(PHV), EV and fuel-cell HV (FCHV). The main sector is being replaced by HV and PHV, but still stands on ICE. In these hybridized ICE systems, unconventional oils, gaseous-, synthetic- and bio-fuels will be inevitably used more besides conventional fuels. EV as a subsidiary sector for short range use imposes additional demands of electricity from new energy sources. FCHV as another subsidiary sector for route bus and large trucks requests new infrastructures for hydrogen from new energy sources besides industrial by-products. Facing to this situation, it has become important for us to obtain concrete experiences on technologies for future energy sources by conducting R&D by ourselves although the production of energy is out of our business category. R&D for bio-ethanol production from cellulose has been already conducted. This paper describes our other typical recent R&D activities on future energy sources including laser nuclear fusion, solar-pumped lasers, solar cells and artificial photosynthesis, with special attentions to ceramics as key materials.

INTRODUCTION

Possible decrease in global supply of conventional oil and greenhouse warming caused by CO_2 emission make demands for reduction of oil consumption and diversification of fuels. In association with this, automotive powertrains are also diversifying from conventional internal combustion engine (ICE) into hybrid vehicles(HV), plug-in hybrid vehicles(PHV), electric vehicles (EV) and fuel-cell hybrid vehicles (FCHV). The main sector of automobiles is being replaced by HV and PHV, but still stands on ICE. For these hybridized ICE systems, unconventional oils, gaseous fuels, synthetic fuels and bio-fuels will be inevitably used more in addition to the conventional fuels. EV as a subsidiary sector for short range use in imposes additional demands of electricity which must be supplied from new energy sources as well as conventional power plants. FCHV as an another subsidiary sector for route bus and large trucks requests new infrastructures for hydrogen from new energy sources as well as from industrial by-products.

Facing to this situation, it has become important for us to obtain concrete knowledge and experiences on technologies for future energy sources by conducting R&D by ourselves although the production of energy is out of our business category. R&D for bio-ethanol production from cellulose has been conducted in TCRDL. This paper describes our other typical recent research activities on future energy sources including laser nuclear fusion, solar-pumped lasers, solar cells and artificial photosynthesis, with special attentions to ceramics as key materials.

COMPACT AND HIGH REPETITION(10HZ) LASER NUCLEAR FUSION

It is believed that nuclear fusion can meet the global energy demand with much less burdens on the environment than other energy sources as well as solar energy. However it is still at an immature stage with a lot of technical challenges remaining to be solved. These are being tackled by big science projects such as ITER(International Thermonuclear Experimental Reactor) Project for plasma fusion power plants and LIFE(Laser Inertial Fusion Energy) project in National Ignition Facility in US for inertial fusion power plants. Under an idea that we may achieve laser fusion power plants in a different approach using only two counter laser beams with additional one beam for fast-ignition in contrast to the NIF's 192 laser beams from all directions, a compact and high repetition(10Hz) laser nuclear fusion project has been started by The Graduate School for the Creation of New Photonic Industries, Hamamatsu Photonics K.K., TOYOTA Motor Corporation and TCRDL. We have already observed reasonable amount of neutron yield from D-D fusion by implosion of double-deuterated polystyrene foils separated by 100 μm by counter- illumination succeeded by fast heating[1]. Figure 1 shows our present status and future plan displayed as neutron yield versus laser energy. Since it is necessary to develop high- repetition MJ lasers before achieving cost and energy pay- back condition as a power plant beyond the break even condition, some intermediate outputs as neutron sources for analysis, medical and industrial applications are envisaged[2]. At present, Nd- doped phosphate laser glass is a key laser medium whether it is NIF's flashlamp- pumped type or our diode-pumped type[3]. It is said that the efficiency can be improved by employing diode- pumped Yb-doped fluorapatite crystals. Moreover, the laser medium may be further scaled up for higher

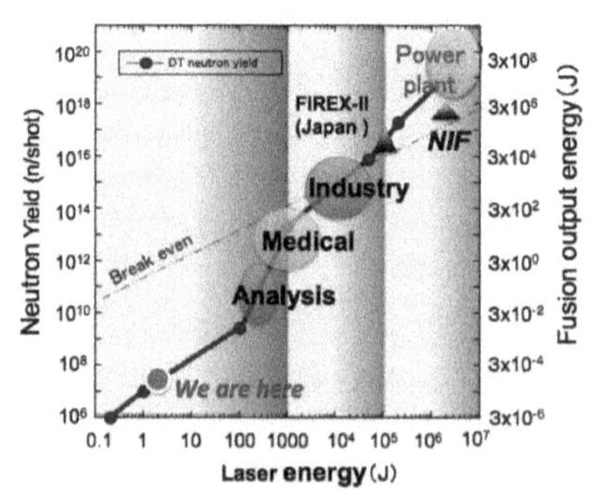

Fig.1. Present status and future plan displayed as neutron yield versus laser energy.

energy output by employing transparent fluorapatite polycrystalline ceramics rather than crystals if micro-domain-control is successful[4,5].

COMPACT SOLAR-PUMPED LASER

High-grade Nd-doped YAG ceramics for laser medium accelerated R&D on solar pumped lasers[6]. Solar pumped lasers usually employ Fresnel lenses or concave mirrors of several meters in diameter to concentrate sunlight into a water-cooled Nd-doped YAG rod of typically 10 mm in diameter and 100mm in length[7]. In contrast, employing an off-axis parabolic mirror of 50mm in diameter and a Nd -doped YAG prismatic rod of 1 x 1 x 5mm in size without water-cooling, we have developed a much more compact solar-pumped laser which can stably emit 1064nm laser light tracking the sun on a commercially available equatorial mounting for amateur astronomers as shown in Fig.2[8]. In the same system, Nd-doped ZBLAN (ZrF_4, BaF_2, LaF_3, AlF_3, NaF) fiber was also successful in laser emission in place of the prismatic rod[9]. The 1064nm laser light or its higher harmonics such as of 532nm is expected to be easily transmitted long distance, to be concentrated into a fine spot to attain high temperature for production of hydrogen from water, and to be converted into electricity using photovoltaic cells with its optical band gap just below the photon energy of the laser light so as to minimize thermal loss.

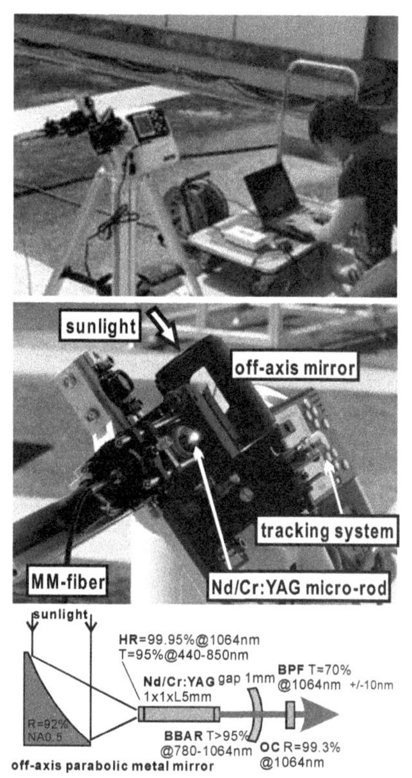

Fig.2 Fabricated compact solar pumped laser mounted on a solar tracker which enables several hours of continuous laser oscillation by tracking the sun.

SOLAR CELLS

Dye-sensitized Solar Cells

R&D on dye-sensitized solar cells(DSSC) comprises of sciences and technologies of ceramics for the optical electrodes typically of sintered nano-porous TiO_2 layers in anatase phase, organic and metal-organic chemistry for sensitizers typically of Ru-complexes and photo-electrochemistry. Conventional solar cell manufacturers working typically on Si p/n junction solar cells have been rather not familiar with such versatile fields but automotive manufacturers are. Taking this advantage, we have been working on DSSC for more than 15 yeaisplayrs with Aisin Seiki Co. Ltd[10,11]. Figure 3 shows outdoor performance test of battery-operated night-lights charged by dye-sensitized solar cells during daylight near the TOYOTA beam line at the synchrotron orbital radiation site "SPring-8" in Japan. Although the top efficiency over 12% has been reported but there remain still tough challenges of outdoor durability mainly because of degradation of organic dyes and liquid electrolytes[12-14]. Although more than 15 years durability was confirmed for small cells, large modules still have different challenges to be durable. In 2012, there came up possible game-changing technologies reported from plural research groups[15,16] in which sensitizers and electrolyte can be replaced by inorganic materials opening the door to all-solid inorganic dye-sensitized solar cells.

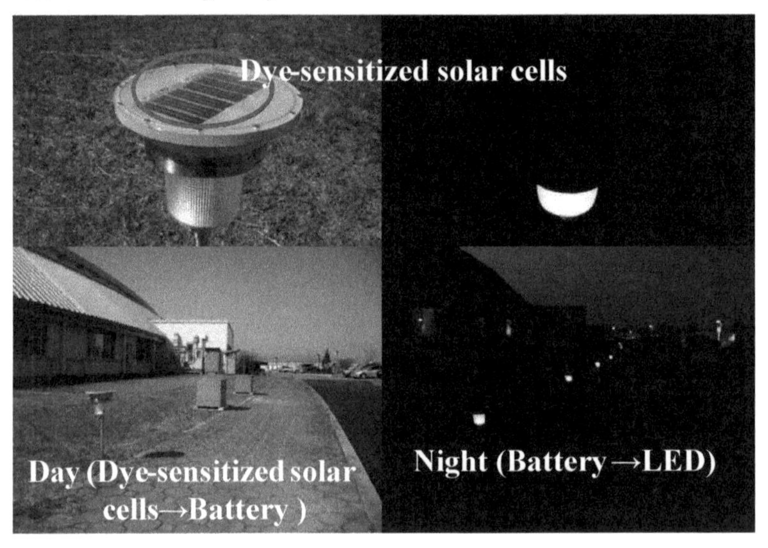

Fig.3 Outdoor performance test of battery-operated LED night-lights charged by monolithic dye-sensitized solar cells in day time (since April, 2009).

Cu_2ZnSnS_4 Thin Film Solar Cells

Inorganic semiconductor solar cells have evolved starting from single-elemental Si, through binary GaAs and CdTe to essentially ternary $Cu(In_{1-x}Ga_x)Se_2$. We have been working on quaternary Cu_2ZnSnS_4 system of kesterite crystal structure since 2006[17-22], because (1) Cu_2ZnSnS_4 is composed of earth-abundant and nontoxic elements, (2) thin-film technologies can reduce both the material cost and the fabrication cost, and (3) the bandgap energy of Cu_2ZnSnS_4 about 1.4eV is optimal for single-junction solar cells leading to the detailed balance limiting efficiency of about 31%[23]. In reactive sintering process, a precursor layer formed on a Mo-coated glass substrate changes into a polycrystalline Cu_2ZnSnS_4 thin film where grain boundaries take important effect on photovoltaic properties. Although this quaternary system can cause difficulties in stable production, the number of R&D reports is increasing rapidly now as shown in Fig.4 since it uses neither rare elements nor environmental pollutants, and since there is an estimation that only this system can make the electricity generation cost lower than that of Si solar cells. More recently, solution-processed kesterite with an additional element of Se has been successful attaining an energy conversion efficiency beyond 11%[24,25]. On the other hand, a quaternary system without Zn but with Ge: $Cu_2Sn_{0.83}Ge_{0.17}S_3$ was found to have a bandgap of 1.0eV and attained energy conversion efficiency of 6%[26]. This also shows a possibility of higher efficiency constructing a multi-junction type solar cell using this $Cu_2Sn_{0.83}Ge_{0.17}S_3$ for a bottom cell[26].

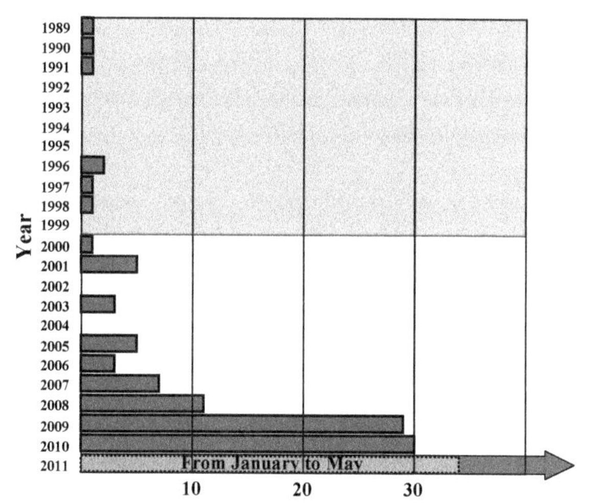

Fig.4 Number of published papers on the topics related to Cu_2ZnSnS_4

The Third Generation Solar Cells

The best strategy to decrease thermal loss of photon energy larger than the bandgap and transmission loss of photon energy less than the bandgap in such a widespread solar spectrum is to construct multi-junction solar cells comprising of plural active materials of different optical absorption edges. To compensate the increase in the fabrication cost to form multi-junction structures, concentration of solar-light onto a solar cell of small area is usually employed. However, since more than half of the sky is shadowed by cloud in average and the percentage of direct sunlight which can be concentrated is lower than 80% even in a typical clear day because of moisture in Japan, the area in which the combination of the solar concentrator and the multi-junction solar cells of small area is applied advantageously is quite limited. The third generation solar cells such as (1) hot-carrier type, (2) multi-exciton generation type, and (3) intermediate-band type have been proposed to decrease the loss mentioned above without using solar concentrators. We have also studied these solar cells both theoretically[27-32] and experimentally [33,34]. Spontaneous formation and arrangement of quantum dots is a key factor in these solar cells. At the present time of writing, however, we are not successful to get any promising perspective for these third generation solar cells.

ARTIFICIAL PHOTOSYNTHESIS OF FORMIC ACID FROM CO_2 AND H_2O ONLY USING SUNLIGHT

Hydrogen production by photo-electrochemical water splitting has been extensively studied all over the world. Hydrogen can be also produced from water electrolysis combined with solar cells. In this situation, syntheses of hydrocarbons using solar energy are much more challenging rather than hydrogen. We had an experience of R&D on visible-light photo-catalysis in nitrogen-doped TiO_2 for photolysis of organic contaminants which has been commercialized as a series of environmental catalysts under the name of VCAT [35]. Based on this experience of photolysis, we have begun R&D on photosynthesis of methanol from CO_2 and H_2O only using sunlight. At present, photo- synthesis of formic acid has been attained[36,37]. Figure 5 schematically shows the process. Here, H_2O molecules are oxidized to yield oxygen using holes formed by photo-excitation in the left electrode and residual electrons are transmitted via conducting wire to the right electrode and reduce CO_2 to form formic acid. To increase selectivity of CO_2 reduction in comparison with hydrogen formation, a Ru-complex layer was formed on the surface of the right electrode. By choosing an appropriate materials combination of the two electrodes, solar-energy conversion efficiency of 0.04% was attained which is 20% of solar-energy conversion efficiency of natural photosynthesis in switchgrass. This was the first achievement of complete artificial photosynthesis from CO_2 and H_2O only using sunlight without applying any bias voltage or additional chemicals in the water. We are continuing effort on further improvement of conversion efficiency and development of electrode materials of lower cost[38].

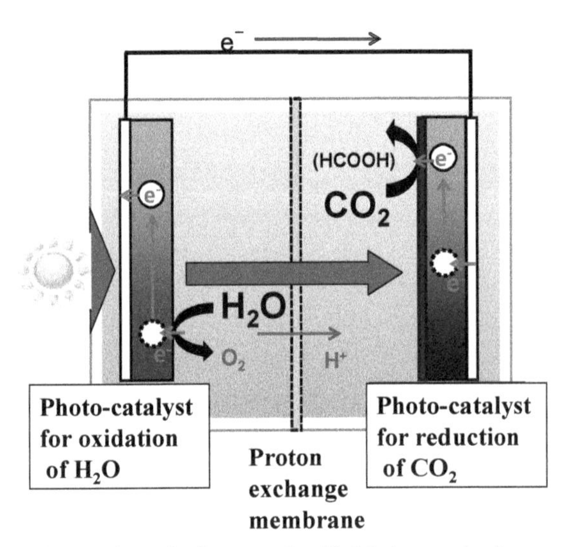

Fig.5 Schematic diagram of artificial photosynhesis of formic acid from CO_2 and H_2O only using sunlight.

SUMMARY

Typical recent research activities on future energy sources including laser nuclear fusion, solar-pumped lasers, solar cells and artificial photosynthesis in TCRDL are reviewed with special attentions to ceramics as key materials. Although every activity introduced here has attained some innovative milestones, we also realize that there remains still a long way to go to make it practically useful. On the other hand, social and economical environment of energy demand and supply fluctuate quickly including the Fukushima nuclear disaster and the shale gas revolution. Flexible but durable activities of R&D must be kept going on for long term goals as well as short term ones.

REFERENCES

[1]Y. Kitagawa, Y. Mori, O. Komeda, K. Ishii, R. Hanayama, K. Fujita, S. Okihara, T. Sekine, N. Satoh, T. Kurita, M. Takagi, T. Kawashima, H. Kan, N. Nakamura, T. Kondo, M. Fujine, H. Azuma, T. Motohiro, T. Hioki, Y. Nishimura, A. Sunahara, and Y. Sentoku, Fusion Using Fast Heatng of a Compactly Imploded CD Core, Phys. Rev. Lett., **108**, 155001 (2012).

[2]O.Komeda, Y.Mori, R.Hanayama, S. Okihara, K.Fujita, K.Ishii, Y.Kitagawa, T.Kawashima, N.Satoh, T.Sekine, M.Takagi, H.Kan, N.Nakamura, T.Kondo, M.Fujine, H.Azuma, T.Hioki, M.Kakeno, T.Motohiro, Y.Nishimura, Neutron generator using a spherical target irradiated with ultra-intense diode-pumped laser at 1.25 Hz, Fusion Science and Technology **63**, 296-300 (2013).

[3]R. Yasuhara, T. Kawashima, T. Sekine, T. Kurita, T. Ikegawa, O. Matsumoto, M. Miyamoto, H. Kan, H. Yoshida, J. Kawanaka, M. Nakatsuka, N. Miyanaga, Y. Izawa, and T. Kanabe, 213 W average power of 2.4 GW pulsed thermally controlled Nd:glass zigzag slab laser with a stimulated Brillouin scattering mirror, OPTICS LETTERS, **33**, 1711-3 (2008).

[4]J. Akiyama, Y. Sato, and T. Taira, Laser Demonstration of Diode-Pumped Nd^{3+} Doped Fluorapatite Anisotropic Ceramics, Appl. Phys. Express, **4**, 022703 (2011).

[5]J. Akiyama, Y. Sato, and T. Taira, Laser ceramics with rare-earth-doped anisotropic materials, Opt. Lett., **35**, 3598-600 (2010).

[6]J. Lu, K. Ueda, H. Yagi, T. Yanagitani, Y. Akiyama, and A. Kaminskii, Neodymium doped yttrium aluminum garnet($Y_3Al_5O_{12}$) nanocrystalline ceramics — a new generation of solid state laser and optical materials —, J. Alloys and Compounds, **341**, 220-5 (2002).

[7]T.Yabe, T. Ohkubo, S. Uchida, K. Yoshida, M. Nakatsuka, T. Funatsu, A. Mabuti, A. Oyama, K. Kakagawa, T. Oishi, K. Daito, B. Behgol, Y, Nakayama, M. Yoshida, S. Motokoshi, Y. Sato, and C. Baasandash, High-efficiency and economical solar-energy-pumped laser with Fresnel lens and chromium co-doped laser medium, Appl. Phys. Lett., **90**, 261120 (2007).

[8]H.Ito, K. Hasegawa, S. Mizuno, and T. Motohiro, A Solar-pumped Micro-rod Laser without Active Cooling, ALPS2p-20 in Conference Program and Abstracts of The 1st Advanced Lasers and Photon Sources sponsored and organized by The Laser Society of Japan, 26th- 27th, April, 2012 at Pacifico Yokohama, Yokohama, Japan

[9]S. Mizuno, H. Ito, K. Hasegawa, T. Suzuki, and Y. Ohishi, Laser emission from a solar-pumped fiber, Opt. Exp., **20**, 5891-5 (2012).

[10]T. Toyoda, T. Sano, J. Nakajima, S. Doi, S. Fukumoto, A. Ito, T. Tohyama, M. Yoshida, T.

Kanagawa, T. Motohiro, T. Shiga, K. Higuchi, H. Tanaka, Y. Takeda, T. Fukano, N. Katoh, A. Takeichi, K. Takechi, M. Shiozawa, Outdoor performance of large scale DSC Modules, J. Photochem. & Photobiol. A:Chem., **164**, 203-7(2004).
[11]Y. Takeda, N. Kato, K. Higuchi, A. Takeichi, T. Motohiro, S. Fukumoto, T. Sano, and T. Toyoda, Monolithically series-interconnected transparent modules of dye-sensitized solar cells, Sol. Ener. Mater. Sol. Cells, **93**, 808-11(2009)
[12]H. Tanaka, A. Takeichi, K. Higuchi, T. Motohiro, M. Takata, N. Hirota, J. Nakajima, and T. Toyoda, Long-term durability and degradation mechanism of dye-sensitized solar cells sensitized with indoline dyes Sol. Ener. Mater. Sol. Cells, **93**, 1143-8(2009).
[13]N. Kato, Y. Takeda, K. Higuch, A. Takeichi, E. Sudo, H. Tanaka, T. Motohiro, T. Sano, and T. Toyoda, Degradation analysis of dye-sensitized solar cell module after lon g-term stability test under outdoor working condition, Sol. Ener. Mater. Sol. Cells, **93**, 8 93-7(2009).
[14]N. Kato, K. Higuchi, H. Tanaka, J. Nakajima, T. Sano, and T. Toyoda, Improvement in long-term stability of dye-sensitized solar cell for outdoor use, Sol. Ener. Mater. Sol. Cells, **95**, 301-5 (2011) .
[15]I.Chung, B. Lee, J. He, R.P. H. Chang, and M. G. Kanatzidis, All-solid-state dye-sensitized solar cells with high efficiency, nature, **485**, 486-9(2012).
[16]M. M. Lee, J. Teuscher, T. Miyasaka, T. N. Murakami, and H. J, Snaith, Efficient Hybrid Solar Cells Based on Meso-Superstructured Organometal Halide Perovskites, SCIENCE, **338**, 643-7 (2012).
[17]H. Katagiri, K. Jimbo, S. Yamada, T. Kamimura, W. S. Maw, T. Fukano, T. Ito, and T. Motohiro, Enhansed conversion efficiencies of Cu_2ZnSnS_4-based thin film solar cells by using preferential etching technique, Appl. Phys. Express, **1**, 041201 (2008).
[18]J. Paier, R. Asahi, A. Nagoya, and G. Kresse, Cu_2ZnSnS_4 as a potential photovoltaic material: A hybrid Hartree-Fock density functional theory study, Phys. Rev. B, 79, 115126 (2009).
[19]A. Nagoya, and R. Asahi, Defect formation and phase stability of Cu_2ZnSnS_4 photovoltaic material, Phys. Rev. B 81, 113202 (2010).
[20]T. Washio, H. Nozaki, T. Fukano, T. Motohiro, K. Jimbo, and H. Katagiri, Analysis of lattice site occupancy in kesterite structure of Cu_2ZnSnS_4 films using synchrotron radiation x-ray diffraction, J. Appl. Phys., **110**, 074511 (2011).
[21]T. Washio, T. Shinji, S. Tajima, T. Fukano, T. Motohiro, K. Jimbo and H. Katagiri, 6% Efficiency Cu_2ZnSnS_4-based thin film solar cells using oxide precursors by open atmosphere type CVD, J. Mater. Chem., **22**, 4021-4 (2012).
[22]S. Tajima, H. Katagiri, K. Jimbo, N. Sugimoto, and T. Fukano, Temperature Dependence of Cu_2ZnSnS_4 Photovoltaic Cell Properties, Appl. Phys. Express, **5**, 082302 (2012).
[23] W. Shockley, and H. J. Queisser, Detailed Balance Limit of Efficiency of *p-n* Junction Solar Cells, J. Appl. Phys. **32**, 510 (1961).

[24]D. B. Mitzi, O. Gunawan, T. K. Todorov, K. Wang, S. Guha, The path towards a high-performance solution-processed kesterite solar cell, Sol. Ener. Mater. and Sol. Cells, **95**, 1421-36 (2011).
[25] T. K. Todorov, J.Tang, S. Bag, O. Gunawan, T. Gokmen, Y. Zhu, D. B. Mitzi, Beyond 11% Efficiency: Characteristics of State-of-the-Art $Cu_2ZnSn(S,Se)_4$ Solar Cells, Advanced Energy Materials, **3**, 34-8(2013).
[26] M. Umehara, Y. Takeda, T. Motohiro, T. Sakai, H. Awano, and R. Maekawa, $Cu_2Sn_{1-x}Ge_xS_3$ (x =0.17) Thin-Film Solar Cells with High Conversion Efficiency of 6.0%, Appl. Phys. Express, **6**, 045501 (2013).
[27]Y. Takeda, T. Ito, T. Motohiro, D. Konig, S. Shrestha, and G. Conibeer, Hot carrier solar cells operating under practical conditions, J. Appl. Phys.,**105**, 074905 (2009)
[28]Y. Takeda, T. Ito, R. Suzuki, T. Motohiro, S. Shrestha and G. Conibeer, Impact ioniz ation and Auger recombination at high carrier temperature, Sol. Ener. Mater. Sol. Cells **93**, 797-802(2009).
[29]Y.Takeda, T. Motohiro, D. Konig, P. Aliberti, Y. Feng, S. Shrestha, and G. Conibeer, Practical Factors Lowering Conversion Efficiency of Hot Carrier Solar Cells, Appl. Phys. Express, **3**, 104301(2010)
[30]Y. Takeda, and T. Motohiro, Requisites to realize high conversion efficiency of solar cells utilizing carrier multiplication, Sol. Ener. Mater. Sol. Cells., **94**, 1399-405(2010).
[31]Y. Takeda, and T. Motohiro, Intermediate-band-assisted hot-carrier solar cells using indirect-bandgap absorbers, Progress in Photovoltaics: Research and Applications, John Willey & Sons, Ltd.,2012.
[32]Y.Takeda, and T.Motohiro, Hot-Carrier Extraction from intermediate-Band Absorbers through Quantum-Well Energy-Selective Contacts Jpn. J. Appl. Phys.,51, 10ND03_1-6(2012).
[33]K.Nishikawa,Y.Takeda, K.Yamanaka, T.Motohiro, D.Sato,J.Ota, N.Miyashita, Y. Okada, Over 100ns intrinsic radiative recombination lifetime in type II InAs/ GaAs1-xSbx quantum dots J.Appl.Phys.,111, 044325_1-6(2012).
[34]K. Nishikawa, Y. Takeda, T. Motohiro, D. Sato, J. Ota, N. Miyashita, Y. Okada, Extremely long carrier lifetime over 200ns in GaAs wall-inserted type II InAs quantum dots, Applied Physics Letters, **100**, 113105_1-3(2012).
[35]R. Asahi, T. Morikawa, T. Ohwaki, K. Aoki, and Y. Taga, Visible-Light Photocatalysis in Nitrogen-Doped Titanium Oxides, Science, **293**, 269-71 (2001).
[36]S. Sato, T. Morikawa, S. Saeki, T. Kajino and T. Motohiro Visible-Light-Induced Selective CO_2 Reduction Utilizing a Ruthenium Complex Electrocatalyst Linked to a p-Type Nitrogen-Doped Ta_2O_5 Semiconductor, Angew. Chem., Int. Ed., **49**, 5101-5 (2010).
[37]S. Sato, T. Arai, T. Morikawa , K. Uemura , T. M. Suzuki , H. Tanaka , and T. Kajino, Selective CO_2 Conversion to Formate Conjugated with H_2O Oxidation Utilizing Semiconductor/Complex Hybrid Photocatalysts , J. Am. Chem. Soc., **133**, 15240-3 (2011).

[38]T. Morikawa, T. Arai, and T.Motohiro, Photoactivity of p-Type Fe_2O_3 Induced by Anionic/Cationic Codoping of N and Zn, Appl. Phys. Express, **6** , 041201 (2013).

Solid Oxide Fuel Cells and Hydrogen Technology

STRUCTURAL AND ELECTRICAL CHARACTERIZATION OF $Pr_xCe_{0.95-x}Gd_{0.05}O_{2-\delta}$ ($0.15 \leq x \leq 0.40$) AS CATHODE MATERIALS FOR LOW TEMPERATURE SOFC

Rajalekshmi Chockalingam[a], Suddhasatwa Basu[a] *and Ashok Kumar Ganguli[b]
[a]Department of Chemical Engineering, [b]Department of Chemistry Indian Institute of Technology New Delhi, 110016, India

ABSTRACT

Structural and electrical properties $Pr_xCe_{0.95-x}Gd_{0.05}O_{2-\delta}$ ($0.15 \leq x \leq 0.40$) have been investigated as cathode materials for low temperature solid oxide fuel cells. Four compositions of $Pr_xCe_{0.95-x}Gd_{0.05}O_{2-\delta}$(PCGO) have been prepared by varying the Pr content. Phase formation, thermal expansion, ionic conductivity, ionic transference number and electronic conductivity have been studied. XRD results indicate that $Pr_xCe_{0.95-x}Gd_{0.05}O_{2-\delta}$ samples crystallize in the fluorite structure, and no other faces has formed. The coefficient of thermal expansion increases with increasing x, and at x = 0.2 shows CTE value of 13 x 10^{-6} K^{-1} comparable to the CTE value of GDC electrolyte. The ionic transference number decrease while electronic conductivity increase with increasing x. Gd^{3+} contributes to ionic conduction by creating oxygen vacancies and Pr^{4+} contributes electronic conduction by decreasing the band gap of CeO_2. A single cell with configuration $Pr_{0.20}Ce_{0.75}Gd_{0.05}O_{2-\delta}$-$Ce_{0.80}Gd_{0.20}O_{2-\delta}$ (cathode) // $Ce_{0.80}Gd_{0.20}O_{2-\delta}$(electrolyte) // NiO-$Ce_{0.80}Gd_{0.20}O_{2-\delta}$(anode) delivered a maximum power density of 167.9 mW cm^{-2} and current density of 372.5 $mAcm^{-2}$ at 650 °C.

INTRODUCTION

Zirconia based high temperature solid oxide fuel cells (SOFC) suffer due to high cost and lower reliability [1]. Reducing the operating temperature of SOFC below 800 °C is very challenging due to poor catalytic activity of existing cathode materials [2]. Conventional cathode material, $La_{1-x}Sr_xMnO_3$ (LSM) performs poor at low temperatures (500 -650 °C [3]. Several groups have investigated perovskites based materials such as $NdBaCo_{2-x}Cu_xO_{5+\delta}$, $GdBaCo_{1.0}Cu_{1.0}O_{5+\delta}$, $La_{1-x}Sr_xCo_{1-y}FeO_{3-\delta}$ (LSCF), $La_{0.9}Sr_{0.1}Ga_{0.8}Mg_{0.2}O_{3-\delta}$ (LSGM), $Ba_{1-x}Sr_xCo_{1-y}Fe_yO_{3-\delta}$ (BSCF) and $SrCo_{1-y}Sb_yO_{3-\delta}$ (SCS) as alternative cathode materials for intermediate temperature SOFC applications [4-8]. Despite improvements in the area-specific resistance (ASR), these materials do not perform well in the temperature range of 500-650 °C. Recently, Takasu et al.,[8]reported high electronic conductivity for $Ce_{0.60}Pr_{0.40}O_{2-\delta}$ above 600 °C. Nauer et al., [6] studied praseodymia and niobia doped ceria and observed an increase in electrical conductivity with increasing praseodymium content. There were also reports of multi doped ceria to improve the electrical conductivity by lowering the association enthalpy of oxygen vacancies, increase of configuration entropy, modification of elastic strain in the crystal lattice and change in the grain boundary composition [6]. Maffei and Kuriakose [7] doped ceria with gadolinium and praseodymium to reduce the electronic conductivity of ceria and found that double doping scheme is not effective in reducing the electronic conductivity of ceria. Most of the published work till date on Pr and Gd co-doped CeO_2 have focused on its application as an oxygen sensor, or for oxygen storage and oxygen permeability devices and no detailed study has been reported about its suitability as a cathode material for low temperature solid oxide fuel cells. In the present work, structural and electrical properties of the mixed ionic and electronic conducting oxides $Pr_xCe_{0.95-x}Gd_{0.05}O_{2-\delta}$ ($0.15 \leq x \leq 0.40$) have been studied for the application of low temperature SOFC cathode.

EXPERIMENTAL DETAILS

Powders of $Pr_xCe_{0.95-x}Gd_{0.05}O_{2-\delta}$ (PCGO) with x=0.15, 0.20, 0.30 and 0.40 were prepared by conventional solid state route. High purity oxides of CeO_2, Gd_2O_3 and Pr_6O_{11} (99.95%, Alfa Aesar, Johnson Matthey Company, USA) in stoichiometric ratio were ball milled in ethanol for 24 h. The resulting powders were dried at 80 °C for 12 h and pre-calcined at 1100 °C for 6 h. The calcined powders were uni-axially die-pressed at 70 MPa into 13 mm diameter pellets. The samples were vacuum-packed and iso-statically pressed at 250 MPa for 5 minutes. The green pellets were sintered at 1600 °C for 10 h in air with a heating and cooling rate of 2 °C per minute. An X'PERT PRO Panalytical X-ray diffractometer was used to determine the phase and crystal structure of PCGO powder samples. CuKα radiation was generated with an accelerating voltage of 40 kV and current of 30 mA. JCPDS database was used to identify phases. Bulk density of the samples was determined using Archimedes principle using de-ionized water. The microstructures of the sintered samples were evaluated using a scanning electron microscopy (SEM). The thermal expansion measurements were performed in air on the sintered rod shaped specimens with a relative density of > 95% using the NETZSCH 402 PC dilatometer over a temperature range 25-800 °C with a heating rate of 2 °C per min. Standard fused silica sample was used for calibration. Complex impedance measurements were performed in the temperature range, 550-700 °C in air using a AUTOLAB PGSTAT30 FRA in the frequency range from 0.1 Hz to 13 MHz with 10 mV signal amplitude. Nyquist plots were then generated using commercially available nonlinear least square fit software NOVA 1.8 and the conductivity in S/cm was determined for each composition. The ionic transference number has been measured by using a homemade set up. The cell configuration used in our study was pO'_1, Ag //PCGO//Ag, pO''_2. Dried and pure oxygen gas was passed on one side (cathode side) as a reference gas and air or mixture of O_2-Ar gases ranging from 5 to 45% O_2 were passed into the anode side of the cell. The average ionic transference number is calculated using the equation [9]

$$t_0 = \frac{E_O}{E_N} \tag{1}$$

where E_O is the e.m.f. measured under open circuit conditions across the MIEC and E_N is the Nernst potential imposed across the sample. The value of E_N was determined by the activity of electroactive species at the electrode-electrolyte interfaces.

The E_N is given by [9]

$$E_N = \frac{RT}{4F} \ln\left[\frac{pO''_2}{pO'_2}\right] \tag{2}$$

Where, R is the universal gas constant, T is the absolute temperature, F is the Faraday's constant pO_2' and pO_2'' are the partial pressures of oxygen at the two interfaces of the cell.

The total conductivity of PCGO samples were obtained by AC impedance spectroscopy. The ionic conductivity is estimated as

$$\sigma_0 = t_0\ \sigma_T \tag{3}$$

The electronic conductivityhas been estimated as

$$\sigma_e = (1 - t_0)\,\sigma_T \tag{4}$$

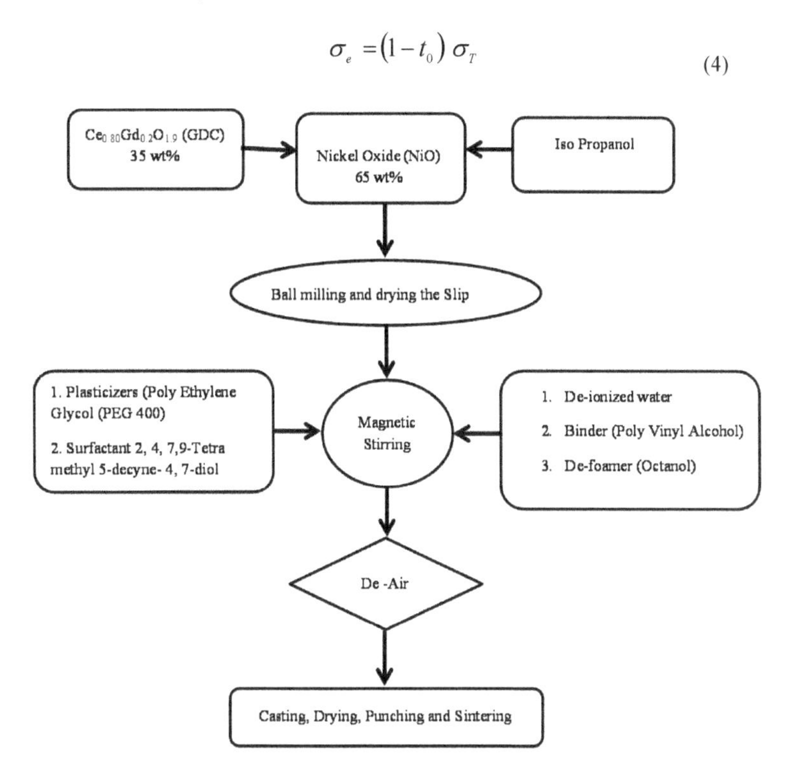

Figure 1. Flow diagram showing various steps involved in the fabrication of NiO-GDC anode tapes via aqueous tape casting.

Figure. 1. shows the flow diagram of various steps involved in the preparation of NiO-GDC anode tapes through an aqueous based tape casting technique [9]. Appropriate amounts of CeO_2 and Gd_2O_3 powders were mixed and ballmilled in ethanol for 24 hours to form a stoichiometric composition of $Ce_{0.8}Gd_{0.2}O_{1.9}$ powder. The resultant powders were dried and calcined at1100 °C for 5 h. The as synthesized $Ce_{0.8}Gd_{0.2}O_{1.9}$ powders were then mixed with NiO powder (99.95%, Alfa Aesar, Johnson Matthey Company, USA) in a weight ratio of 35:65 and again ball milled in isopropanol for 6 h and the slip was dried at 50 °C. The anode materials, NiO-GDC mixture were mixed with deionized water and poly acrylic acid (dispersant) using magnetic stirring for 24 h followed by another 24 h of stirring after adding other ingradients like binder, plasticizer, surfactant etc. The slip was then tape casted and the cast tapes were then allowed to dry over night at room temperature before casting the bilayer of GDC electrolyte. The electrolyte slurry was prepared by mixing appropriate amount of GDC powder, poly vinyl butyral (binder) and menhaden fish oil as dispersant in iso propanol using magnetic stirring for 12 h and the resultant slip was casted on the alreadyprepared anode layer. The thickness of the anode layer was 1mm and the electrolyte layer was 500 μm. This green anode/electrolyte bilayer was punched into 13 mm diameter disks and cofired at 1500 °C for 10 h hold at heating rate of 1°C per minute. Cathode slurry was prepared by mixing PCGO nano powder and GDC electrolyte powder in a weight ratio 70:30 and appropriate amounts of ethyl cellulose and poly vinyl butyral were added as a pore former and binder respectively and mixed in α-terpineol and made into a fine ink on a magnetic stirrer. This cathode ink was brush painted on the electrolyte side of the sintered anode/electrolyte bilayer, dried and again sintered at 1350 °C for 2h. The sintered single cells consist of 0.5-0.6 mm thick electrolyte layer and 0.3-0.5 mm thick anode and cathode layers respectively. Finally both anode and cathode surfaces were covered with nickel gauze and silver gauze respectively as current collectors, which were then coated with silver conductive ink as a sealant. Silver wires were connected as the connection leads and heat treated at 400 °C for 1 h for a better electrical contact. The as prepared cell was then mounted in between the air and fuel chambers of an indigenously developed stainless steel test station. The cells were tested between 500, 550 and 650 °C with humidified hydrogen gas as the fuel and zero air was used as oxidant. Both gas flow rates were controlled between 40 and 100 mL min^{-1} at 1 atmosphere pressure. The performance analysis of the fuel cells was carried out using Potentistat/Galvanostat (Autolab, PGSTAT 30).

RESULTS AND DISCUSSION

The XRD patterns of $Pr_xCe_{0.95-x}Gd_{0.05}O_{2-\delta}$ ($0.15 \leq x \leq 0.40$) samples sintered at 1600 °C for 10h are shown in Figure. 2. All the samples show fluorite crystal structure and no extra peaks are observed, indicating that no secondary phases are formed. Figure 3 shows the SEM analysis

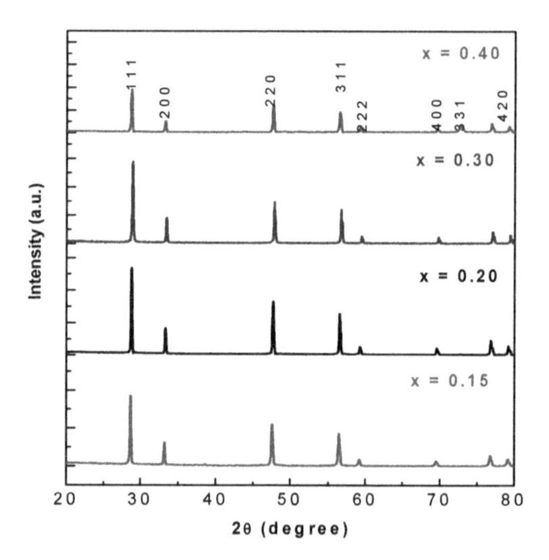

Figure 2.X-ray diffraction analysis of $Pr_{1-x}Ce_xGd_{0.05}O_{2-\delta}$ ($0.15 \leq x \leq 0.40$) cathode samples sintered at 1600 °C for 10 h

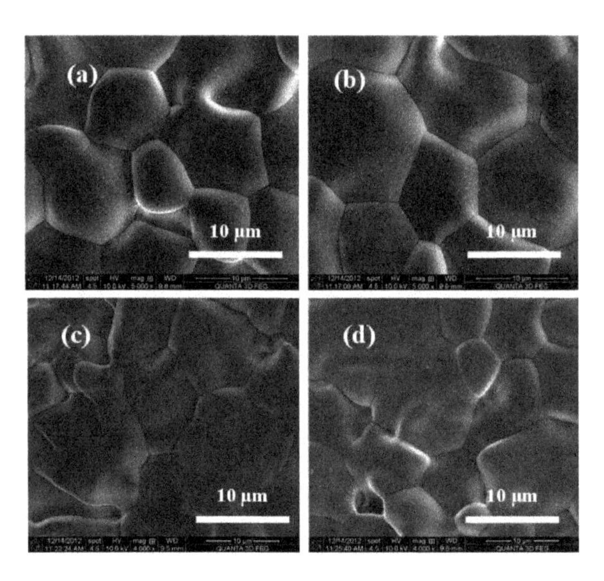

Figure 3.SEM analysis of $Pr_{1-x}Ce_xGd_{0.05}O_{2-\delta}$ ($0.15 \leq x \leq 0.40$) cathode samples (a) x = 0.15 (b) x = 0.20 (c) x = 0.30 and (d) x = 0.40 sintered at 1600 °C for 10 h.

of the fracture surfaces of the cathode samples sintered at 1600 °C for 10h, which indicates well sintered and fully densified samples with a grain size ranging from 5-10 μm size.The density of the samples were estimated by Archemedes principle using deionized water and the percentage density of the samples were shown in figure 4 (a). The percentage density increases with increasing Pr content. The x=0.40 sample exhibited a maximum density of 98% whereas the sample with x =0.15 exhibited 88 %. Figure 4 (b) shows the shrinkage vs Pr content of samples sintered at 1600 °C for 10 h and the shrinkage decreases with increasing Pr content. The results show that sinterability of the sample improved with addition of Pr.

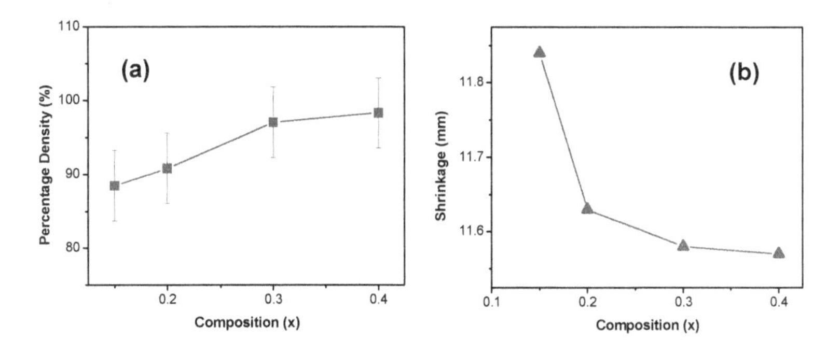

Figure 4. (a) Percentage density and (b) Shrinkage vs composition of $Pr_{1-x}Ce_xGd_{0.05}O_{2-\delta}$ ($0.15 \leq x \leq 0.40$) cathode samples sintered at 1600 °C for 10 h.

Thermal expansion coefficients (TEC) of PCGO calculated from the thermal expansion curves by linear regression in the temperature range 25-800 °C as a function of Pr content is shown in the Figure. 5. The PCGO sample, x=0.15 exhibits TEC value of 12.8 x 10^{-6} K^{-1} and x=0.20 exhibits a value of 13.6 x 10^{-6} K^{-1}, which is comparable to the value of GDC substrate 11.6 x 10^{-6} K^{-1} at 550 °C. On the other hand, the TEC values of the compositions x=0.30 and x=0.40 exhibit higher TEC values of 27x 10^{-6} and 19.8x 10^{-6} K^{-1} respectively. The larger increase

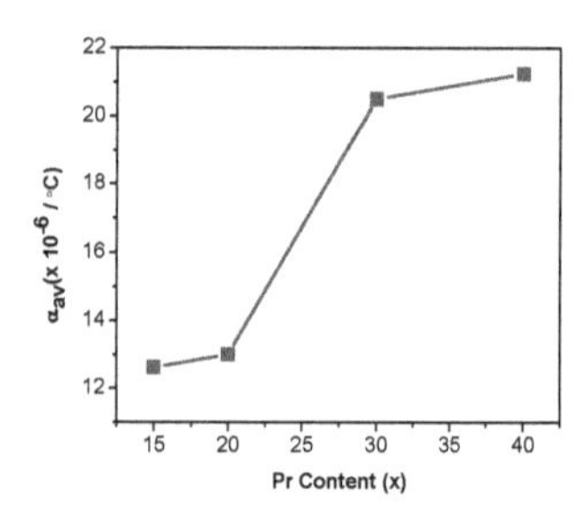

Figure 5. Coefficient of thermal expansion (CTE) of $Pr_xCe_{0.95-x}Gd_{0.05}O_{2-\delta}$ ($0.15 \leq x \leq 0.40$) as a function of Pr content measured from 25 to 800 °C.

in the thermal expansion coefficient of the compositions x=0.30 and x=0.40 is due to the loss of oxygen and associated reduction of Pr^{4+} to Pr^{3+} and Ce^{4+} to Ce^{3+}.

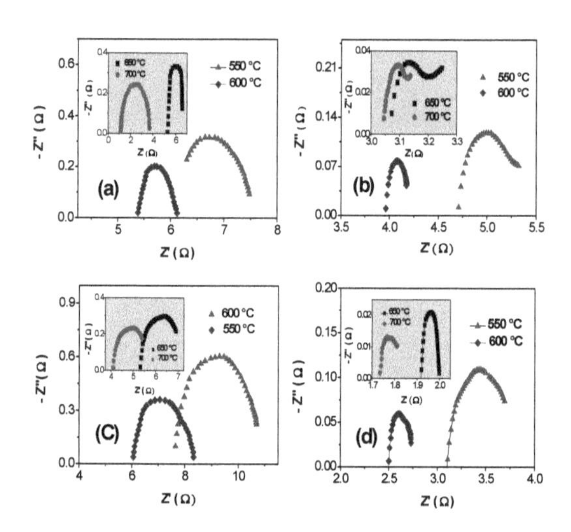

Figure 6. AC impedance plots of $Pr_xCe_{0.95-x}Gd_{0.05}O_{2-\delta}$ as a function of Pr content measurement. (a) x= 0.15, (b) x= 0.20, (c) x=0.30 and (d) x=0.40. The inset shows the impedance spectra at 650 and 700 °C.

AC impedance measurements were carried out in the temperature range of 550-700 °C with symmetric half-cell configuration, PCGO-GDC//GDC//PCGO-GDC in air and results are presented in Figure.6. The typical impedance spectra show single arc for all compositions at all temperatures except for the composition x=0.20 at 650 and 700 °C, indicates single reaction process takes place with specific relaxation time during oxygen reduction reaction (ORR).

Nauer et al., [6] reported smilar observations in praseodymia and niobia doped ceria. They observed single semicircle in the impedance spectra when the electronic contribution to the total conductivity is higher than that of ionic contribution and two semi circles appear when the ionic contribution to the total conductivity is appreciable. Based on the above arguments, it appears that the conductivity mechanism in all PCGO samples is dominated by electronic conductivity as evidenced by the presence of a single arc in the impedance plot. Stefanik et al., [10] observed that more than 20% Pr doping with CeO_2 improves the electronic conductivity at high oxygen partial pressures due to the formation of impurity band with in the band gap between the O-2p valence band and the Ce-4f conduction band of CeO_2. The total conductivity of PCGO compositions is presented in Figure.7. as a function of temperature. The composition x = 0.20 exhibited highest total conductivity of 0.02 Scm^{-1} at 550 °C. Both Gd and Pr contribute oxygen vacancies as described by the defect equation[11];

$$Pr_2O_3 \xleftrightarrow{CeO_2} 2Pr'_{Ce} + V_O^{\bullet\bullet} + 3\,O_O \tag{5}$$

$$Gd_2O_3 \xleftrightarrow{CeO_2} 2Gd'_{Ce} + V_O^{\bullet\bullet} + 3O_O \tag{6}$$

At high oxygen partial pressures and temperatures $[Pr'_{Ce}] \gg [Ce'_{Ce}]$ and $[GD'_{Ce}] \gg [Ce'_{Ce}]$

The electro neutrality condition is [11];

$$[Pr'_{Ce}] + [Gd'_{Ce}] = 2\,[V_O^{\bullet\bullet}] \tag{7}$$

To sparate the electronic contribution to total conductivity transference nunmber measurement has been performed and results are presented in Figure. 8. It is evident from the figure that the ionic transference number (t_{ion}) decreases with increasing Pr content. The electronic contribution to the total conductivity decreses with increasing temperature. In other words, the ionic contribution to the total conductivity increases with increasing temperature because vacancy formation is a thermally activated process which is responsible for increases the ionic conductivity. The electronic contribution is 60, 68, 70 and 75 % of the total conductivity for the compositions x = 0.15, 0.20, 0.30 and 0.40 respectively at the temperature 550 °C. One can clearly see that the electronic contribution to the total conductivity varied from 50 to 85 %. The compositions x = 0.15 and 0.20 shows mixed ionic and electronic conducting behavior and x = 0.40 shows more electronic conduction behavior . It is well known that a good cathode material should have mixed ionic and electronic conductivity along with comparable CTE with electrolyte and current collectors. It is evident from the results that the composition x = 20 could be a promising material for low temperature cathode applications due to its high conuctivity as well as comparable CTE value with that of electrolyte comparied to other composiotions.

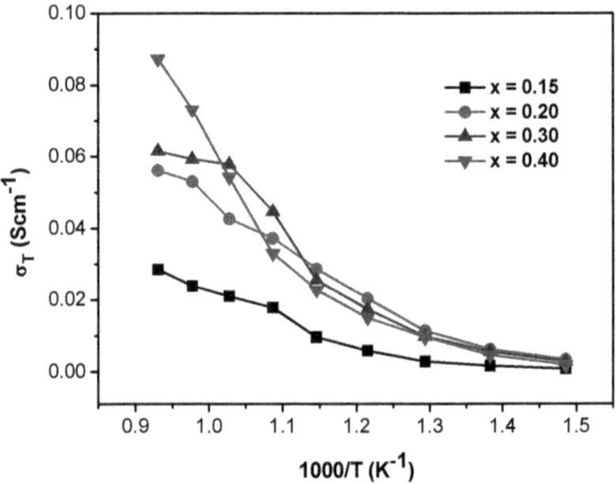

Figure 7. Total conductivity of $Pr_xCe_{0.95-x}Gd_{0.05}O_{2-\delta}$ as a function of Pr content measurement. (a) x= 0.15, (b) x= 0.20, (c) x=0.30 and (d) x=0.40.

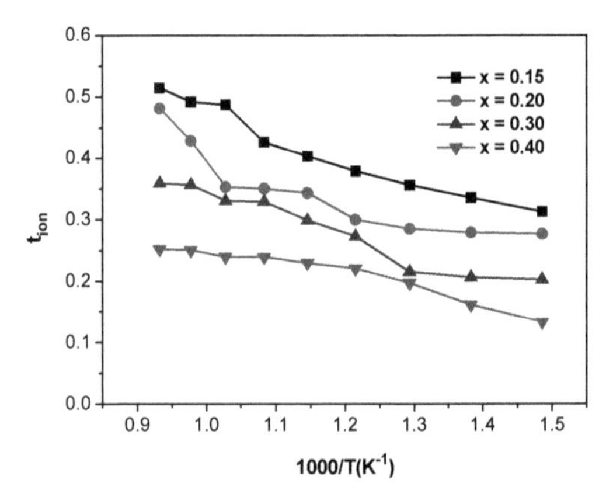

Figure 8. Ionic transference number of $Pr_xCe_{0.95-x}Gd_{0.05}O_{2-\delta}$ as a function of Pr content measurement. (a) x= 0.15, (b) x= 0.20, (c) x=0.30 and (d) x=0.40.

The phase stability of NiO-GDC anode tapes were evaluated using XRD. Figure. 9.(a) and (b) shows the X-ray patterns of NiO-GDC green tapes and tapes were sintered at 1250 °C for 15 h. The X-ray pattern revealed characteristic peaks of NiO and GDC and no unwanted phases were observed even after a prolonged heat treatment, indicates that NiO-GDC anodes are stable.

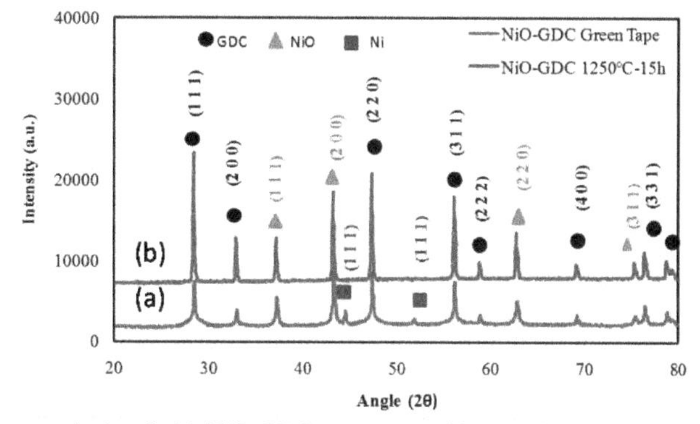

Figure 9. X-ray analysis of (a) NiO-GDC green tape (blue) and (b) NiO-GDC tape sintered at 1250 °C for 15 h.

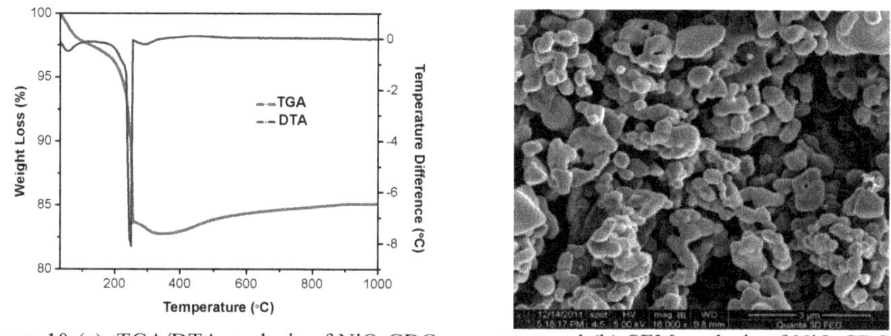

Figure 10 (a). TGA/DTA analysis of NiO-GDC green tape and (b) SEM analysis of NiO-GDC tape sintered at 1250 °C for 15 h.

Figure. 10 (a) shows the thermo gravimetric and differential thermal analysis (TGA-DTA) of NiO-GDC green tape. The first weight loss from room temperature to 100 °C is associated with an endothermic peak at ~70 °C in the DTA curve caused by the loss of molecular water. One can also observe a second weight loss from 200 °C to 280 °C and an endothermic peak at ~250 °C. This may be due to the decomposition of organic residues [13]. Figure 10 (b) shows the microstructure of NiO-GDC anode laminates sintered at 1250 °C for 15 h. It is evident from the micrograph that there is a uniform distribution of porous Ni particles surrounded by hexagonal GDC particles. These types of connected porosity is very much favorable for effective fuel transportation, which in turn enhances anodic reactions.

The performance analyses of a single cell with a configuration $Pr_{0.20}Ce_{0.75}Gd_{0.05}O_{2-\delta}$ - $Ce_{0.80}Gd_{0.20}O_{2-\delta}$ (cathode) // $Ce_{0.80}Gd_{0.20}O_{2-\delta}$ (electrolyte) // NiO-$Ce_{0.80}Gd_{0.20}O_{2-\delta}$ (anode) at three different temperatures 500, 550 and 650 °C is presented in Figure 11. The power density of the cell increased with increasing temperature. Similar observation has previously been reported by Lalanne et al. [15] for single cell using Neodymium nickelate as cathode material. It was found that the single cell delivered a maximum power density of 167.9 mW cm^{-2} and current density of 372.5 $mA.cm^{-2}$ at 650 °C. The better performance of the cell may be due to higher concentration of oxygen ion vacancies, which in turn accelerates the dissociation of oxygen molecules into oxygen atoms and increases the oxygen exchange kinetics and oxygen reduction reaction [16].

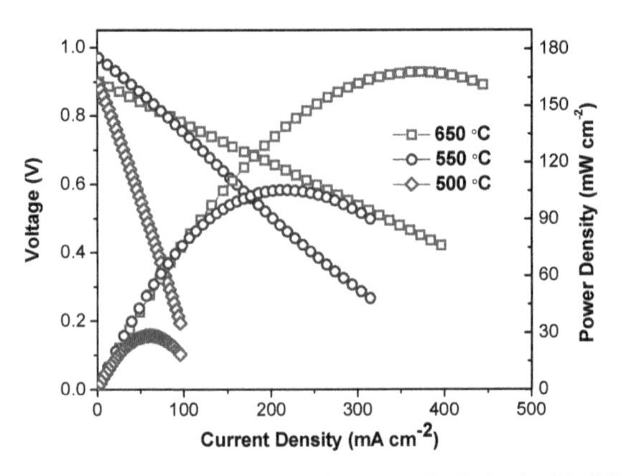

Figure 11. Performance analysis of $Pr_{0.20}Ce_{0.75}Gd_{0.05}O_{2-\delta}$ –GDC (cathode) //GDC (electrolyte) //NiO-GDC (anode) single cell using hydrogen as fuel and air as oxidant.

CONCLUSIONS

The cathode composition, $Pr_{0.20}Ce_{0.75}Gd_{0.05}O_{2-\delta}$ showed comparable TEC value of 13 $x10^{-6}$ K^{-1} with that of GDC and highest total conductivity of 0.02 Scm^{-1} at 550 °C. The ionic transference number for the composition x = 0.2 varied from 0.5 to 0.3 in the temperature range of 400-800 °C indicates mixed ionic and electronic conduction behavior. The electronic conductivity of $Pr_xCe_{0.95-x}Gd_{0.05}O_{2-\delta}$ increased with increase in Pr due to the formation of an impurity band in between O-2p valence band and the Ce-4f conduction band, which in turn reduces the band gap there by increasing the electronic conductivity. Both Gd and Pr contributed to the improvement in ionic conduction. The results signify that the PCGO sample with x=0.2 could be a promising cathode material for low temperature SOFC applications. A single cell with configuration $Pr_{0.20}Ce_{0.75}Gd_{0.05}O_{2-\delta}$-$Ce_{0.80}Gd_{0.20}O_{2-\delta}$ (cathode) // $Ce_{0.80}Gd_{0.20}O_{2-\delta}$(electrolyte) // NiO-$Ce_{0.80}Gd_{0.20}O_{2-\delta}$(anode) delivered a maximum power density of 167.9 mW cm^{-2} and current density of 372.5 $mAcm^{-2}$ at 650 °C. The improved performance of PCGO cathode composition is attributed to higher concentration of oxygen ion vacancies, which in turn accelerates the dissociation of oxygen molecules into oxygen atoms and increases the oxygen exchange kinetics and oxygen reduction reaction.

ACKNOWLEDGMENTS

We gratefully acknowledge partial support of this research by Council of Scientific and Industrial Research (CSIR), New Delhi, India Project No: 22(0524)/10/EMR-II. We also acknowledge IIT, New Delhi for the financial assistance for graduate studies.

REFERENCES

1. M. L. Perry and T. F. Fuller, A Historical Perspective of Fuel Cell Technology in the 20th Century, *J. Electrochem. Soc.*, **149**, S59-S67 (2002).
2. J. W. Fergus, Electrolytes for solid oxide fuel cells, *J. Power. Sources.*,**162**,30-40 (2006).
3. B. C. H. Steele, Materials for IT-SOFC stacks: 35 years R&D: the inevitability of gradualness?*Solid State Ionics*, **134**, 3-20 (2000).
4. P. Bance, N. P. Brandon, B. Girvan, P. Holbeche, S .O'Dea, and B. C. H. Steele, Spinning-out a fuel cell company from a UK University—2 years of progress at Ceres Power, *J. Power Sources,* **131,** 86-90 (2004).
5. Z. Shao, and S. M. Haile, A High Performance Cathode for the Next Generation Solid-Oxide Fuel Cells, *Nature,* **431**, 170-173 (2004).
6. M. Nauer, C. Ftikos, and B. C. H. Steele, An evaluation of Ce-Pr oxides and Ce-Pr-Nb oxides mixed conductors for cathodes of solid oxide fuel cells: Structure, thermal expansion and electrical conductivity, *J. Eur. Ceram., Soc.*,**14** 493-499 (1994).
7. N. Maffei and A. K. Kuriakose, Solid oxide fuel cells of ceria doped with gadolinium and praseodymium, *Solid State Ionics*, **107**, 67-71(1997).
8. Y. Takasu, T. Sugino, Y. Matsuda, Electrical conductivity of praseodymia doped ceria, *J. Appl. Electrochem*. **14**, 79-81 (1984).
9. V. V. Kharton, and F. M. B. Marques, Interfacial effects in electrochemical cells for oxygen ionic conduction measurements: I. The e.m.f. method, *Solid State Ionics*, **140,** 381-394 (2001).
10. C. Fu, S. H. Chan, Q. Liu, X. Ge and G. Pasciak, Fabrication and evaluation of Ni-GDC composite anode prepared by aqueous-based tape casting method for low-temperature solid oxide fuel cell, *Int. J. of Hyd. Energy*, **35**, 301-307 (2010).
11. T. S. Stefanik and H. L. Tuller, Nonstoichiometry and Defect Chemistry in PraseodymiumCerium Oxide *J. Electro ceramics*, **13,** 799-803 (2004).
12. L. Navarro, F. Marques and J. Frade, n-Type Conductivity in Gadolinia-Doped Ceria
    a. *J. Electrochem. Soc.*, **144**, 267-273 (1997).
13. Fu YP, Chen SH. Preparation and characterization of neodymium-doped ceria electrolyte materials for solid oxide fuel cells, *Ceramic International*, **36**(2): 483-490 (2010).
14. Hench LL, West JK. Chemical processing of advanced materials. John Wiley & Sons, In., 288-289 (1992).
15. Lalanne C, Mauvy F, Siebert E, Fontaine ML, Bassat JM, Ansart F, Stevens P, Grenier J C. Intermediate temperature SOFC single cell test using $Nd_{1.95}NiO_{4+\delta}$ as cathode. *J. European Ceramic Society*, **27**, 4195-4198 (2007).
16. Zhu B. Proton and oxygen ion-mixed-conducting ceramic composites and fuel cells. *Solid State Ionics*, **145**,371-380 (2001).

SOLID OXIDE METAL-AIR BATTERIES FOR ADVANCED ENERGY STORAGE

Xuan Zhao, Yunhui Gong, Xue Li, Nansheng Xu, Kevin Huang*
Department of Mechanical Engineering
University of South Carolina, Columbia, SC 29201

ABSTRACT

Metal-air batteries have the potential to achieve high specific energy without the need for storage of oxygen. The majority of conventional metal-air batteries utilize liquid electrolytes, invoking severe chemical reactions with electrode materials and limiting the reversibility of the battery. We report the progress made in a new type of metal-air battery developed from reversible solid oxide fuel cell (RSOFC) and redox couple chemical looping technologies. The paper first discusses the principle of the new battery, followed by the unique features inherited by the battery. Finally, a few examples of solid oxide metal-air operated at both 800 and 550°C are given to show the promises of the new battery as an advanced energy storage mechanism.

INTRODUCTION

Metal-air batteries have attracted a great deal of attention recently due to their high intrinsic specific energy and needless storage for oxygen. The known metal-air batteries can be generally classified into two categories based on the type of the electrolytes employed: cation conductor, *e.g.* Li^+ in Li-air batteries, and anion conductor, *e.g.* OH^- in Zn-air batteries. A major issue facing these batteries is, however, the very poor reversibility caused by the chemical reaction between liquid electrolyte and electrode materials, congestion of air pathway, decomposition and evaporation of electrolytes. In addition, the reversibility becomes worsened at higher rates of cycling, severely limiting the practical use of these batteries.

An all solid-state metal-air battery seems to be a promising solution to cope with the aforementioned problems arisen from the use of liquid electrolyte. Here we demonstrate a new all solid-state metal-air battery based on a reversible solid oxide fuel cell and a redox couple. The former fulfills the discharge and charge cycles by utilizing its dual functionality of "fuel cell" and "electrolyzer", respectively, while the latter stores the electrical energy into the redox couple though a H_2/H_2O mediated redox reaction. [1-4] In this paper, we show the progress made in the past two years about this new battery technology.

A BRIEF REVIEW ON THE RSOFCS AND HYDROGEN CHEMICAL LOOPING

The first functional solid oxide fuel cell (SOFC) was demonstrated in 1940s. [5-6] However, it was not until 1980s that the first RSOFC was operated as both power and hydrogen generators. [7] A RSOFC can efficiently convert the stored chemical energy in hydrocarbon fuels including H_2 to electricity and split H_2O into H_2 with DC electricity in a reversed mode. In other words, a RSOFC can work in both fuel-cell and electrolysis modes. Fig. 1 shows a typical RSOFC configuration. It appears that the reversible conversion between chemical and electrical energy in RSOFCs can be utilized as a means of storing energy provided that the consumed and generated fuel (e.g., H_2) by RSOFCs are properly stored during the charge-discharge cycles. However, an operational SOFC demands for a large amount of H_2, which requires a large-capacity hydrogen separation and storage facility to support it. Similarly, a RSOFC operating under "electrolysis" mode requires a large and constant supply of H_2O and heat to sustain its operation. A bulky and complex facility for H_2 storage (often at lower efficiency) and generation of steam would make the RSOFC for energy storage commercially less competitive. Therefore, there is an urgent need to develop an efficient way to store the H_2 generated during the electrolysis in order for RSOFC to be applicable for energy storage.

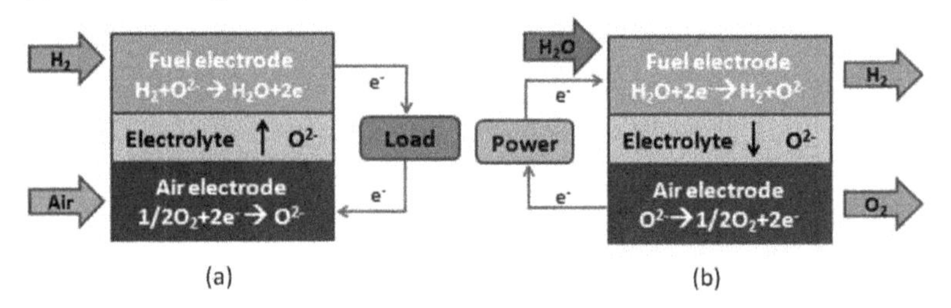

Fig.1 RSOFC operated in (a) fuel cell mode; (b) electrolysis mode

Storage of H_2 in solids such as metal hydrides is an attractive means of achieving high volumetric storage capacity. However, combining these metal hydrides with RSOFC for *in-situ* H_2 storage presents a formidable challenge: instability of metal hydrides at elevated temperatures (500°C-1000°C) where a RSOFC must operate. On the other hand, storing H_2 in chemically stable metal/metal oxide redox couples at high temperatures has been demonstrated as a promising approach in today's metal-steam hydrogen chemical looping processes. [8-10] During such a process, pure H_2 is generated by the steam-metal reaction shown in eq. (1), where M represents metal and MO_x is the corresponding metal oxide. After all the metal is converted into oxide, a reverse reaction will be needed to regenerate the metal by introducing H_2.

$$M + xH_2O = MO_x + xH_2 \quad (1)$$

E*q*. (1) involves reactions between a gas and solid phase easy to be integrated into a practical chemical looping system. The unique advantage of such a chemical looping process lies in its ability to indirectly store hydrogen in the form of reduced solid metals that are stable at high temperatures.

The new solid oxide metal-air battery combines a RSOFC with a metal/metal-oxide redox couple to fulfill discharge-charge cycles with *in-situ* storage and release of H_2 in a closed system [1].

WORKING PRINCIPLE AND KEY FEATURES OF THE NEW BATTERY

Fig.2 shows the working principle of the solid oxide metal-air battery with a series of enabling electrochemical and redox reactions. The free-standing energy storage unit (ESU) is situated in the fuel-electrode chamber where a mixture of H_2O/H_2 mediates the electrical-chemical energy conversion. The operation of a charge and discharge cycle can be described as follows. During the discharge, the interaction between H_2O and metal (Me) produces H_2 locally in the ESU via the following chemical reactions

$$Me(s) + xH_2O_{(g)} = MeO_{x(s)} + xH_{2(g)} \tag{2}$$

The generated H_2 then diffuses to the fuel-electrode surface of the RSOFC operating in fuel-cell mode by which H_2 is electrochemically oxidized, producing electricity and H_2O via the following electrochemical reactions

$$H_{2(g)} + O^{2-} = H_2O_{(g)} + 2e^- \tag{3}$$

The product H_2O then diffuses back to the surface of Me and further reacts to produce more H_2 to keep reaction (2) active. When all (or a controlled portion of) the Me phase is oxidized and no (or a controlled portion of) H_2 is produced, the discharge cycle is stopped and the battery needs to be recharged. For the charge cycle, the H_2O in the mediation gas H_2-H_2O is electrochemically split into H_2 at the fuel-electrode of the RSOFC operating under the electrolysis mode

$$H_2O_{(g)} + 2e^- = H_{2(g)} + O^{2-} \tag{4}$$

The generated H_2 then migrates to the ESU where metal oxide (MeO_x) is chemically reduced to Me by

$$MeO_{x(s)} + xH_{2(g)} = Me_{(s)} + xH_2O_{(g)} \tag{5}$$

The product H_2O is then further split by reaction (4) to produce more H_2 for reaction (5). When all (or a controlled portion of) MeO_x is reduced to Me by H_2 and no (or a controlled portion of) H_2O is produced, the charge cycle is completed. The freshly reduced and chemically active Me is then ready for the next discharge cycle as described by reactions (2) and (3). According to Gibbs phase rule, the residual freedom Fr=0 for a dual-phase (P=2) binary system (C=2) under isothermal and isobaric conditions, which implies that the intensive quantity-electrode potential- is independent of the state-of-charge.

The overall chemical reaction from individual reactions (2)-(5) occurring inside the battery can be written by

$$Me_{(s)} + x/2O_{2(g)} = MeO_{x(s)} \tag{6}$$

which suggests the nature of a metal-air battery.

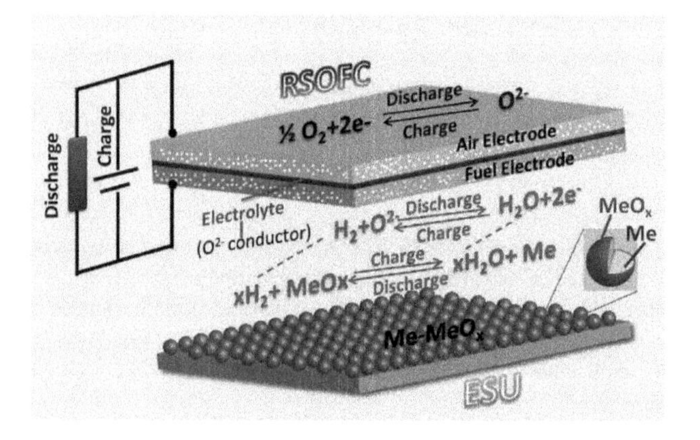

Fig.2 Schematic of the working principle of the solid oxide metal-air battery. The metal/metal-oxide redox couple is decoupled from the fuel-electrode of the RSOFC

The new battery differs from the conventional metal-air batteries in several ways:

1) All of the battery components are in solid-state, which avoids the strong chemical reactions occurred between liquid electrolyte and solid electrodes.
2) The ionic conduction in the solid electrolyte is divalent, promising a higher rate of discharge/charge and higher specific energy.
3) The redox couple, which is physically separated from the electrodes, takes the responsibility of energy storage, preventing the electrodes from detrimental volume changes. In addition, the decoupling design offers the flexibility of experimenting different chemistries for better performance.
4) The high electrical efficiency and high power density of a conventional RSOFC can be naturally inherited by the new metal-air battery.
5) Higher operating temperature can promote faster electrode reactions, ion transport and redox kinetics, allowing speedy charge/discharge cycles.
6) The gas composition in the battery is controlled by the thermodynamic equilibrium between Me and MeO_x phases, yielding state-of-the-charge independent Nernst Potential.
7) The overall power (kW) and energy capacity (kWh) of the new battery system can be separately designed to meet practical storage requirements by varying the number of cells (and/or the size of electrodes) or the amount of redox material in the ESU, respectively.

SELECTION OF REDOX COUPLE ENERGY STORAGE UNIT

According to eq.(6), the overall performance of the new battery depends on the chemistry of metal/metal-oxide redox couples. Therefore, the selection of redox couple is as important as that of RSOFC. Studies of RSOFCs have been widely documented. However, studies of the use of a metal/metal-oxide redox couple in a battery system are rarely reported in the literature. One focus of this paper is to show how to improve the battery performance from the perspective of redox couple based ESU. To facilitate the understanding of fundamentals, Fe-based redox couples are used as an example.

The first step to select a chemically stable redox couple is to examine the phase diagram. Taking Fe-O system as an example, Fig.3 indicates that there are two sets of redox couple existed in Fe-O system: Fe-FeO above 600°C and Fe-Fe_3O_4 below 600°C. Once the redox couple is determined, the maximum theoretical specific energy density (MTSE) and Nernst potential (EMF) can then be calculated from

$$MTSE = -\Delta G_f^o/kg \cdot metal \quad (7)$$

$$EMF = -\Delta G_f^o /nF \quad (8)$$

Here $-\Delta G_f^o$ and n are Gibbs free energy of formation of metal oxide and number of electrons transferred during the reaction, respectively. Fig.4 shows the comparison of MTSE and EMF of Fe-FeO and Fe-Fe_3O_4 redox couples. It is evident that the battery with Fe-Fe_3O_4 as the ESU possesses a higher EMF and MTSE than the battery with Fe-FeO ESU.

Another important factor needing consideration during the selection of a redox couple is the kinetics of the H_2-H_2O mediated redox reactions. An ideal situation is that the rate constant for metal oxidation and oxide reduction is identical, resulting in a reversible redox reaction. The Fe-FeO redox couple offers rather reversible redox kinetics due to the high concentration of point defects available in FeO that promote the diffusion of Fe and O. In contrast, Fe-Fe_3O_4 may not possess redox kinetics as well as that of Fe-FeO for the reason of less point defects available in Fe_3O_4. Therefore, the thermodynamic advantage may not be fully realized due to the kinetic limitation.

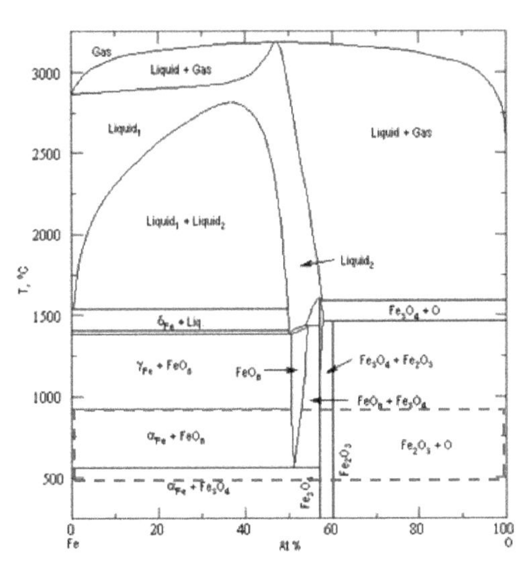

Fig. 3 Phase diagram of Fe-O system. The red dash square indicates the operating temperature of the new battery

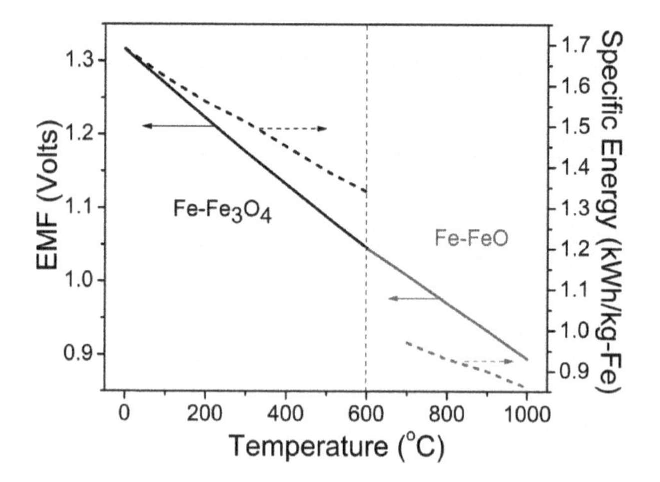

Fig.4 Plots of Nernst potential (EMF) and specific energy of all solid state iron air battery as a function of temperature. Solid lines: EMF; dashed lines: specific energy

CYCLING PERFORMANCES AT 800°C AND 550°C

The testing results as examples illustrating the performance of solid oxide metal-air battery are shown in Fig.5 at 800°C and 550°C. The details for materials preparations and experimental procedures can be found in our recent publications. [2, 11] The 800°C battery (Fig.5 (a)) yielded a discharge specific energy (DSE) of 876 Wh/kg·Fe, which is up to 94.2% of the MTSE at 800°C (MTSE=930Wh/kg Fe), and a round trip efficiency of 89.1%, when compared with the charge specific energy (CSE) =983 Wh/kg·Fe. The 550°C battery (Fig.5 (b)) yielded a DSE of 1,237 Wh/kg·Fe, which is up to 91.0% of the MTSE at 550°C (=1,360Wh/kg Fe), and a round trip efficiency of 82.5%, when compared with the CSE=1,500 Wh/kg·Fe. While these results show the promising features of this new battery concept with stable cycling performance and the specific energy achieved at 800°C and 550°C are close to their respective theoretical values, it is to be noted that the 800°C battery was cycled at J=100 mA/cm^2 whereas the 550°C battery was cycled at J=10mA/cm^2. Lower temperature obviously limited the rate capacity of the battery. Lower round-trip efficiency at lower temperature is an indication of slower redox kinetics. However, lower operating temperature can reduce the cost and increase the reliability of the battery. Therefore, improving the performance of RSOFC and redox couple at intermediate temperatures is a rational approach to realize a high performance, low cost and robust battery.

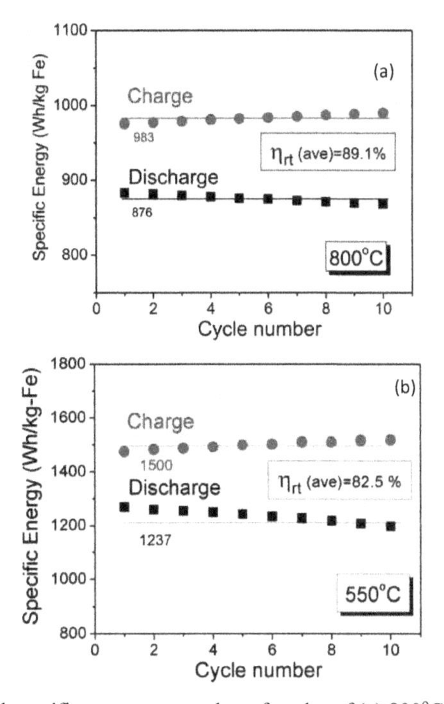

Fig. 5 The averaged specific energy vs number of cycles of (a) 800°C and (b) 550°C. The specific energy is normalized to the amount of iron equivalent to oxygen flux (current) to support the redox reaction

CONCLUSION

A solid oxide metal-air battery combining reversible solid oxide fuel cell and hydrogen chemical looping technologies has been demonstrated as a new energy storage mechanism. The performance tested at 800 and 550°C showed the potential of the new battery to be a high rate capacity and high specific energy battery. Along with many of its unique features, the new solid oxide metal-air battery is expected to play a role in large-scale stationary energy storage applications.

REFERENCES

1. N. Xu, X. Li, X. Zhao, J.B. Goodenough, and K. Huang, *Ener. & Environ. Sci.*, **4** (12), 4942 (2011).
2. X. Zhao, N. Xu, X. Li, Y. Gong, and K. Huang, *RSC Adv.*, **2** (27), 10163(2012).
3. X. Zhao, N. Xu, X. Li, Y. Gong, K. Huang, accepted, to be published in *ECS Trans.* (Volume 45), 2013.
4. X. Zhao, N. Xu, X. Li, Y. Gong, K. Huang, accepted, to be published in *ECS Trans.* (Volume

50), 2013.
5. E. Bauer and H. Preis, *Z. Elektrochem.*, **43**, 727-732 (1937).
6. C. Wagner, *Naturwissenschaften*, , **31**, 265-268 (1943).
7. W. Doenitz and R. Schmidberger, *Intl. J. Hydro Ener.*, **7**, 321-330 (1982).
8. M. Thaler and V. Hacker, *Intl. J. Hydro Ener.*, **37**, 2800-2806 (2012).
9. E. Lorente, J. A. Pena, J. Herguido, *Intl. J. Hydro. Ener.*, **33**, 615-626 (2008).
10. E. Lorente, J. A. Pena, J. Herguido, *Journal of Power Sources*, **192**, 224-229 (2009).
11. Xuan Zhao, Yunhui Gong, Xue Li, Nansheng Xu, Kevin Huang, submitted to *J. Electrochem. Soc.*.

FABRICATION OF CeO_2/Al MULTILAYER THIN FILMS AND THE THERMAL BEHAVIOR

Shumpei Kurokawa, Takashi Hashizume, Masateru Nose, Atsushi Saiki
University of Toyama
Toyama, Toyama, Japan

ABSTRACT

Cerium dioxide (CeO_2) thin film is an attractive material with multiple applications, such as electrolytes in solid oxide fuel cells. In this study, CeO_2 and Al multilayer thin film were fabricated by using the differential pumping co-sputtering system which had multi-chamber and each layer was deposited in divided atmospheres and depositing condition independently. Their morphologies and the effect of annealing were investigated. As a result, as-deposited samples showed a peak of XRD at $2\theta = 38.5°$ corresponding to the (111) plane of Al. However annealed samples did not show the peaks of Al.

INTRODUCTION

Yittria Stabilized zirconia(YSZ) is used as an electrolyte of the SOFC[1-3]. Working temperature of SOFC is very high such as 900℃[2,4].Two yttrium ions can create disorderly one oxygen ion vacancy inside of the YSZ electrolyte, since some of the 4+ zirconium ions are replaced to 3+ yttrium ions[3,4]. At high temperature, oxygen ions can move easily through these oxygen ion vacancies[5,6]. SOFC brings high power generation efficiency by steam generation using exhaust heat close to 1000 ℃[6,7]. In addition, since the carbon monoxide with hydrogen can be used as a fuel, in combination with coal gasification plant, power generation reaction proceeds easily by high operation temperature[8,9]. Because YSZ electrolyte is solid state, it is hard to evaporate or move away even at high temperature. And then, it is possible to extend the life of the fuel cell. However there is a drawback that constructional materials of fuel cell are likely to deteriorate by high operating temperatures[10]. So many engineers have researched about an electrolyte which can be operated at lower temperature everyday[2,3]. The working temperature of SOFC using CeO_2 as electrolyte is lower than YSZ , When CeO_2 are used as the electrolyte, the operating temperature can be lowered to 700 ℃[12,13], because a resistance of CeO_2 is less than that of YSZ. At operating temperature below 700 ℃, low-cost stainless steels could be used as the components of fuel cell systems. Ce is the most present the element in the rare-earth element[11].We consider Al may reduce CeO_2 and oxygen vacancy will be created in CeO_2. In this study, we made multilayer thin films of CeO_2 and Al, we examined the effect of the existence of Al layer in the CeO_2 layer and effect of annealing.

Table I. Experimental condition.

	Chamber(A)	Chamber(B)
Target	Al	CeO_2
Substrate	Si	Si
Power	RF	RF
RF Power	20(W)	200(W)
Ar gas flow rate	20(sccm)	20(sccm)
Substrate temperture	Room Temperture	Room Temperture
Base pressure	6×10^{-4}(pa)	6×10^{-4}(pa)

Step time	15(min)	30(min)	1(h)
Sputtering time	1(h)	2(h)	3(h)
Layer	4(layer)	4(layer)	3(layer)

EXPERIMENTAL

CeO_2/Al multilayer thin films in this study were deposited by differential pumping co-sputtering system. Commercially available Si(100) substrate were used as substrate. The substrate was cut to 25.0×24.5 mm. The substrate were cleaned ultrasonically in ethanol, acetone and butanol to remove the loosely bonded deposit and dried in Ar gas. Argon gas was used as the sputter gas. CeO_2 and Al were used as the sputtering targets. The sputter chamber was pumped down first from atmospheric pressure to a base pressure of 6.0×10^{-4} Pa. Then argon gas was flow into the sputtering chamber at 20 sccm and the CeO_2 target and Al target were pre-sputtered at 100W in pure argon atmosphere for 30 min to remove any oxide layer or contaminant on the target surface. The DC power of CeO_2 and Al supply was tuned up to 200 W and 20 W respectively to start the multilayer thin film deposition. The sputter conditions are shown in the table.1. After the sputtering the samples were annealed at 800°C for 1,2,3 h in He atmosphere in order to prevent oxidation of the samples by air.

The cross section morphology of the deposited thin film was observed by the field emission scanning electron microscope (JEOL JSM−6700F). The crystal structure of the deposited thin film was analyzed by the X-ray diffraction (Bruker's AXS D8 DISCOVER system). The composition depth profile of thin films was observed by the glow discharge spectrometry (Horiba GD-Profiler2). The chemical bounding state information for Al element in the Al layer was studied using the X-ray photoelectron spectroscopy (Thermo Fisher Scietific ESCALAB250X). Specific resistance of thin film was measured by the 2-terminal method.

RESULT AND DISCUSSION

Fig.1, Fig.2 shows FE-SEM micrographs of the cross section of CeO_2/Al multilayer thin film deposited by the differential pumping co-sputtering system. It can be seen that the CeO_2/Al multilayer thin films were well deposited. Boundary was not clear between Al layer and CeO_2 layer. However, differences in contrast of layers were seen. Shape of cross section of CeO_2 was irregularity in the as-deposited samples. However, cross section of CeO_2 showed the shape of edge, and voids were observed in Al layer when the sample was annealed for 1h, 2h. On the other hand, shape of cross section of the sample of annealed for 3h showed unevenness. Film thickness of CeO_2 layer in sputtering time for 30 min step was about 0.80μm and in sputtering time for 1h step was about 1.15 μm. Film thickness of Al layer in sputtering time for 30 min step was about 60 nm and in sputtering time for 1h step was about 265 nm. From this result, it was found that the film thickness was not proportional to the time. This reason was thought to be as follows that deposition was inhibited due to position of the substrate and the target by preliminary experiment device characteristics.

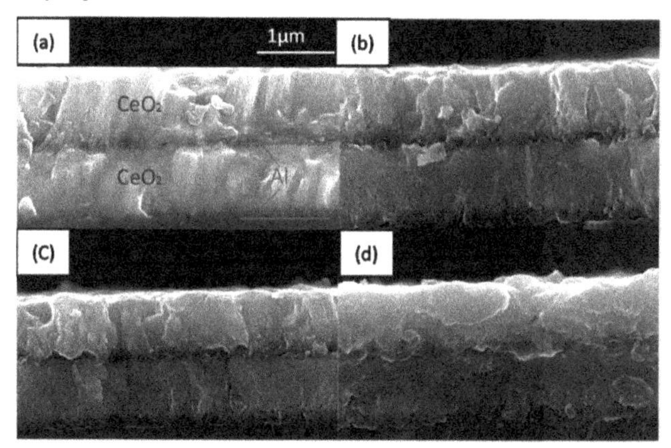

Fig.1 Cross sectional SEM images of the 4-layer films depsited by the differential pumping co-sputtering system(sputtered for 30 min×4(2h CeO_2/Al/CeO_2/Al/Si)) (a):as-deposited, (b):annealed 800°C, 1 h, (c):annealed 800°C, 2 h, (d):annealed 800°C, 3 h.

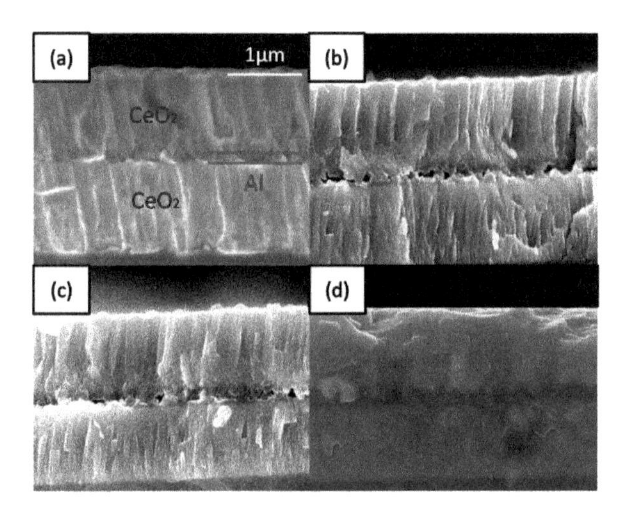

Fig.2 Cross sectional SEM images of the 4-layer films depsited by the differential pumping co-sputtering system(sputtered for 1 h×3(3h CeO_2/Al/CeO_2/Si)).
(a):as-deposited, (b):annealed 800°C, 1 h, (c):annealed 800°C, 2 h, (d):annealed 800°C, 3 h.

Fig.3, Fig.4 shows the XRD patterns of the CeO_2/Al multilayer thin film. As-deposited samples showed a peak at $2\theta = 38.5°$ corresponding to the (111) plane of Al and all another peaks corresponding to the CeO_2. However, annealing samples did not show the peaks of Al. From this result, metal Al was changed to Al oxide by annealing. However, the peaks of the products were not identified in annealing samples. It was thought that the product concentration was too low to detected by XRD and the product was not crystallized or Al melted into the CeO_2. In addition, peaks of CeO_2 were shifted to the higher angle about 0.3~0.4° in the annealing samples. They were more close to the ICDD card data and peak intensity was increased, peaks of CeO_2 became sharper than the as-deposited samples. This indicated that the compressive stress of CeO_2 crystal was released and growth of the CeO_2 crystal grain and lattice spacing of CeO_2 crystal perpendicular to the substrate decreased smaller than that of as-deposited sample. However, intensity of several peaks was decreased by an increase in heat treatment time. This was considered was that the CeO_2 crystal was distorted due to Al diffusion as the increase at the annealing time. In Fig.3 and Fig.4 seen the difference in orientation, this was consider due to difference of the elements of the film immediately above the substrate.

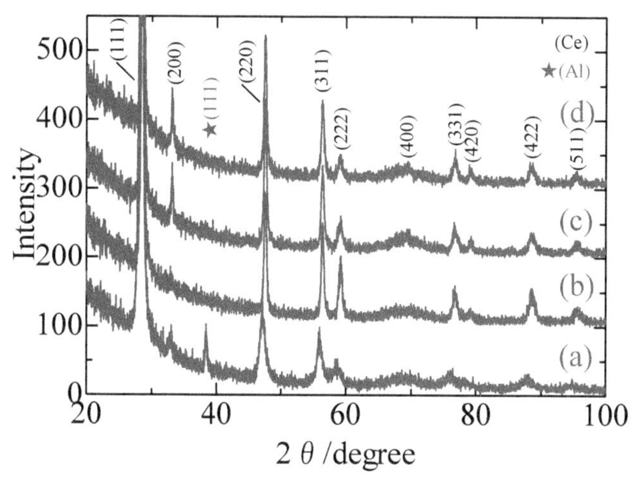

Fig.3 XRD profiles of the layer structured films (sputtered for 30 min×4(2 h CeO_2/Al/CeO_2/Al/Si)) (a):as－deposited, (b):annealed at 800°C, 1 h, (c): annealed at 800°C, 2 h, (d):annealed at 800°C, 3 h.

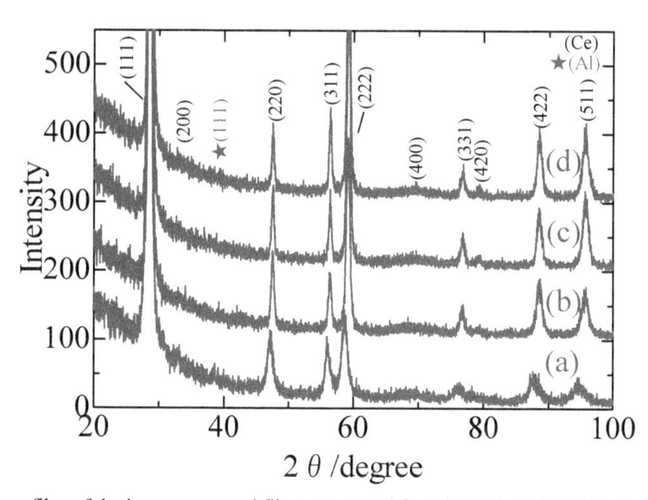

Fig.4 XRD profiles of the layer structured films (sputtered for 1 h×3(3 h CeO_2/Al/CeO_2/Si)) (a):as－deposited, (b):annealed at 800°C, 1 h, (c): annealed at 800°C, 2 h, (d):annealed at 800°C, 3 h.

Fig.5, Fig.6 shows GDS measurements result of CeO_2/Al multilayer thin film. It can be seen abundance of oxygen has been increase in the Al layer in the sample after annealing. In addition, presence of aluminum range was spreaded in the sample after annealing. This indicated that the Al was diffused into CeO_2 layer and was oxidized by annealing. Which was consistent with our conclusion from XRD analysis and SEM images. Abundance of Ce was increased in surface of samples due to reaction with air.

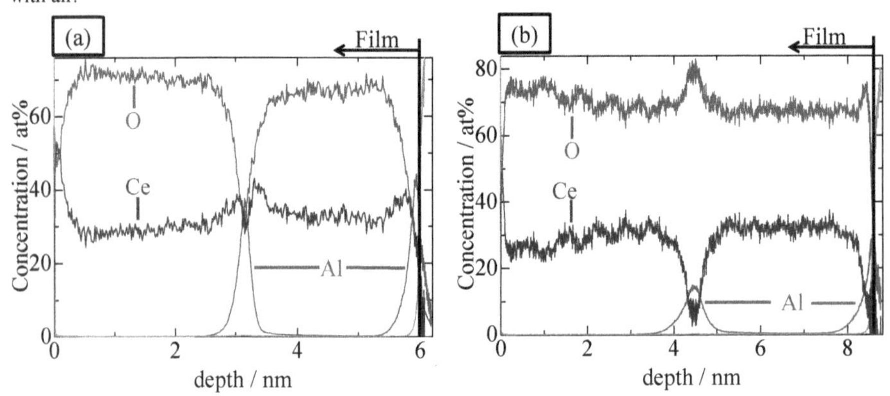

Fig.5 GDS profiles of the layer structured films(CeO_2/Al/CeO_2/Al/Si).
(a):as-deposited, (b):annealed at 800°C, 3 h.

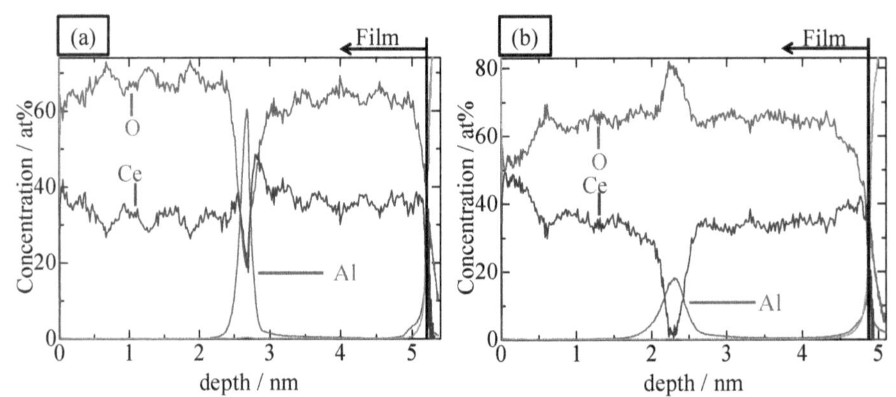

Fig.6 GDS profiles of the layer structured films(CeO_2/Al/CeO_2/Si).
(a):as-deposited, (b):annealed at 800°C, 3 h.

Fig.7, Fig.8 shows XPS spectra for the CeO_2/Al multilayer thin film. Fig.7 shows XPS spactra for the Al 2p core level. Al-Al bond was observed consistently in the sample before the annealing, Al-O bond was observed at the surface of Al layer due to presenting the oxide layer. However, intensity of Al-O bond was increased nearby the CeO_2 layer in the sample after the annealing. Fig.8 shows XPS spactra for the Ce 3d core level. CeO_2 bond of annealing sample was markedly lower than before the annealing sample. In addition, all CeO_2 bond and Ce_2O_3 bond was degrees in the sample after annealing, This indicated that the CeO_2 was reduced by oxidation reaction of Al.

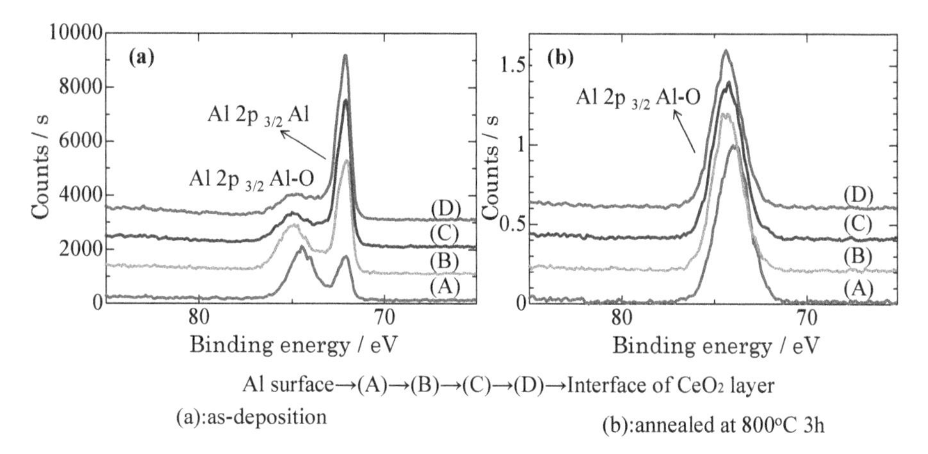

Fig.7 XPS profiles (Al 2p) of the layer structured films (Al/CeO_2/Si)
(a):as-deposited, (b):annealed at 800°C, 3 h.

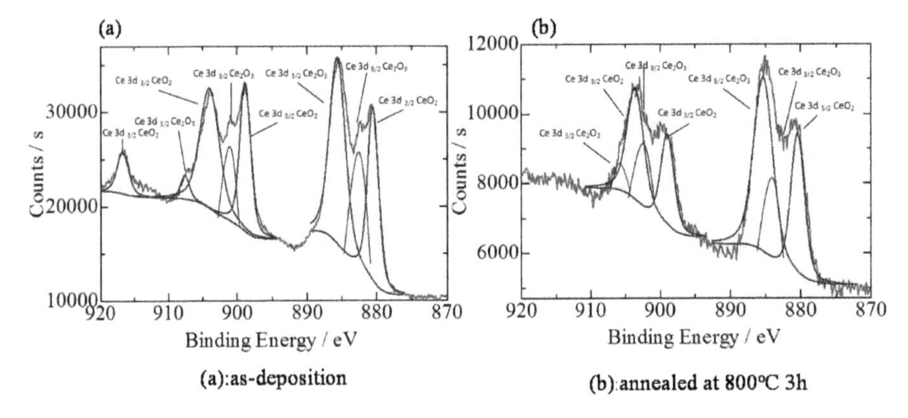

Fig.8 XPS profiles (Ce 3d) of the layer structured films (Al/CeO_2/Si)
(a):as-deposited, (b):annealed at 800°C, 3 h.

Table.2 shows specific electrical resistance for the CeO_2/Al multilayer thin film. Specific electrical resistance decreased after the annealing. It was thought that the crystals of CeO_2 were well crystallized and the Al was diffused into CeO_2 layer, the CeO_2 was reduced by oxidation reaction of Al after annealing from XRD and GDS, XPS result. We Indicated below the reaction equation.

$$CeO_2 + \frac{2}{3}xAl \rightarrow CeO_{2-x} + Al_{2/3x}O_x \ (0<x<0.5)$$

or

$$CeO_2 + 1/3Al \rightarrow 1/2Ce_2O_3 + 1/6Al_2O_3$$

$$Ce_{Ce} \rightarrow Ce_{Ce}^{\cdot} + e^{\prime}$$

We discussed specific electrical resistance was reduced by electron becomes a carrier in this manner.

Table 2 Specific electrical resistance of CeO_2/Al multilayer thin film.

	As-deposition sample	Annealing sample
Specific electrical resistance	$6.41\times10^2(\Omega\cdot m)$	$6.02\times10^1(\Omega\cdot m)$

CONCLUSIONS

CeO_2/Al multilayer thin films were deposited on Si substrate by differential pumping co-sputtering system. In annealed samples, voids were observed locally in Al layer. Annealed samples did not show the peaks of Al. Peaks of CeO_2 were shifted to the high angle about 0.3~0.4°. Abundance of oxygen has been increase in the Al layer and presence of aluminum range is spreading in the sample after annealing. Al-O bond was increased nearby the CeO_2 layer in the sample after the annealing. It was also shown CeO_2 was reduced by oxidation reaction of Al. Specific electrical resistance was decreased after the annealing.

ACKNOWLEDGMENTS

We acknowledge the Japan society for the promotion of Science (JSPS) for the partial support from a Grant – in – Aid for Scientific Research (B) (No,21360359)

This study was supported by Center for Research and Development in Natural Science, Instrumental Analysis Laboratory, University of Toyama.

REFERENCES

[1] Joonho Kim,Youngsus Park,Dae jin sung,Sangjin Moon,Ki Bongh Lee,Suk-In Hong, 27 (2008)pp.136-701

[2] Kai YAN,Qiang ZHEN,Xiwen Song,Rare Metals 26(2007)pp.311-316

[3] V.Thangadurai,P.Kopp, 25(2007)pp.178-183

[4] A.E. Rakshani, Solid-state Electron. 29 (1986) 1.

[5] H. Derin, K. Kantarli, Appl. Phys. A 75 (2002) 391.

[6] H.C. lu, C.L.Chu, C.Y. Lai, Y.H. Wang, Thin Solid Films 517 (2009) 4408.

[7] T. Ghodselahi, M.A. Vesaghi, A. Shafiekhani, A. Baghizadeh, M. Lameii, Appl. Surf. Sci. 225 (2008) 2730.

[8] K. Borgohain, N. Murase, S. Mahamuni, J. Appl. Phys. 92 (2002) 1292.

[9] C.C. Chusuei, M.A. Brookshier, D.W. Goodman, Langmuir 15 (1999) 2806.

[10] M. Fox, Optical Properties of Solid, Oxford University Press, New York, NY, 2001

[11] S. Kanakaraju, S. Mohan, A.K. Sood, Thin Solid Films 305 (1997) 191

[12] C.J. Fu, Q.L. Liu, S.H. Chan, Hydrogen energy 35 (2010) 112000-112007

[13] Y.S. Hong, S.H. Kim, W.J. Kim, H.H. Yoon, Current Applied Physics 11 (2011) 163-168

Direct Thermal to Electrical Energy Conversion Materials and Applications

REDUCED STRONTIUM TITANATE THERMOELECTRIC MATERIALS

Lisa A. Moore and Charlene M. Smith
Corning Incorporated
Corning, NY, USA

ABSTRACT

The efficiency of a thermoelectric material for the conversion of heat into electricity is a function of its Seebeck coefficient (S), electrical conductivity (σ), thermal conductivity (K), and the operating temperature (T) as expressed by a figure-of-merit, $ZT = \sigma S^2T/K$. The strong coupling between these parameters poses a significant challenge for achieving high ZT. One approach is to maintain high electrical conductivity while simultaneously decreasing thermal conductivity. In this study, the effects of co-doping and reduction on the high temperature thermoelectric properties of strontium titanate-based materials were investigated. Bulk, polycrystalline $(La_x,Y_ySr_{1-x-y})_{1-z}Ti_{1+z}O_{3-\delta}$ materials were prepared by powder processing and spark plasma sintering. Reduction of the powders prior to sintering increased the power factor (σS^2) and ZT over the unreduced material. The addition of a reductant (TiC, TiN, TiB_2) further increased the power factor while suppressing increases in thermal conductivity. The fraction of thermal conductivity due to the lattice component was reduced from over 97 percent to less than 75 percent. ZT values up to 0.33 at 1050K, were obtained in co-doped, reduced strontium titanate with a reductant.

INTRODUCTION

Thermoelectric materials can provide a source of "green energy" through the conversion of waste heat into electricity. Numerous sources of waste heat exist today including industrial waste heat generated in chemical reactors, incineration plants, iron and steel melting furnaces, and in automotive exhaust. The maximum efficiency of a thermoelectric material depends on temperature and material properties: Seebeck coefficient (S), electrical conductivity (σ), and thermal conductivity (K). The efficiency of a thermoelectric material is described by its power factor (PF) and a dimensionless figure-of-merit (ZT), where:

$$ZT = \sigma S^2T/K = (PF)T/K. \tag{1}$$

The strong coupling between the material parameters makes high ZT values difficult to achieve.[1] In particular, low thermal conductivity is difficult to obtain in conjunction with high electrical conductivity since a portion of the heat is transported by electrons:

$$K = k_{elec} + k_{latt} \tag{2}$$

where k_{elec} is the electronic thermal conductivity, given by the Wiedemann-Franz law:

$$k_{elec} = L\sigma T \tag{3}$$

The Lorenz number (L) = $2.4 \times 10^{-8}\ V^2K^{-2}$ for metals.[19] The second term in Equation (2) represents the portion of the heat transported by phonons and is referred to as the lattice thermal conductivity. Minimization of k_{latt} has been proposed as a route for maximizing ZT.[2]

Efficient thermoelectric materials that can operate at high temperatures (i.e., 800-1000K) in air and are low cost and composed of non-toxic elements are of interest. High efficiency (ZT > 1) has been achieved in bulk $(Bi, Sb)_x(Te,Se)_y$ alloys, but due to a lack of chemical stability these materials have use temperatures limited to less than about 700K and they contain elements considered to be environmentally unfriendly.[2] Complex oxide materials have been the subject of much research.[3] Doped, oxygen-deficient $SrTiO_3$ materials have received particular attention because of their high Seebeck coefficients and electrical conductivity.[4] However, high thermal conductivity has inhibited the realization of bulk, high ZT strontium titanate materials and various methods for reducing thermal conductivity have been investigated.[3,5-7]

In the present study, the effects of La and Y co-doping and reduction on the high temperature thermoelectric properties of strontium titanate materials were investigated. Bulk, polycrystalline materials with the composition $(La_xY_ySr_{1-x-y})_{1-z}Ti_{1+z}O_{3-\delta}$ where x=0-0.15, y=0-0.1, x+y=0.05-0.25, and z=0-0.11 were prepared by powder processing and spark plasma sintering. Reduction of the powders was achieved by two methods: (1) heat treatment of the powders in a graphite bed prior to sintering, or (2) addition of TiC, TiB_2, or TiN and heat treatment of the powders in a graphite bed prior to sintering. The use of these additives was found to be the most effective at improving the high temperature ZT by increasing the electrical conductivity without causing large increases in thermal conductivity.

EXPERIMENTAL PROCEDURE

Dense, polycrystalline ceramics were prepared from mixtures of dry $SrCO_3$, La_2O_3, Y_2O_3, TiO_2 and Ti_2O_3 powders. The powders were turbula-mixed and cold-pressed into pellets which were calcined at 1200 °C for 12 hours in air, milled, then re-calcined under the same conditions and re-milled. In some cases, an additive – TiC (Sigma Aldrich, <200nm), TiN (Sigma Aldrich, <3 microns) or TiB_2 (NaBond, ~60nm) -- was mixed into the calcined powder. Cold-pressed pellets were buried in a graphite bed consisting of a silica crucible filled with graphite powder. The assembly was heated to 1400 °C in an air atmosphere and held for 6 hours. The reduced materials were milled and sieved to -325 mesh. Powders were sintered in a 20mm-diameter graphite die using spark plasma sintering. The sintering schedule and a representative densification curve are shown in Fig. 1. The maximum temperature and hold time were 1500 °C/15 sec with an applied force of 11 kN.

Polished samples for measurements were prepared from the sintered parts. Bulk densities were calculated from the dimensions and weights of the polished samples and were typically within 95% of the density of single crystal $SrTiO_3$ (5.117 g.cm^{-3}).[20] Electrical conductivity and Seebeck coefficients were measured from room temperature to 800 °C in a helium atmosphere using an ULVAC-ZEM3 instrument. Thermal conductivity was measured by laser flash method on a TA Instruments Flashline 4000 using a Corning Code 9606 standard.

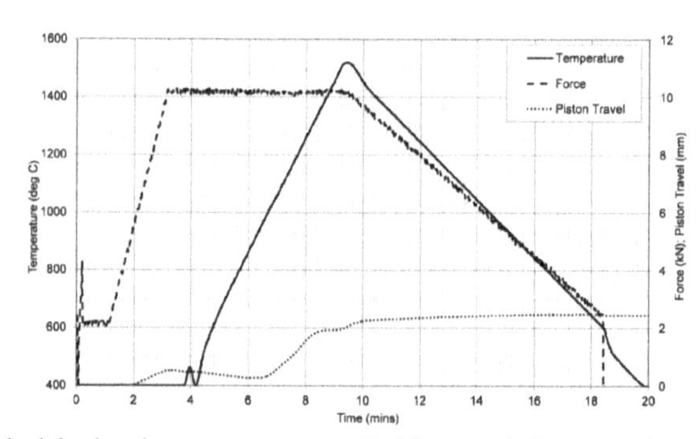

Figure 1. SPS schedule showing temperature, applied force, and piston travel as a function of time.

Scanning electron microscopy (SEM) images and energy dispersive x-ray (EDX) spectra of polished samples of the sintered materials were taken using a Hitachi SU-70 FESEM with an OXFORD Inca EDX system.

RESULTS

Table I is a list of sample compositions and ZT values at 1050K for the sintered materials prepared from as-made (non-reduced) powders, graphite-bed reduced powders, and powders with 5 wt.% TiC addition. Several observations can be made from the data. Considering Compositions A-G, reduction of the powders in the graphite bed clearly produced an increase in ZT over that of materials prepared from the as-made powders. The addition of TiC produced a further significant increase in ZT. Amongst the samples with TiC, ZT increased with the total dopant level (x+y) up to 0.2, then decreased at the higher level. This trend is not apparent in the as-made and reduced-only materials. The proportion of La to Y in these materials does not have a large effect on ZT, although co-doping with La and Y produced higher ZT values than the individual dopants alone (Compositions B-D). The highest ZT at 1050K of 0.30 was obtained with Composition E, $La_{0.1}Y_{0.1}Sr_{0.8}TiO_{3-\delta}$ + 5 wt.% TiC, and Composition F, $La_{0.15}Y_{0.05}Sr_{0.8}TiO_{3-\delta}$ + 5 wt.% TiC. Compositions H-L are Composition F with excess titanium oxide batched before the calcination process step (z > 0). The maximum ZT at 1050K=0.25 was obtained at z=0.09, Composition K.

Table I. Sample compositions (nominal) and ZT at 1050K.

	(LaxYySr1-x-y)1-zTi1+zO3				ZT at 1050K		
No.	(x+y)	x	y	z	As Made	Reduced	5wt% TiC + Reduced
A	0.05	0.03	0.03	0.00	0.06	0.14	0.17
B	0.10	0.10	0.00	0.00	0.01	0.16	0.19
C	0.10	0.00	0.10	0.00	0.05	0.14	0.20
D	0.10	0.05	0.05	0.00	0.04	0.11	0.28
E	0.20	0.10	0.10	0.00	0.02	0.10	0.30
F	0.20	0.15	0.05	0.00	0.07	0.15	0.30
G	0.25	0.15	0.10	0.00		0.24	0.24
H	0.20	0.15	0.05	0.04		0.01	
I	0.20	0.15	0.05	0.06		0.05	
J	0.20	0.15	0.05	0.08		0.21	
K	0.20	0.15	0.05	0.09		0.25	
L	0.20	0.15	0.05	0.11		0.18	

Figure 2 shows thermoelectric properties as a function of temperature for Composition F: as-made, reduced, and with 5 wt.% TiC addition. Electrical conductivity increased as a consequence of the reduction step and still further with the addition of TiC. The electrical conductivity of the latter peaked around 560K, then decreased with increasing temperature. At 1050K, the electrical conductivity of the TiC material was two times greater than that of the material reduced without TiC. Seebeck coefficients were negative for all three materials, indicating n-type conduction, and decreased monotonically with increasing temperature. Reduction increased the Seebeck coefficient at 1050K from -269 μV/K to -195 μV/K, while the addition of TiC produced a relatively small further increase to -179 μV/K. Thus, the power factor of the material with TiC was greater than that of the reduced-only and as-made materials, reaching a maximum of 1.3×10^{-3} W/m.K^2 at 660K and

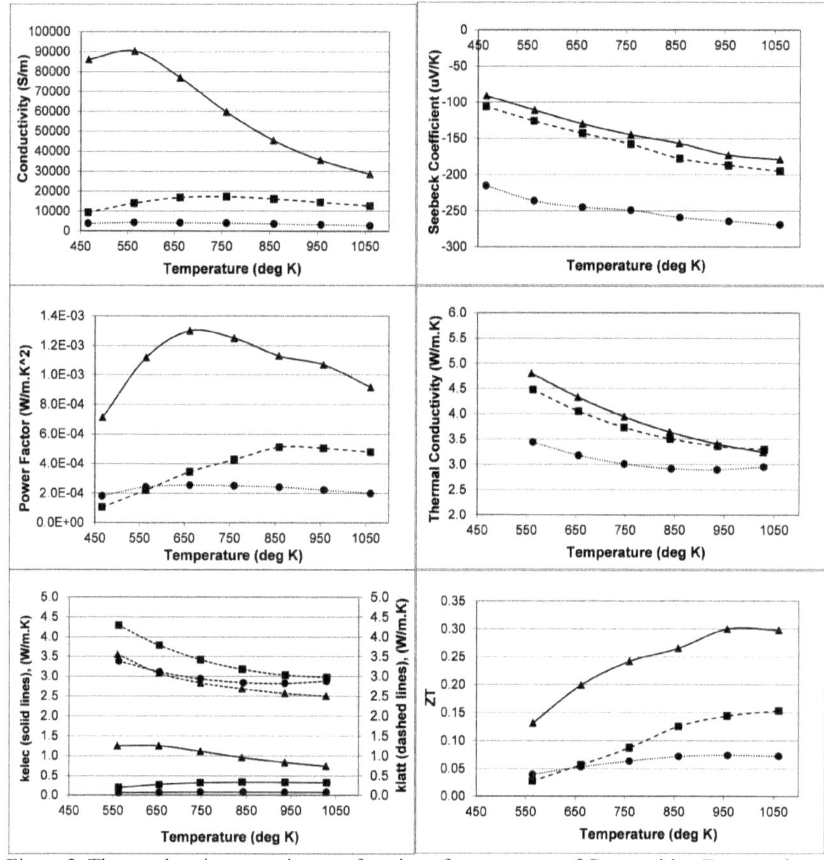

Figure 2. Thermoelectric properties as a function of temperature of Composition F: as-made (circles); reduced (squares); and 5 wt.% TiC addition (triangles).

0.9×10^{-3} W/m.K^2 at 1050K. Thermal conductivity increased with reduction from 3-3.4 W/m.K to 3.3-4.5 W/m.K. Significantly, the thermal conductivity of the TiC material was similar to that of the reduced material without TiC addition. To understand this effect, k_{elec} and k_{latt} were calculated using Equations (2) and (3). Indeed, k_{latt} of the TiC material was lower than that of the reduced material without TiC addition. ZT of both reduced materials was higher than that of the as-made material and increased with temperature up to 1050K. The highest ZT values between 550 and 1050K were obtained with TiC addition.

Another way of expressing the effect of reduction and TiC addition on thermal conductivity is by the ratio of the lattice thermal conductivity to the total thermal conductivity: $k_{latt}/K = k_{latt}/(k_{elec}+k_{latt})$. Figure 3 shows plots of k_{latt}/K as a function of temperature for Composition F as-made, reduced, and with different levels of TiC added: 1.5, 3.5, 5, and 7 wt.%. For the as-made material, k_{latt}/K was close to one. The ratio decreased with reduction and with TiC additions of 3.5-7

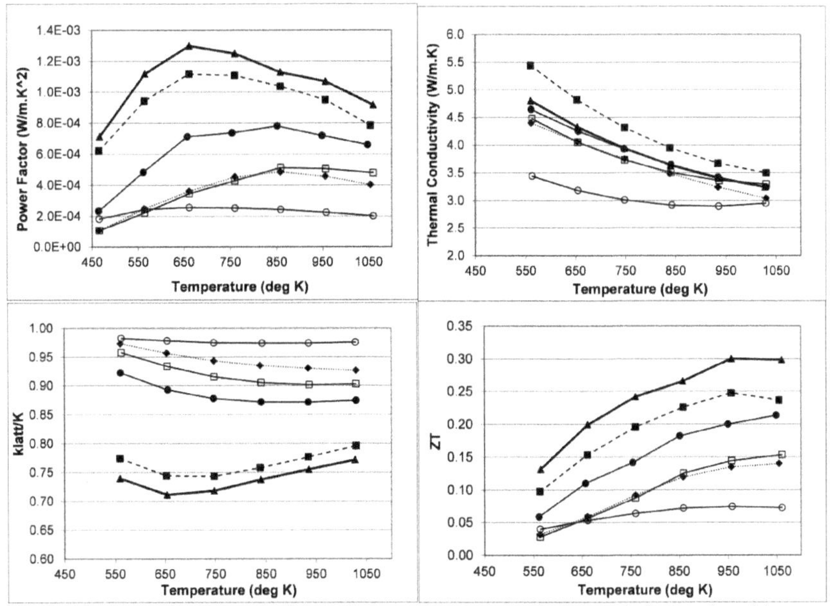

Figure 3. Thermoelectric properties of Composition F: as-made (open circles); reduced (open squares); 1.5 wt.% TiC (diamonds); 3.5 wt.% TiC (squares); 5 wt.% TiC (triangles); and 7 wt.% TiC (circles).

wt.%. The ratio was lowest for 5 wt.% TiC over the entire temperature range explored. At 1050K, $k_{latt}/K=0.77$. ZT and PF were maximized at this composition as well with ZT(1050K) = 0.30.

TiB_2 and TiN additions were found to have an effect similar to TiC. Composition F was prepared with 1.7, 3, 6, and 8 wt.% TiB_2 and with 1.5, 5, and 7 wt.% TiN. Thermoelectric properties as a function of temperature are shown in Figures 4 and 5, respectively. Decreases in k_{latt}/K and increases in PF and ZT over that of the Composition F reduced material from 550 to 1050K were obtained with both additives. This effect was obtained with 5wt.% TiN, but only 1.7wt.% TiB_2 was needed. The highest ZT at 1050K achieved with TiB_2 was 0.33 (3 wt.%) and with TiN was 0.30 (7 wt.%). These compositions also gave the lowest k_{latt}/K at 1050K, 0.75.

Figure 6 shows the thermoelectric properties of Compositions H-L as a function of temperature. A decrease in k_{latt}/K over that of the stoichiometric Composition F and increases in PF and ZT were observed from 550 to 1050K for Compositions K and L. The lowest k_{latt}/K, 0.81, and the highest PF and ZT at 1050K were obtained for Composition K (z=0.09). For this material, ZT (1050K) = 0.25.

The microstructures of the ceramics were examined by SEM and EDX. Figure 7 shows backscattered electron images of Composition F reduced, Composition K reduced, and Composition F with TiN, TiC, and TiB_2 additions. Both the addition of excess titania (Fig. 7b) and the TiX additives (Figs 7c-e) produced a large increase in grain size, from ≤ 5 microns to ≤ 20 microns, compared to the reduced stoichiometric material without additives (Fig. 7a). All of the materials were compositionally inhomogeneous as shown by the variations in gray level in the images. The matrix grains were determined to be La,Y-strontium titanate (see EDX spectrum, Fig. 7f). The gray

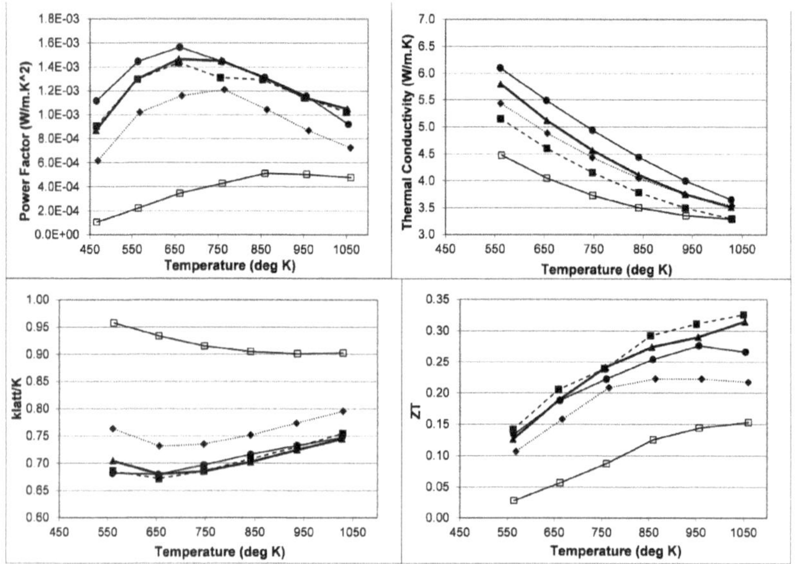

Figure 4. Thermoelectric properties of Composition F: reduced (open squares); 1.7 wt.% TiB_2 (diamonds); 3 wt.% TiB_2 (squares); 6 wt.% TiB_2 (triangles); and 8 wt.% TiB_2 (circles).

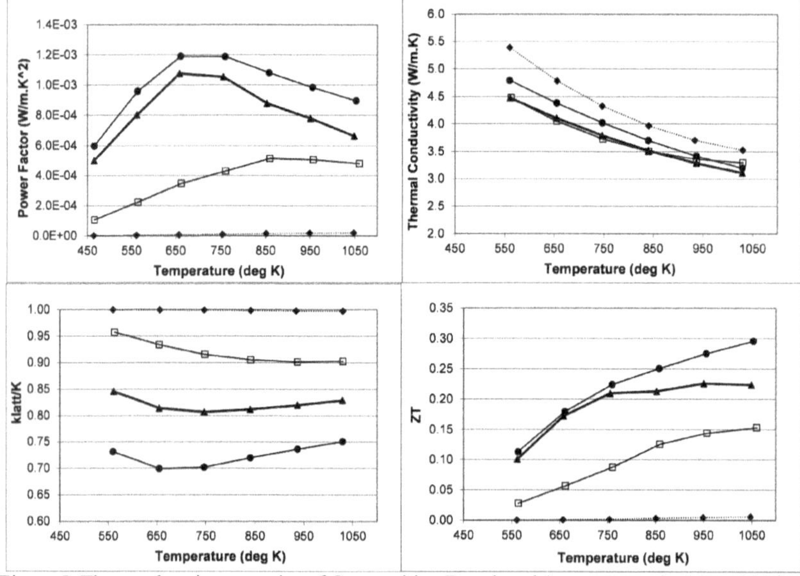

Figure 5. Thermoelectric properties of Composition F: reduced (open squares); 1.5 wt.% TiN (diamonds); 5 wt.% TiN (triangles); and 7 wt.% TiN (circles).

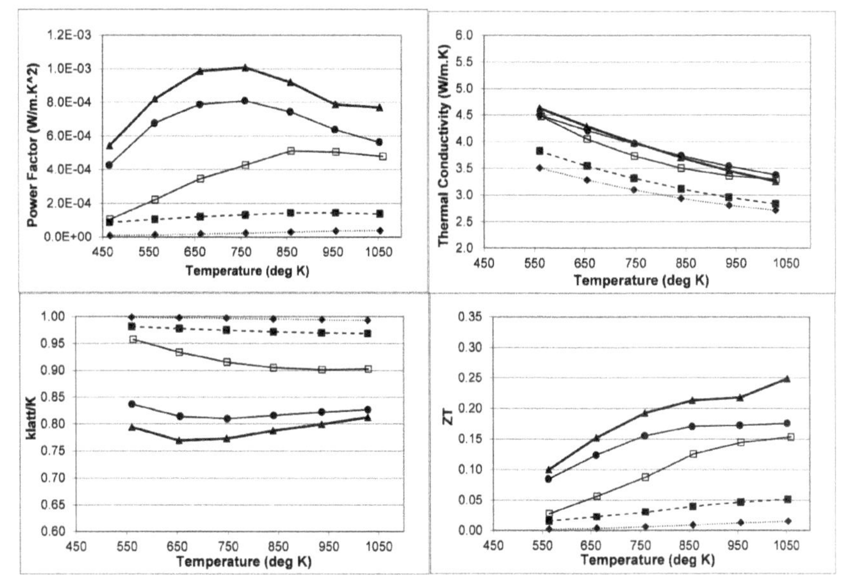

Figure 6. Thermoelectric properties of Compositions F (open squares); H (diamonds); I (squares); K (triangles); and L (circles). All composition reduced.

level gradation within the grains from dark gray to white indicates an increasing level of La enrichment. Areas of Y enrichment were also present. The black intergranular phase was identified as a titania-rich phase. In the excess titania (Fig 7b) and TiC (Fig. 7c) materials, this phase has been identified as a titanate phase containing lower levels of strontium (see EDX spectrum, Fig. 7g). The TiN material (Fig 7d) had a similar microstructure with more of the dark intergranular phase. In contrast, the TiB_2 material (Fig. 7e), appeared much more homogeneous, exhibiting less gray level variation within the matrix grains. No TiC, TiN, or TiB_2 was detected by XRD.

DISCUSSION

The effects of single dopants, Y and La, and oxygen partial pressure (pO_2) on the electrical conductivity and thermoelectric properties of polycrystalline $SrTiO_3$ ceramics have been reported in the literature.[8-10] La^{+3} and Y^{+3} substitute for Sr^{+2} in the perovskite crystal lattice. La^{+3} doping up to x=0.4 in $La_xSr_{1-x}TiO_3$ and Y^{+3} doping up to x=0.1 in $Y_xSr_{1-x}TiO_3$ were demonstrated in ceramics with a single cubic structure as determined by XRD.[9,10] Electrical conductivity increased with increasing dopant concentrations and decreasing pO_2 and is believed to occur through the formation of electrons localized on Ti sites, Ti^{+3}.

In the present study, La+Y co-doping with the introduction of a TiX (X=N, C, B_2) additive was shown to be more effective at improving the thermoelectric properties of strontium titanate than single element doping or co-doping with reduction alone. All of the additives produced increased electrical conductivities and power factors while suppressing lattice thermal conductivity.

A key role of the additives is to lower pO_2 within the graphite bed during the reduction step. This effect can be shown through thermodynamic calculations. Here, the reaction between TiO_2,

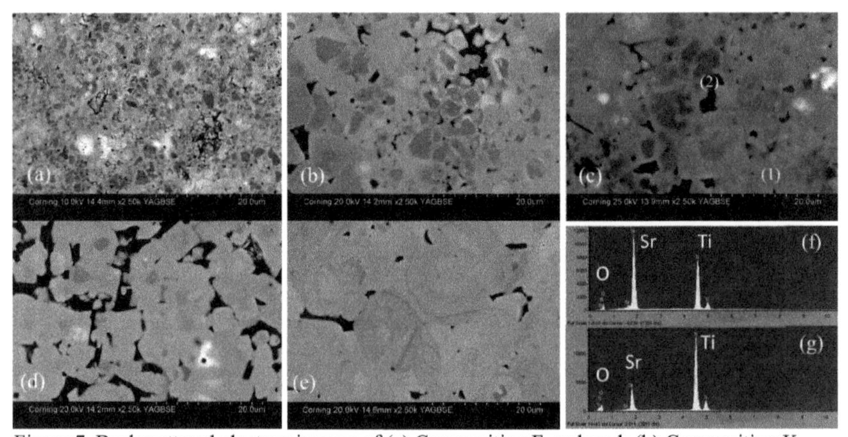

Figure 7. Backscattered electron images of (a) Composition F, reduced; (b) Composition K, reduced; (c) Composition F with 5 wt.% TiC; (d) Composition F with 7 wt.% TiN; and (e) Composition F with 6 wt.% TiB_2. EDX spectrum of (f) area 1 in (c) (matrix phase); (g) area 2 in (c) (black phase).

TiN, TiC, TiB_2 and pure CO(g) at 1400 °C are considered. Factsage 6.2[11] was used to calculate pO_2 for different molar CO:TiX ratios. Only pure phases (gases, liquids, solids) were included in the calculations. The results are shown in Table II. A difference of several orders of magnitude in pO_2 is observed between TiO_2 and the TiX compounds, for which pO_2 hardly varies with CO:TiX. Thus at 1400 °C:

$$pO_2\ (TiB_2) \leq pO_2\ (TiC) < pO_2\ (TiN) \ll pO_2\ (TiO_2).$$

The additives create a more reducing environment than the graphite bed alone and can, therefore, be termed "reductants". The electrical conductivity of $SrTiO_3$ has been shown to increase with decreasing pO_2 through the formation of oxygen vacancies and Ti^{+3}.[12] The addition of these reductants produced higher electrical conductivities and power factors than the graphite bed atmosphere alone, consistent with lower pO_2 during synthesis.

Table II. Thermodynamic calculations of pO_2 for TiO_2, TiN, TiC, and TiB_2 in equilibrium with CO(g) at T=1673K, P(total)=1 atm.[11]

INPUTS (moles)	0.1 CO(g)	1 CO(g)	5 CO(g)	10 CO(g)
1 TiO2	2.4E-11	6.6E-12	1.2E-12	5.9E-13
1 TiN	1.6E-16	1.6E-16	1.6E-16	1.6E-16
1 TiC	7.3E-17	7.3E-17	7.3E-17	7.3E-17
1 TiB2	3.5E-17	3.5E-17	7.3E-17	7.3E-17

This increase in electrical conductivity was not accompanied by a proportional increase in thermal conductivity predicted by Equation (2). The additives, therefore, play a second role of suppressing the lattice component of thermal conductivity as shown in Figures 2-5. It is possible that oxygen vacancies induced by reduction increase the phonon scattering in the crystal lattice.[13] Alternatively, the additives could act as a source of excess titania. Excess TiO_2 dissolved in the strontium titanate crystal would produce oxygen vacancies and Sr-site vacancies which, again, could act as phonon scatterers.[12] When excess titania was batched into the La+Y co-doped strontium titanate, lower k_{latt}/K and higher power factors were obtained (Fig. 6). Since the Ti_2O_3 was batched before the calcination step, it was most likely oxidized and would not have contributed to the reduction of the material during subsequent process steps as the TiX reductants would. The actual compositions of the strontium titanate materials were not measured. Komornicki et. al., however, observed that the grain size of sintered polycrystalline $Sr_xY_{0.02}TiO_3$ (x=0.94-1.02) ceramics varied strongly with composition.[14] Smaller grains were observed for Ti/(Sr+Y) < 1, while larger, more regular-shaped grains were observed for Ti/(Sr+Y) > 1. In the present study, the grain sizes of materials with TiN, TiC, TiB_2 and excess titania additions that showed improved thermoelectric properties all had significantly larger grain sizes than the "stoichiometric" reduced material.

In addition to creating point defects, excess titania can lead to the formation of other phases. The Sr-Ti-O phase diagram shows a region of limited Ti solubility in $SrTiO_3$.[15] Subsequent studies indicated a TiO_2 solubility in $SrTiO_3$ of 100-1000 ppm ($z=5x10^{-5}-5x10^{-4}$).[12,16] Beyond this solubility range, equilibrium phase assemblages of $SrTiO_3$(ss), TiO_2, Magneli phases (Ti_nO_{2n-1}), and other reduced titania compounds are predicted to form depending on the level of oxygen deficiency. SEM micrographs of materials in the present study did reveal multiphase microstructures including inhomogenous grains of La, Y strontium titanate and a titania-rich phase containing strontium. A low-strontium titanate phase does appear on the phase diagram, however the composition of the phase in the present material cannot be assigned at this time. Nevertheless the materials are composites of primarily La,Y-doped strontium titanate and at least one other phase. Bergman et. al., showed that while the power factor of a composite thermoelectric material could exceed that of its components, the figure-of-merit could not.[17,18] Thus, it is suggested that higher ZT might be obtained if more homogeneous materials were produced using the reduction process with added reductants described here.

The lack of TiN, TiC or TiB_2 detectable by XRD in the materials was curious at first. This observation can be explained, however, using the thermodynamic calculations described above. In addition to producing lower pO_2, the calculations show that all three additives can be partially or fully converted to reduced titania, TiC, and carbon in the presence of CO(g) during the reduction step. In the case of TiB_2, gaseous and liquid boron oxide phases may also form. While no TiC or carbon was apparent in the SEM images, TEM examination of Composition F with 5 wt.% TiC did reveal small regions of TiC at triple junctions between grains. Such nanophases would not be detected by XRD, but could act as phonon scatterers, thereby reducing the lattice thermal conductivity.

Finally, some mention of the differences in the amounts of the reductant additives needed to achieve beneficial effects should be made. Addition levels which were effective at lowering k_{latt}/K (and increasing PF and ZT) were lower for TiB_2 than TiC which were in turn lower than TiN. This minimum addition level correlates with their ability to decrease pO_2 (see Table II). That is, TiB_2 is more effective at reducing pO_2 than TiC which is more effective than TiN. An alternative explanation can be made based on the difference in particle sizes between the additives. The TiN powder was <3 microns, TiC <0.3 microns, and the TiB_2 ~60nm. These size differences may have affected the extent of reaction and interdiffusion within the powder compacts during the reduction and sintering steps. Indeed, SEM images of the TiB_2 material did show a more homogeneous microstructure than the other materials at all addition levels. This observation recommends the use of very small particle size reductants (i.e. nanopowders).

CONCLUSION

The addition of a reductant during powder processing has been shown to improve the high temperature (550-1050K) thermoelectric properties of bulk, polycrystalline La,Y-doped strontium titanate ceramics. TiC, TiN, and TiB_2 were all shown to be effective for increasing the power factor while simultaneously suppressing the lattice thermal conductivity. Their effect may be attributed to increased charge carrier concentrations due to lower pO_2 during processing, the generation of point defects and secondary phases. Using these additives, the lattice component of the thermal conductivity was reduced from 97% of the thermal conductivity to 75% at 1050K. Materials with ZT up to 0.33 at 1050K were produced using this process.

ACKNOWLEDGMENTS

The authors thank Dr. Todd St. Clair and Dr. Monika Backhaus for their contributions to this work; Ms. Teresa McDermott and Mr. Andrew Russell for materials preparation; Ms. Kim Work, Ms. Michelle Wallen, Dr. Robert Fretz, and Mr. Ron Davis for materials characterization.

REFERENCES

[1]T.M. Tritt and M.A. Subramanian, Thermoelectric Materials, Phenomena, and Applications: A Bird's Eye View, *MRS Bulletin*, **31**, 188-198 (2006).
[2]G.J. Snyder and E.S. Toberer, Complex Thermoelectric Materials, *Nature Materials,* **7,** 105-114 (2008).
[3]K. Koumoto, I. Terasaki, and R. Funahashi, Complex Oxide Materials for Potential Thermoelectric Applications, *MRS Bulletin,* **31**, 206-210 (2006).
[4]W. Wunderlich, H. Ohta, and K. Koumoto, Enhanced Effective Mass in Doped $SrTiO_3$ and Related Perovskites, *Physica B*, **404,** 2202-2212 (2009).
[5]W. Wunderlich and K. Koumoto, Development of High-Temperature Thermoelectric Materials Based on $SrTiO_3$-layered Perovskites, *Int. J. Mat. Res.*, **97**, 657-662 (2006).
[6]H. Ohta, Thermoelectrics Based on Strontium Titanate, *Materials Today*, **10** [10], 44-48 (2007).
[7]R-Z Zhang, C-L Wang, J-C Li, and K. Koumoto, Simulation of Thermoelectric Performance of Bulk $SrTiO_3$ with Two-Dimensional Electron Gas Grain Boundaries, *J. Am. Ceram. Soc.*, **93**, 1677-1681 (2010).
[8]S. Hui and A. Petric, Electrical Properties of Yttrium-Doped Strontium Titanate under Reducing Conditions, *J. Electrochem. Soc.*, **149**, J1-J10 (2002).
[9]O. Marina, N.L. Canfield, and J. W. Stevenson, Thermal Electrical, and Electrocatalytical Properties of Lanthanum-doped Strontium Titanate, *Solid State Ionics*, **149**, 21-28 (2002).
[10]H. Muta, K. Kurosaki, and S. Yamanaka, Thermoelectric Properties of Rare Earth Doped $SrTiO_3$, *J. Alloys and Compounds,* **350**, 292-295 (2003).
[11]C.W. Bale, E. Bélisle, P. Chartrand, S.A. Decterov, G. Eriksson, K. Hack, I.-H. Jung, Y.-B. Kang, J. Melançon, A.D. Pelton, C. Robelin and S. Petersen, FactSage Thermochemical Software and Databases – Recent Development, *Calphad*, **33**, 295-311 (2009); www.factsage.com.
[12]N.-H. Chan, R.K. Sharma, and D.M. Smyth, Nonstoichiometry in $SrTiO_3$, *J. Electrochem. Soc.*, **128**, 1762-1769 (1981).
[13]C. Yu, M.L. Scullin, M. Huijben, R. Ramesh, and A. Majumdar, Thermal Conductivity Reduction in Oxygen-Deficient Strontium Titanates, *Appl. Phys. Lett.*, **92,** 191911-1-3 (2008).
[14]S. Komornicki, S. Kozinski, and M. Rekas, The Influence of Cationic Ratio on Dielectric Properties of Yttrium Doped $SrTiO_3$ Based Ceramics, *Solid State Ionics*, **39**, 159-162 (1990).

[15]G.J. McCarthy, W.B. White, and Rustum Roy, Phase Equilibria in the 1375°C Isotherm of the System Sr-Ti-O, *J. Am. Ceram. Soc.*, **52**, 463-467 (1969).
[16]S. Witek, D.M. Smyth, and H. Pickup, Variability of the Sr/Ti Ratio in $SrTiO_3$, *J. Am. Ceram. Soc.*, **67**, 372-375 (1984).
[17]D. J. Bergman and O. Levy, Thermoelectric Properties of a Composite Medium, *J. Appl. Phys.*, **70**, 6821-6833 (1991).
[18]D.J. Bergman and L.G. Fel, Enhancement of Thermoelectric Power Factor in Composite Thermoelectrics, *J. App. Phys.*, **85**, 8205-8216 (1999).
[19]C.A. Wert and R.M. Thomson, *Physics of Solids*, 2nd edition, pp. 243-244. McGraw-Hill, New York, 1970.
[20]W.J. Tropf, M.E. Thomas, and T.J. Harris, "Properties of Crystals and Glasses," *Handbook of Optics*, Vol. II, 2nd edition, p. 33.42. Edited by M. Bass, E.W. Van Stryland, D.R. Williams, and W.L. Wolfe. McGraw-Hill, New York, 1995.

Photovoltaic Materials and Technologies

DENSIFICATION AND PROPERTIES OF FLUORINE DOPED TIN OXIDE (FTO) CERAMICS BY SPARK PLASMA SINTERING

Meijuan Li [1,2], Kun Xiang[1], Qiang Shen [1], Lianmeng Zhang [1]
[1]. Wuhan University of Technology
Wuhan, Hubei, China
[2]. University of California, Davis
Davis, California, U S

ABSTRACT:

Fluorine doped tin oxide (FTO) has been recognized as a very promising transparent conductive material which is used in a wide range of devices. For the preparation of high quality FTO thin film by magnetron sputtering or pulse laser deposition (PLD) technique, full dense FTO target materials are fabricated. The FTO powders (fluorine concentration is set according to the mole ratio of [F]/[Sn]=0.1~0.3) are synthesized and then annealed at 400°C to ensure the doping fluorine into the tin oxide lattice. And nearly full-densified FTO ceramics are sintered via spark plasma sintering (SPS). During the rapid sintering process, the limited vaporization of fluorides is beneficial to the sintering of tin oxide by vapor transport, while it brings about only few pores in the sinters. The relative density of SPS sintered FTO ceramics can be higher than 98.5% of the theoretical density when the samples are sintered at 850~900°C. The microstructures observed by scanning electronic microscope illustrate that the grain growth in SPS as-sintered FTO ceramics is obvious. These results contribute to a relative low electrical resistivity of 0.012Ω·cm and high hall mobility of 15 $cm^2/(V \cdot s)$ for FTO ceramics.

INTRODUCTION

Transparent conducting oxides (TCOs) are a class of electrical conductive materials with comparably low absorption of electromagnetic waves within the visible region of the spectrum. They are usually prepared with thin film technologies and used in opto-electrical devices. Fluorine doped tin oxide (FTO) has been recognized as a very promising TCO material which is used in a wide range of devices, including applications as opto-electronics, touch screen displays, thin film photovoltaics, energy-saving windows, RFI/EMI shielding and other electro-optical applications [1]. Outstanding optical characteristics have been provided by tin-, indium- and zinc oxides. And indium tin oxide (ITO) is the most widely used transparent conducting oxide because of its high electrical conductivity and optical transparency [2]. However, the constricted supply of indium has induced that TCO materials formed from inexpensive and abundant starting materials become the focus of many researches. FTO is a possible alternative to ITO because it combines good electrical conductivity and high visible transmission, and it is inexpensive as well as chemically and thermally stable [3, 4].

In the recent years, various techniques have been developed for TCO thin film deposition. The various CVD techniques [5-8], rf sputtering [9, 10], sol-gel [11], pulsed laser ablation [12], and magnetron sputtering [13-16] are some of the processing methods for the deposition of the SnO_2-based thin films, and the FTO thin film coating is mainly deposited by CVD techniques [6-8]. In recent years, other techniques, such as inkjet printing [17] and RF magnetron sputtering [18] have been developed to prepare FTO thin film. Among these deposition techniques, magnetron sputtering is a widely used method in industry because it can perform the coating of large area substrates at a high rate with competitive costs. To prepare a homogeneous and dense film by magnetron sputtering or pulse laser deposition (PLD) technique, high quality targets are needed. As far as we know, most of F sources for magnetron sputtering is fluorides, such as

HF[10], NH_4F[17] and SnF_2 [18], and there is no report about the preparation of FTO ceramic targets. In this present work, we synthesize FTO powders, and consolidate FTO ceramics via SPS technique. The sintering behavior of FTO powders and the effects of fluorine concentration on the phase compositions, density, microstructure and electrical conductivity of as-sintered FTO ceramics are investigated.

EXPERIMENTAL PROCEDURES

The initial materials are commercially available tin(□) oxide (SnO_2, purity≥99.5%, average diameter is about 0.2μm) from Sinopharm Chemical Reagent Co., Ltd, Shanghai and Tin(II) fluoride (SnF_2, purity≥99%) from Alfa Aasar, Tianjin. To synthesize the F-doped tin oxide powders, appropriate amounts of SnO_2 and SnF_2 powder were well mixed by grinding and the fluorine concentration is set according to the mole ratio of [F]/[Sn]=0.1~0.3. The mixture was pressed into discs of 30 mm diameter, and then calcined in furnace at 400°C for 2h.

FTO samples were consolidated in a spark plasma sintering system (SPS-1050). The FTO powders were loaded in a 20mm inner diameter graphite die, and then the die was transferred into the SPS machine. The pulse pattern was kept constant and consisted of twelve pulses (with a pulse duration of 3.3 ms) followed by two periods of zero current. The applied axial pressure on the punches was 40MPa during sintering. All samples were sintered at 850°C or 900°C with a heating rate of 100 °C/min and dwell durations up to 3min in vacuum (~20 Pa).

The density of the as-sintered FTO samples was measured by Archimedes displacement method with distilled water as the immersion medium. The phase compositions of the FTO powders and the as-sintered ceramics were identified by X-ray diffraction (XRD, Rigaku Ultima III, Japan) with Cu Kα radiation (λ=0.15418nm). Microstructures of FTO ceramics were observed by scanning electrons microscopy (SEM, S-3400, Hitachi, Japan) and the grain size was reckoned by image-j software. The electrical conductivity of sintered FTO samples was measured at room temperature by the Hall effect measurement system (HL 5500PC, Accent Optical, UK), and the sintered FTO samples were processed into plates with $10\times10\times1$ mm^3 size for the measurement.

RESULTS AND DISCUSSIONS

Densification and Sintering Behaviors of FTO powders

Figure 1 shows the Z-axis displacement and displacement rate (dZ/dt) of FTO samples during the SPS process when their sintering temperature is set at 900°C. Because the displacement includes the expansion-shrinkage of the graphite die, it is not the exact expansion-shrinkage of FTO samples. However, these curves can reflect the initial sintering temperature and densification procedure of the FTO powders. The densification procedure begins when the displacement rate changes from negative value to positive value ($dZ/dt\geqslant0$ means that the sample is in the shrinkage stage). For FTO powders, the densification procedure starts at about 800°C and the shrinkage stage slightly begins earlier and ends later with increasing fluorine concentration. There is no obvious shrinkage at the dwell duration of sintering temperature. Figure 2 shows the relative density of FTO samples as-sintered at the temperature of 850°C and 900°C respectively, and the theoretical density of FTO is taken to be 6.993 g/cm^3 according to the density of cassiterite SnO_2 (PDF No. 41-1445). The results indicate that the consolidation of FTO powders with small fluorine content needs a higher sintering temperature, while the sample with more fluorine concentration ([F]/[Sn]=0.3) obtains higher density when it is sintered at the lower temperature of 850°C. Fluorine concentration remarkably affects the densification of FTO ceramics. The FTO sample with the mole ratio of [F]/[Sn]=0.2 obtains the highest relative density of 98.6% when it is consolidated at the sintering temperature 900°C.

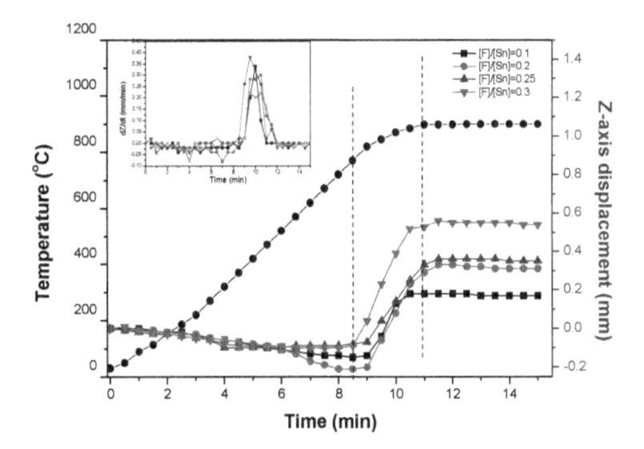

Figure 1. Z-axis displacement and displacement rate curves of FTO samples during the SPS process (consolidated at 900°C)

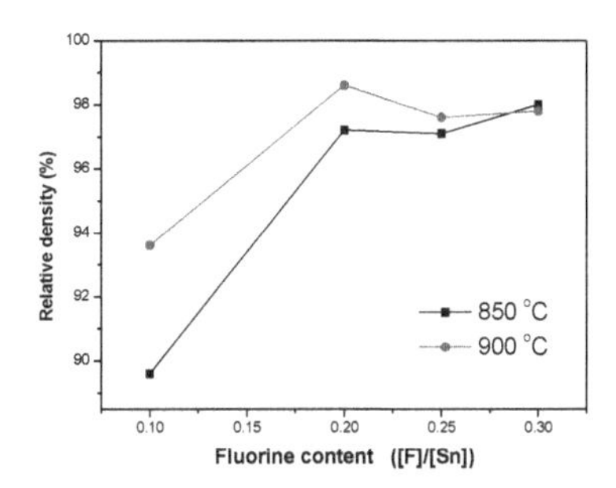

Figure 2. Relative density of SPS as-sintered FTO ceramics

Phase Compositions and Microstructure of FTO ceramics

It is well known that SnF_2 is a fluoride with low melting point and boiling point. Figure 3a shows the XRD patterns of initial materials for FTO synthesis and the as-synthesized FTO powder ([F]/[Sn]=0.2). It can be seen that the synthesized FTO powder consists of cassiterite SnO_2 (PDF No. 41-1445), and tin(II) fluoride (PDF No. 15-0744) which diffraction peaks are different from that of raw material SnF_2. Figure 3b illustrates the XRD patterns of FTO ceramics sintered at the temperature of 900 °C. In the FTO samples with small fluorine concentration ([F]/[Sn]=0.1), there is no tin fluoride phase exist. However, the diffraction peaks of tin fluorides($Sn_{10}F_{34}$, PDF No.30-1371 and SnF_2, PDF No. 15-0744) become obviously with increasing fluorine content. The difference is the tin oxidation state in the fluorides. This indicates that only a part of the doped fluorine ions is cooperated into the tin oxide lattice through substituting the oxygen ions, and the other exists as tin fluorides. Some Sn^{2+} ions are oxidized and converted into higher valence ions during the SPS process. Because the radius of F^- ion (0.133nm) is very similar to that of O^{2-} ion (0.140nm), the F^- substitution for O^{2-} can not lead to an obvious change of the SnO_2 lattice parameters. Therefore, the diffraction peaks of cassiterite SnO_2 do not appear any shift despite some fluorine ions are doped into the SnO_2

lattice.

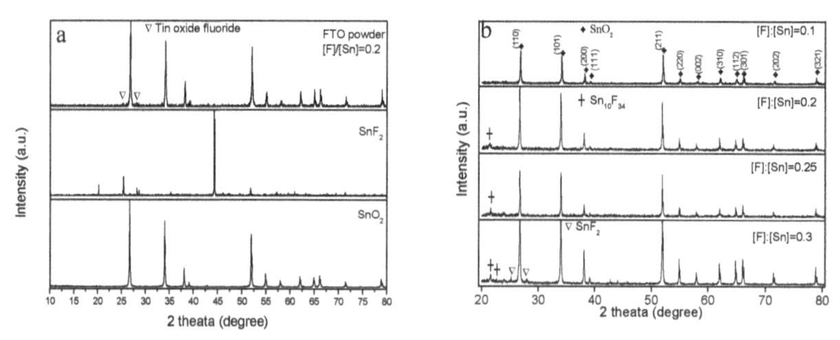

Figure 3. XRD patterns of (a) initial materials, FTO powder and (b) as-sintered FTO ceramics

The SEM micrographs of the consolidated FTO ceramics with different fluorine doping concentrations are provided in Figure 4, and the sintering temperature of FTO samples is 900°C. It is observed that the as-sintered FTO ceramics are nearly full-densified. In the microstructures of FTO ceramics, the tin oxide grains adjoin closely and only few pores exist. More than 200 grains were measured to reckon the grain size in FTO microstructures. The grain sizes are 1.08±0.60μm, 1.15±0.57μm, 0.99±0.66μm and 1.31±0.64μm for the FTO samples with [F]/[Sn]=0.1, 0.2, 0,25 and 0.3 respectively. The grain size in FTO sinters is larger than the particles' size of initial SnO_2 powder (average diameter is about 0.2μm). These results indicate that the grains of tin oxide grow in the annealing and spark plasma sintering process. Because tin fluoride melts at low temperature (its melting point is 213 °C) and vaporizes at higher temperature, the generation of liquid phase or vaporization is beneficial to mass transfer by liquid and/or vapor transport. Therefore, the grain growth in SPS sintered FTO samples is obvious, and the region of grain size distribution broadens with increasing fluorine concentration.

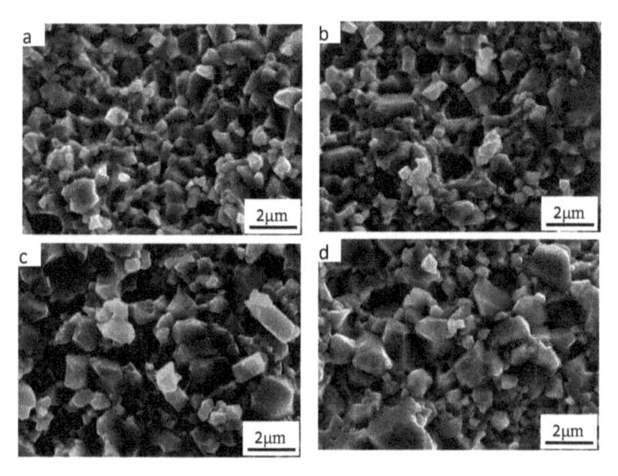

Figure 4. SEM micrographs for FTO ceramics with different fluorine content
(Sintered at 900 °C and the ratio of [F]/[Sn] is:a 0.1, b 0 .2, c 0.25 and d 0.3)

Electrical Properties of FTO Ceramics

The electrical properties of the as-sintered FTO ceramics with different F doping concentration are characterized by applying the van der Pauw method via a Hall effect measurement system. FTO ceramic is an n-type semiconductor. In the F-doped SnO_2 lattice, F^- ions substituted for O^{2-} ions act as donors and create electrons. The resistivity (ρ), Hall mobility (μ) and carrier concentration (n) for FTO samples sintered at 900 °C are listed in Table I. It can be seen that their carrier concentration is on the order of $10^{19}cm^{-3}$, and the carrier concentration slightly increases with increasing F doping concentration. Their resistivity is on the order of $10^{-2}\Omega\cdot cm$, it is superior to that of the SnO_2:F thin films (with similar F concentrations) deposited by RF magnetron sputtering[18]. Meanwhile, their Hall mobility is related to the density and microstructure of FTO samples. The highest Hall mobility is about 15 $cm^2/(V\cdot s)$ for the samples with fluorine doped ratio of [F]/[Sn]=0.2-0.25. The lowest bulk resistivity is 0.01224 Ω·cm, which is obtained when the F doping concentration is [F]/[Sn]= 0.25.

Table I. Electrical Conductivity Properties of FTO Ceramics (Sintered at 900 °C)

No	Mole Ratio of [F]/[Sn]	Bulk Resistivity (Ω·cm)	Hall Mobility ($cm^2\cdot V^{-1}s^{-1}$)	Bulk Carrier Concentration ($10^{19}cm^{-3}$)
1	0.10	0.0909	4.1	1.67
2	0.20	0.0227	14.4	1.92
3	0.25	0.01224	18.8	2.72
4	0.30	0.0583	3.7	2.89

CONCLUSIONS

FTO powders with fluorine doping concentration of [F]/[Sn]=0.1~0.3 are synthesized, and FTO ceramics with high density and high electrical conductivity are fabricated by spark plasma sintering. The highest relative density of SPS as-consolidated FTO ceramics is higher than 98.5% of the theoretical density when the samples are sintered at 850~900°C. The microstructure of as-sintered FTO ceramics indicates that SnO_2 grains grow obviously in the rapid SPS process. The grain growth is attributed to the liquefaction and vaporization of fluorides, which is beneficial to the mass transfer and diffusion. These results contribute to a relative low electrical resistivity of 0.012Ω·cm and high hall mobility of 15 $cm^2/(V\cdot s)$ for FTO ceramics.

ACKNOWLEDGMENTS

This work is financially supported by the Research Fund of International Science and Technology Cooperation Program of China (No.2011DFA52650) and the State Scholar Fund of China.

REFERENCES

[1]P. Gerhardinger and D. Strickler, Fluorine Doped Tin Oxide Coatings – Over 50 Years and Going Strong, *Key Engineering Materials,* **380**, 169-78 (2008).

[2]A. Stadler, Transparent Conducting Oxides – An Up-To-Date Overview, *Materials*, **5**, 661-83 (2012).

[3]C. G. Granqvist, Transparent Conductors as Solar Energy Materials: A Panoramic Review, *Sol. Energy Mater. Sol. Cells*, **91**, 1529-98 (2007).

[4]W. H. Baek, M. Choi, T. S. Yoon, H. H. Lee, and Y. S. Kim, Use of Fluorine-doped Tin Oxide Instead of Indium Tin Oxide in Highly Efficient Air-Fabricated Inverted Polymer Solar Cell, *Appl. Phys. Lett.,* **96**, 133506-1-3 (2010).

[5]J.R. Brown, P.W. Haycock, L.M. Smith, A.C. Jones, and E.W Williams, Response Behaviour of

Tin Oxide Thin Film Gas Sensors Grown by MOCVD, *Sensors and Actuators B: Chemical,* **63**, 109-14 (2000).
[6]C. G. Borman and R. G. Gordon, Reactive Pathways in the Chemical Vapor Deposition of Tin Oxide Films by Tetramethyltin Oxidation, J. Electrochem. Soc., 136, 3820-8 (1989)
[7]M. A. Olopade, O. E. Awe, A. M. Awobode, and N. Alu, Characterization of SnO_2:F Films Deposited by Atmospheric Pressure Chemical Vapour Deposition for Optimum Performance Solar Cells, *The African Review of Physics*, **7**, 177-81 (2012).
[8]N. Noor and I. P. Parkin, Enhanced Transparent-conducting Fluorine-doped Tin Oxide Films Formed by Aerosol-Assisted Chemical Vapour Deposition, *J. Mater. Chem. C,* **1**, 984-96 (2013).
[9]S. Boycheva, A. K. Sytchkova, M. L. Grilli, and A. Piegari, Structural, Optical and Electrical Peculiarities of R. F. Plasma Sputtered Indium Tin Oxide Films, *Thin Solid Films*, **515**, 8469-73 (2007).
[10]A. N. Banerjee, S. Kundoo, P. Saha, and K. K. Chattopadhyay, Nanocrystalline Fluorine Doped SnO2 Thin Films by Sol-gel Method, *J. Sol–Gel Sci. Technol.*, **28**, 105-10 (2003).
[11]J. H. Kim, K. A. Jeon, G. H. Kim, and S. Y. Lee, Electrical, Structural, and Optical Properties of ITO Thin Films Prepared at Room Temperature by Pulsed Laser Deposition, *Appl. Surf. Sci.*, **252**, 4834-7 (2006).
[12]J. Ma, X. T. Hao, H. L. Ma, and X. G. Xu, RF Magnetron Sputtering SnO_2:Sb Films Deposited on Organic substrates, *Solid State communications*, **121**, 345-9 (2002).
[13]B. Yoo, K. Kim, S. H. Lee, W. M. Kim, and N. G. Park. ITO/ATO/TiO_2 Triple-Layered Transparent Conducting Substrates for Dye-sensitized Solar Cells, *Solar Energy Materials & Solar cells*, **92**, 873-7 (2008).
[14]J. Boltz, D. Koehl, and M. Wuttig, Low Temperature Sputter Deposition of SnO_x:Sb Films for Transparent Conducting Oxide Applications, *Surface & Coating Technology*, **205**, 2455-60 (2010).
[15]E. Medvedovski, N. Alvarez, O. Yankov, and M. K. Olsson, Advanced Indium-tin Oxide Ceramics for Sputtering Targets, *Ceramic International*, **34**, 1173-82 (2008).
[16]D. J. Kwak, B. H. Moon, D. K. Lee, C. S. Park and Y. M. Sung, Comparison of Tansparent Conductive Indium Tin Oxide, Titanium-Doped Indium Oxide, and Fluorine-Doped Tin Oxide Films for Dye-Sensitized Solar Cell Application, *Journal of Electrical Engineering & Technology*, **6**(5), 684-7 (2011).
[17]W. Z. Samad, M. M. Salleh, A. Shafiee, and M. A. Yarmo, Structural, Optical and Electrical Properties of Fluorine Doped Tin Oxide Thin Films Deposited Using Inkjet Printing Technique, *Sains Malaysiana*, **40**(3), 251-7 (2011).
[18]F. de Moure-Flores, A. Guillen-Cervantes, K. E. Nieto-Zepeda, J. G. Quinones-Galvan, A. Hernandez-Hernandez, M. de la L. Olvera, and M. Melendez-Lira, SnO_2:F Thin Films Deposited by RF Magnetron Sputtering: Effect of the SnF_2 Amount in the Target on the Physical Properties, *Rev. Mex. Fis.*, **59**, 335-8 (2013).

INTERFACIAL CHARACTER AND ELECTRONIC PASSIVATION IN AMORPHOUS THIN –FILM ALUMINA FOR Si PHOTOVOLTAICS

L.R. Hubbard[1], J.B. Kana-Kana[1], and B.G. Potter, Jr.[1, 2]

[1]Department of Materials Science and Engineering
[2]Optical Sciences Center
University of Arizona
Tucson, AZ 85721

ABSTRACT

The development of Si photovoltaic architectures using n-type base elements has prompted the investigation of alumina thin films as alternative passivation coatings for p-type Si to enhance photocarrier extraction and improve overall energy-conversion efficiency. The relationship between interfacial chemistry and nanostructure and electronic passivation performance was examined in amorphous alumina films, grown using a high-throughput plasma enhanced chemical vapor deposition (PECVD) method onto p-type Si wafers. The specimens were subjected to a range of post-deposition isothermal annealing treatments. Minority carrier lifetime (τ) was measured using resonance-coupled photoconductive decay (RCPCD) and was related to the evolution of interfacial roughness as well as near-interface oxygen-aluminum ratio throughout the iterative thermal treatments. An annealing time of 6 minutes at 500°C under a nitrogen atmosphere produced the greatest enhancement in both fixed space charge at the interface and carrier lifetime observed in this study, consistent with a field-based passivation response. From the correlation established between passivation performance and interfacial structure and chemistry, a mechanistic interpretation of the relationship between thermal processing, nanostructure, and passivation-related properties is offered in the context of an alumina passivation coating produced using an industrial-scale synthesis method.

INTRODUCTION

Passivation of p-type Si in alternative Si photovoltaic architectures incorporating an n-type Si base represents a key challenge in retaining the energy conversion efficiency of such devices. The use of an n-type Si base, offers reduced sensitivity to Si impurities (e.g. Fe) [1] and increased resistance to light induced degradation [2] within the device. A major challenge in the realization of n-type Si base photovoltaics is the identification of suitable passivation coatings for the p-type Si top layer that will successfully provide enhanced charge recovery and improved PV conversion efficiency while remaining industrially viable.

Commonly used passivation coatings for n-type silicon include a-SiN_x:H and thermally grown oxide (SiO_x–H). However, these options are not as effective as amorphous alumina (a-alumina) for the passivation of p-type Si [3-11]. In this case, passivation is associated with the development of a fixed space charge, i.e. field-based passivation. Alumina furnishes a local negative fixed space charge at the alumina-Si interface, resulting in the repulsion of minority carriers (electrons) in the p-type Si emitter layer [4]. The resulting charge separation within the near interface region of the Si leads to a reduced probability for recombination at interfacial trapping sites and an increased minority carrier lifetime. The increased efficiency of charge extraction thereby improves energy-conversion efficiency in the photovoltaic device [12].

Highly uniform, a-alumina ceramic thin films exhibiting superior p-type Si passivation performance have been successfully produced via high throughput, plasma-enhanced chemical vapor deposition (PECVD) [13]. This continuous-deposition method, employing a linear plasma source, was previously shown to produce alumina passivation films with a minority carrier lifetime in excess of 1000 μs on p-type Si [13].

Previous work examining lab-scale chemical vapor deposition (CVD) and atomic layer deposition (ALD) deposition for the production of amorphous alumina films on c-Si have been pursued to investigate processing methods and mechanisms contributing to the passivation effect [12, 14]. These efforts have provided some insight into the underlying structural and chemical parameters contributing to the behavior, primarily associated with modification in point defect and/or coordination environments of species within the alumina films [15-17]. Moreover, these studies typically associate a significant contribution to passivation behavior with a SiO_x interlayer often formed at the interface [12].

In the present work, an iterative isothermal annealing processing routine is used to follow the evolution of the Si-alumina interfacial nanostructure and chemistry in conjunction with measurements of minority carrier lifetime and fixed space charge. Through this approach, the investigation identifies significant correlations between nano-scale interfacial structural and chemical character and the development of passivation-related electronic properties in systems that do not exhibit a SiO_x interfacial layer. Subsequent discussion endeavors to develop a mechanistic link between new evidence for nanoscale roughening and the resulting passivation performance thus providing an opportunity to optimize deposition and processing methods for the development of improved energy conversion efficiency in Si PV cells. Further, the study is performed in the context of production-scale deposition and subsequent thermal processing methods, providing important insight into the potential for material processing to contribute to the development of industrially relevant, high performance passivation coatings.

EXPERIMENTAL METHODS

Material Deposition:

Amorphous alumina thin films (thickness = 20±1.3 nm) were deposited using a linear PECVD plasma source (Linear PECVD™ Reactor, General Plasma, Inc.). The aluminum precursor, tri-methyl aluminum (TMA), was reacted with an oxygen plasma at a substrate temperature of 325 C. The reactant flow rate ratio was maintained at 4:1 (TMA: O_2), consistent with previous reported work [13], producing a process pressure of 20 mT. Additional details of the deposition approach and conditions used can be found in Reference 13. The in-line deposition rate was 50 nm-m/min for the continuous deposition process. The reaction resulted in a solid alumina layer with a mixture of hydrocarbon effluent gases. The alumina films were deposited onto 4-inch diameter, float-zone (FZ), p-type (boron-doped) silicon wafers, (wafer thickness = 280-300 μm, double-side polished). Si substrate resistivity was 3-5 Ω-cm. Prior to deposition, the substrates were subjected to a dilute HF (1 mM, 2 min) etch and deionized water rinse to remove the native oxide on the silicon surface [13].

Thermal Processing:

After deposition, 1x1" samples were cut from the as-deposited wafers and subjected to an iterative, isothermal annealing series at 500°C under flowing N_2 (99.999 % purity, 10 ml/min). Using a 2" diameter fused-silica-lined muffle tube furnace. The annealing temperature and atmosphere are consistent with optimized annealing conditions previously demonstrated for these materials [13] and in previously reported studies [18-24]. The present study thus intends to elucidate the underlying materials-related processes contributing to the passivation behavior of films produced under these conditions. Specimens, supported on a fused silica boat during annealing, were extracted for characterization at 2 minute intervals to a maximum annealing time of 12 minutes. The finite thermal mass of the specimen and fused silica boat, coupled with disturbances in the thermal environment associated with insertion of the specimen into the silica muffle tube furnace, dictated a minimum annealing time step that could be reliably employed while still assuring reasonable time for specimen temperature equilibration. Based on the temperature recovery times observed near the specimen under the conditions used, a minimum annealing time step of two minutes was determined.

Electronic, structural, and chemical analyses were performed after each 2 minute annealing period to characterize the electronic passivation performance and corresponding interfacial characteristics. Observation of significant modification in electronic behavior during the iterative isothermal processing regimen was used to focus more involved interfacial structural and chemical examination on specimens after specific annealing times. For these cases, smaller specimens for transmission electron microscopy and associated chemical analysis were cut from the primary annealing samples as needed to provide corresponding nanostructural information. The annealing schedule and associated characterizations were performed on a minimum of six specimens produced from multiple separate depositions. Additional detail regarding sampling method is provided below.

Electronic Characterization:

Resonance-coupled photoconductive decay (RCPCD) was employed to measure minority carrier lifetime (τ) in the near interface region of the specimens [18]. Specimens were excited with a pulsed solid-state laser (λ = 532 nm, 1064 nm; pulse length = 1.2 ns (synchronous pulses), photodensity at sample = $5.5*10^{13}$photons/cm^2 (per wavelength)). The time-decay (lifetime) of the resulting photoexcited carrier population was obtained by a single-term exponential fit of the RF impedance decay curve [13, 18, 25-30].

Capacitance-voltage (C-V) measurements [20, 21] of the fixed space charge at the interface were performed at each anneal iteration using a mercury probe (Model 802B-150 MDC™ Inc.). A mercury work function of 4.53eV was used for data analysis. Measurements were taken for each sample over 5 days and then averaged to determine the fixed space charge at each annealing interval.

Structural Characterization:

Nanostructural characterization of the a-alumina-Si interface in selected samples was obtained using high-resolution, cross-sectional transmission electron microscopy (HRTEM) (JEOL JEM 4000EX). With a point-to-point imaging resolution of 0.17 nm, the technique allowed an analysis of atomic-scale interfacial roughness evolution with annealing time. Cross-sectional TEM samples were produced via a gallium-based focused ion beam processing. The resulting sample dimensions were 100 μm by 100 μm with a thickness of approximately 100 nm. Image processing software (ImageJ™) was used to measure the root-mean-square (rms) interfacial roughness observed in the HRTEM cross-section analysis.

Electron-beam chemical analysis was pursued using scanning transmission electron microcopy (STEM) with energy dispersive x-ray spectrometry (EDS) (JEOL ARM200F). The instrument has a spot-size of 0.08 nm. Specimen thickness and associated interaction volume provide an EDS spatial resolution of approximately 1 nm. EDS point scans on the cross-section samples were taken at 2 nm intervals along an 8 nm line normal to the alumina-Si interface to provide a limited compositional depth profile. Each depth profile thus included two point scans within the Si, two within the alumina and a point centered over the interface itself for each specimen analyzed. All EDS spectra were reproduced 3 times varying the location along the Si-alumina interface on each sample. EDS peak areas were used to monitor modifications in local chemical stoichiometry with anneal time. Multiple (3-4) specimens from different silicon wafer substrates were analyzed at each annealing time to examine the reproducibility of trends observed.

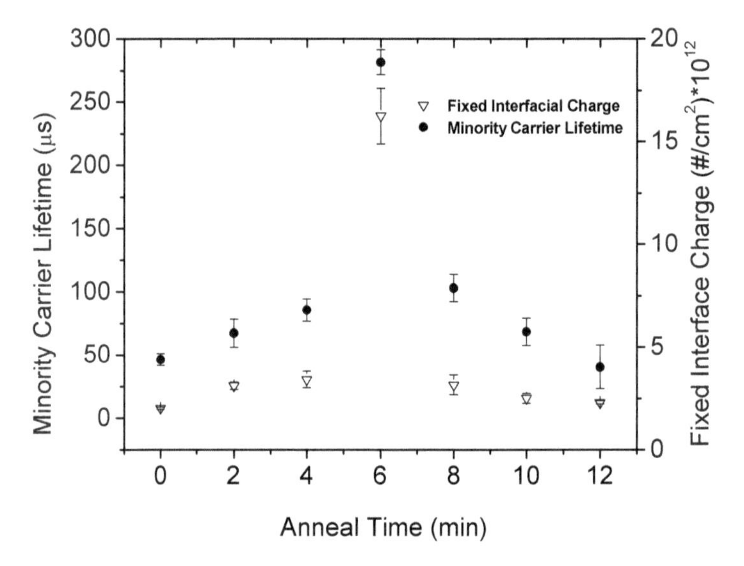

Figure 1. Minority carrier lifetime (τ) (from RCPCD) and fixed interface charge density (from the C-V measurement) as functions of annealing time at 500°C for FZ Si substrate with a ~20 nm thick PECVD a-alumina thin layer. The plot depicts the minority carrier lifetime obtained via RCPCD (square symbols) and the fixed interface charge (circles). Error bars represent the 95% confidence levels for 12 specimens (from 6 independent depositions).

RESULTS

Prior to PECVD a-alumina deposition, measured carrier lifetimes for the HF washed wafers were 6±0.8 μs consistent with that anticipated for an unpassivated wafer. Post deposition, the lifetimes improved to 15±1.3μs while after a 6min anneal at 500°C, a carrier lifetime of 238±25 μs was achieved. For longer annealing times, carrier lifetimes decreased with a 12 min total anneal time resulting in a lifetime value of only 18±0.7 μs. The data thus shows an

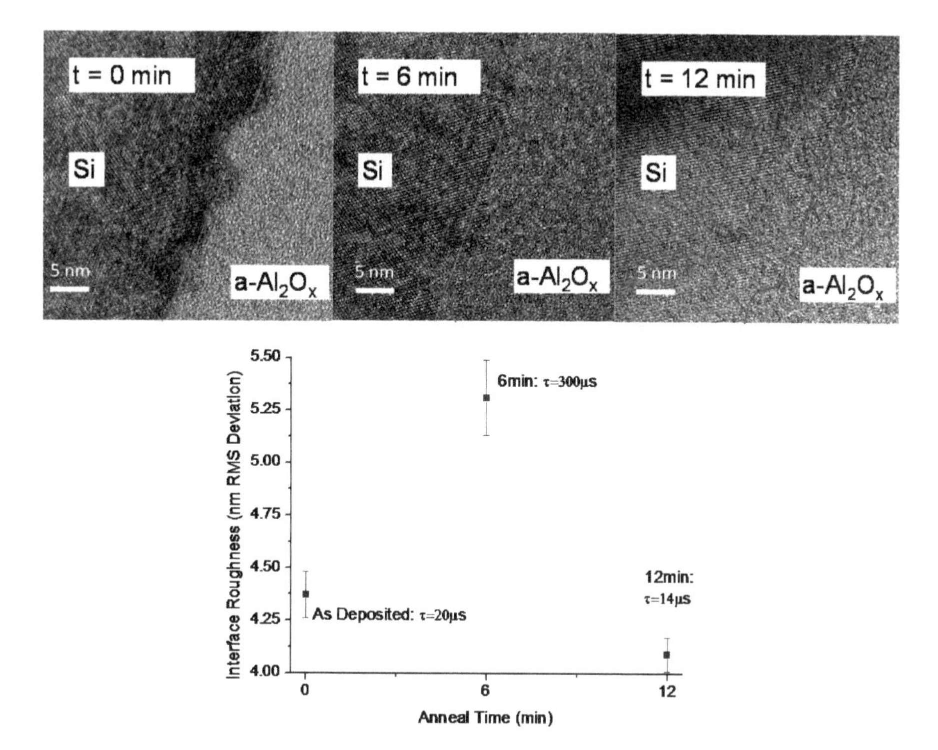

Figure 2. (Upper) Representative HRTEM images showing interfacial evolution with annealing time, t, at 500°C. Annealing times are indicated on the images. The straight lines present on the images denote the interface average position. The location of the interface separating the ordered Si substrate and the amorphous alumina thin film is also identified in each image (wavy lines). These were used to calculate the rms deviation of the interface, providing a metric for interfacial roughness comparison. (Lower) Interfacial rms roughness values obtained from analysis of the HRTEM images as a function of anneal time. The corresponding RCPCD-derived minority carrier lifetime is also provided next to each data point. Error bars represent the data ranges observed for 12 measurements taken over specimens from four independent depositions.

approximately 3-fold increase in carrier lifetime upon PECVD alumina deposition, with the maximum value for carrier lifetime of annealed samples increasing to nearly 15 times that of the as-deposited value and subsequently falling back to the as-deposited range with further anneal. Figure 1 depicts the trends in both the minority carrier lifetime (as measured using RCPCD) and the fixed interfacial charge density (C-V measurement) with isochronal annealing time at 500°C. Both the magnitude of the carrier lifetime and the fixed interface charge density increase with anneal time, reaching a maximum after 6 minutes at temperature and then decrease again with further anneal time.

The enhancement of both the fixed interface charge and the minority carrier recombination time within the sample at the same anneal time is consistent with a field-based passivation mechanism [1, 3, 24]. Of interest is that the magnitude of the EP appears to scale

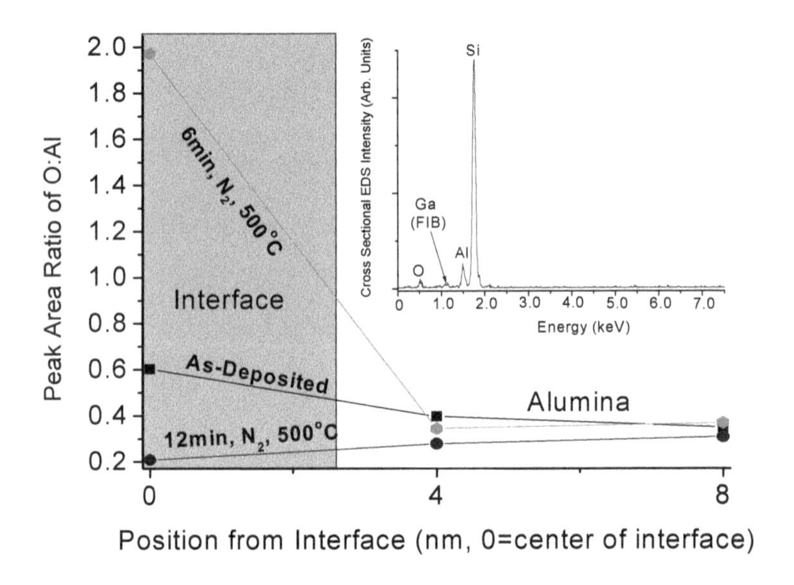

Figure 3. O:Al peak area ratios obtained from EDS point scans with distance from the a-alumina-Si interface (along interface normal, into the alumina layer). An enhancement of the peak area ratio at the interface is observed after a 6 minute anneal at 500°C. Inset: Representative EDS spectrum from an a-alumina:Si sample, ~4 nm into the alumina thin layer

with the square of the fixed charge magnitude, consistent with other, field-based alumina passivation results [12].

Figure 2 contains a representative HRTEM image of the c-Si/amorphous alumina interface (Figure 2a) after a 6 minute anneal at 500°C. Lines included in the image highlight the average interface position and the boundary between amorphous alumina and crystalline Si. No interlayer structure is observed to separate the alumina and Si phase regions that could be attributed to the presence of SiOx. Figure 2b contains a plot detailing the change in interface roughness as measured from cross-sectional images obtained from the specimens subjected to different annealing times. A maximum interfacial root mean square (rms) roughness value is observed after the 6 min anneal. The enhancement in rms roughness after the 6 min anneal is associated with the appearance of ~1 nm diameter Si clusters on the silicon side of the interface.

Both as-deposited films and those annealed > 6 min exhibit a smoother interface and reduced cluster development. Referring to Figure 1, the roughest interface is then correlated with the maximum minority carrier lifetime and fixed space charge density for this system. It may also be noted that longer anneal times lead to a smoother interface than that of the as-deposited measurements

STEM-EDS point scans were taken on the same samples used for the HRTEM imaging. Responses associated with O, Al, Si are clearly observed (inset, Figure 3). In addition, Ga associated with the sample preparation (i.e. Ga-beam used for focused ion beam (FIB) specimen processing)) are also present. Oxygen and aluminum peak areas were determined after a linear background subtraction using a nonlinear least squares fitting routine employing Gaussian peak functions. The ratio of O:Al peak areas with position relative to the alumina-Si interface are provided in Figure 3. An enhancement in this ratio is observed at the interface in specimens subjected to the six minute isothermal anneal. This three-fold increase in O:Al peak area ratio also coincides with the maximum observed minority carrier lifetime, fixed interfacial charge density and interface roughness (Figures 1, 2). Specimens subjected to either lower or higher annealing times consistently exhibited lower O:Al peak area ratios at the interface.

DISCUSSION

The maximum minority carrier lifetime, an indicator of passivation performance, coincides with a 500°C, 6 min anneal under the ultra-high purity N_2 atmosphere used. While the temporal resolution available in the isochronal annealing study (see discussion in the Experimental section) limits an precise identification of the optimum annealing time (and indeed, the maximum carrier lifetime attainable) under these conditions, there was clearly a maximum in the carrier lifetime (Figure 1) that occurred between 4 and 8 minutes into the isothermal annealing sequence. Consistent with a field-based passivation approach, the maximum observed carrier lifetime is also correlated with a maximum in the measured fixed interfacial space charge at this point in the anneal (Figure 1). In addition, the measurement lifetime appears to be scale with the square of the fixed charge density (10^{12} carriers/cm^2), consistent with Hoex et al., and indicative of a field-based passivation mechanism in this system [12].

The increase in both the carrier lifetime and fixed space charge has been further correlated to a increase in the observed nanoscale interface roughness as well as an enhancement of the local O:Al ratio (as indicated by an increase in EDS peak area ratio) in the near interface region. For annealing times examined longer than 6 minutes, degradation in carrier lifetime and the evolution of interfacial structure and chemistry characteristics similar to those of the as-deposited specimens is observed. These observations, and associated correlations with the electronic behavior of the a-alumina:Si specimens with isothermal annealing, can be used to infer a relationship between the thermally induced nanostructural and interfacial chemistry evolution observed and these passivation-relevant electronic characteristics.

Reviewing the trends in carrier lifetime observed in the present work, the bare silicon substrate after HF pre-deposition wash exhibits a minority carrier lifetime of approximately 5-7 μs. The addition of the as-deposited alumina thin layer to the surface of the sample increases the measured carrier lifetime into the range of 25-30 μs. As recognized by others [6-12] the presence of Al-O polar bonds near the interface after PECVD deposition contributes to a net negative interface charge associated with the highly polarizable oxygen species [20]. This increase in the presence of Al-O polar bonds results in the screening of minority carriers (electrons) within the Si and increases the average time to carrier recombination [21] (Figure 3), consistent with a field-based passivation mechanism. Clearly, however, modifications in the magnitude of this effect (and its impact on the resulting carrier lifetimes observed) are observed with thermal processing that produces variations in the nanostructural and chemical characteristics of the alumina-Si interface. A potential link between the observed passivation-relevant electronic characteristics and these factors is now proposed.

During deposition of the amorphous alumina thin layer onto the silicon substrate, O^+ species present in the plasma etch the surface of the substrate leaving a high-free energy Si surface with Si present in a variety of coordination environments [24]. This structural damage would be analogous to the damage layer of ion-beam implantation to form a p-n junction with the exception that the damage in PECVD is implemented on the surface and not within the bulk; both types of damage can be annealed out to reform a lower free-energy crystalline substrate layer [19]. Simultaneous deposition of the alumina film during exposure to the plasma, however, limits the degree of Si structural rearrangement as high-energy Si bond conformations are stabilized through coordination with the newly deposited alumina film [20, 21].

Upon annealing for 6 min at 500°C, sufficient thermal energy allows the high-energy surface-damaged Si atoms to return to a more stable coordination environment (Si defect anneal anticipated at ~485°C [20-22]). The increase in thermal energy enables the formation of an intermediate Si nanomorphology [21] where the surface Si atoms participate in higher energy, more spatially distinct, higher-interfacial-area, structural configurations at the Si-alumina interface associated with increased interfacial roughness [20] (Figure 2). Such Si features can exhibit an environmentally enhanced electron affinity [20-23], further concentrating the local negative fixed charge electron density at the Si interface with the resulting negative field from the stored charge density increasing the average time to carrier recombination. This increased propensity for a negative space charge field is likely further amplified by an overall increase in the relative O:Al ratio at the interface, as observed via EDS (Figure 3). While corresponding modifications in local chemical bond environment associated with of the O:Al ratio change must be assessed to fully understand its contribution to the interfacial fixed space charge, the local increase in O:Al EDS peak area ratio and its correlation with minority carrier lifetime supports such an interpretation.

Upon further annealing, reduction in interfacial energy is anticipated to drive a smoothing of the interfacial Si nanostructure, thus also producing a decreased environmental electron affinity enhancement associated with local Si interfacial nanomorphology. The decrease in interface area as an increasing number of Si atoms join the bulk also results in a reduction in the charge density that can be accommodated at the smoother interface. RCPCD measured carrier lifetimes in this case approach the as-deposited values upon annealing for 8-12 minutes (Figure 1). Further annealing is shown via EDS to produce a decrease in the O:Al ratio (as indicated by a reduction in the EDS peak area ratio (Figure 3)), reducing the opportunity for local, high-polarizability species to contribute to the negative space charge density. These combined nanostructural and local chemical modifications with increased annealing time contribute to a diminished minority carrier (electron) density near the alumina-Si interface and a decreased minority carrier lifetime.

To summarize, the enhanced the electronic passivation effect of a-Al_2O_x PECVD thin layers on c-Si p-type semiconductors subjected to a post-deposition isothermal anneal is associated with the evolution of interfacial nanostructure and local chemistry. Both an increased interfacial roughness, associated with the formation of an Si nanocluster morphology, and an enhanced O:Al ratio near the interface over that found in volume of the alumina film were found to be correlated to the development of an enhanced negative fixed space charge and a corresponding ~15 fold increase in electronic passivation via a field-based passivation effect. An interpretation of these results, in terms of the combined contributions of the interface nanostructure and near-interfacial chemistry to fixed space charge, has been offered to explain both the enhancement and subsequent degradation of passivation performance with time at temperature.

CONCLUSIONS

New electronic passivation layer options for alternative Si solar cell designs incorporating p-type Si emitters are needed [1-11]. Industrial scale deposition of a-alumina PECVD coatings has demonstrated strong p-type Si passivation performance. The present work established clear correlations between passivation-related electronic behavior and thermally induced modifications in a-alumina:Si interfacial chemistry and nanostructure within production scale PECVD-derived films. The analysis suggests that a localized increase in O:Al ratio at the alumina-Si interface, coupled with nanoscale roughening of Si interfacial structure, resulting from the relaxation of high-energy, plasma-damaged Si surface structures, promotes an enhancement in a negative space charge field leading to more effective screening of minority carriers (electrons) from the Si surface thereby increasing carrier lifetime. Additional investigation using alternative film deposition conditions and isothermal annealing atmospheres and temperatures will provide further insight into the energetics and kinetics of the mechanisms proposed. Correlations observed to-date; however, can provide a basis for post-deposition process optimization.

ACKNOWLEDGMENTS

We acknowledge the funding support of the Science Foundation of Arizona (SRG 0408-08). The authors wish to thank General Plasma, Inc. for the thin film materials and for helpful discussions concerning the work. We are also grateful to the LeRoy Eyring Center for Solid State Science at Arizona State University for electron imaging and microanalytical assistance.

REFERENCES

[1] D. MacDonald, and L. Geerligs, Recombination activity of interstitial iron and other transition metal point defects in p- and n-type crystalline silicon. *Appl. Phys. Lett.* **85**, 4061 (2004).
[2] S. Glunz, S. Rein, J. Lee, and W. Warta, Minority carrier lifetime degradation in boron-doped Czochralski silicon. *J. Appl. Phys.* **90**, 2397 (2001).
[3] H. Momida, S. Nigo, G. Kido, and O. Takahisa, Minority carrier lifetime degradation in boron-doped Czochralski silicon. *Appl. Phys. Lett.*, **98**, 042102 (2011).
[4] J. Lelievre, E. Fourmond, A. Kaminski, O. Palais, D. Ballutaud, and M. Lemiti: Study of the composition of hydrogenated silicon nitride SiNx:H for efficient surface and bulk passivation of silicon. *Sol. Energy Mat. Sol. Cells* **93**, 1281 (2009).
[5] J. Benick, B. Hoex, M. van de Sanden, W. Kessels, O. Schultz, and S. Glunz, High efficiency n-type Si solar cells on Al2O3-passivated boron emitters. *Appl. Phys. Lett.* **92**, 253504 (2008).
[6] P. Lu, Z. Wang, A. Lennon, A. Lennon and S. Wenham, Enhanced passivation for silicon solar cells by anodic aluminum oxide. *IEEE Xplore*, **6**, 1491 (2011).

[7] M. Christophersen, V. Fadeyev, B. Phlips, H. Sadrozinski, C. Parker, S. Ely, and J. Wright, Alumina and silicon oxide/nitride sidewall passivation for p-and n-type sensors. *Phys. Rev. A*, **4**, 77 (2012).
[8] J. Cruz-Campa, M. Okandan, P. Resnick, P. Clews, T. Pluym, R. Grubbs, V. Gupta, D. Zubia, and G. Nielson, Microsystems enabled photovoltaics: 14.9% efficient 14 mm thick crystalline silicon solar cell. *Sol. Energy Mat. Sol. Cells*, **95**, 551 (2011).
[9] S. Ktifa, M. Ghrib, F. Saadallah, H. Ezzaouia, and N. Yacoubi, Photothermal deflection spectroscopy study of nanocrystalline si (nc-si) thin films deposited on porous aluminum with PECVD. *Int. J. Photoenergy*, 1, 1155 (2012).
[10] P. Prathap, O. Tuzun, D. Madi, and A. Slaoui, Thin film silicon solar cells by AIC on foreign substrates. *Sol. Energy Mat. Sol. Cells*, **95**, 44 (2011).
[11] J. Bedinger, M. Moore, R. Hallock, K. Alavi, and T. Kazior, Passivation layer for a circuit device and method of manufacture. US Patent No.: US7,902,083 B2, Mar. 8 (2011).
[12] J. Hoex, J. Gielis, M. van de Sanden, and W. Kessels, On the c-Si surface passivation mechanism by the negative-charge dielectric Al2O3. *J. Appl. Phys*. **104**, 113703 (2008).
[13] Q. Shang, W. Seaman, M. Whitney, M. George, J. Madocks, and R. Ahrenkiel, N-type and p-type c-si surface passivation by remote pecvd AlOx for solar cells. Photovoltaic Spec. Conf., 20-25 June (2010).
[14] Y.-I. Ogita, M. Tachihara, Y. Aizawa, N. Saito, Ultralow surface recombination in p-Si passivated by catalytic-chemical vapor deposited alumina films. *Thin Solid Films* **519**, 4469 (2011).
[15] K. Kimoto, Y. Matsui, T. Nabatame, T. Yasuda, T. Mizoguchi, I. Tanaka, and A. Toriumi, Coordination and interface analysis of atomic-layer-deposition Al2O3 on Si(001) using energy-loss near-edge structures. *Appl. Phys. Lett.* **83**, 4306 (2003).
[16] K. Matsunaga, T. Tanaka, T. Yamamoto, Y. Ikuhara, First-principles calculations of intrinsic defects in Al2O3. *Phys. Rev. B* **68**, 085110 (2003).
[17] G. Lucovsky,: Transition from thermally grown gate dielectrics to deposited gate dielectrics for advanced silicon devices: A classification scheme based on bond iconicity. *J. Vac. Sci. Technol.* **A 19**, 1553 (2001).
[18] R. Ahrenkiel, and S. Johnston, Recombination lifetimes using the RCPCD technique: comparison with other methods, NREL/BK-520-32717 08/2002.
[19] M. Kundu, N. Miyata, and M. Ichikawa, Interface stability during the growth of Al2O3 films on Si(001). *J. Appl. Phys*. **93**, 1498 (2003).
[20] S. Ossicini, F. Degoli, E. Iori, E. Luppi, and R. Magri, Simultaneously B- and P-doped silicon nanoclusters: Formation energies and electronic properties. *Appl. Phys. Lett.* **87**, 173120 (2005).
[21] A. Puzder, A. Williamson, J. Grossman, and G. Galli, Surface control of optical properties in silicon nanoclusters. *J. Chem. Phys*., **117**, 6721 (2002).
[22] Z. Zhou, L. Brus, and R. Friesner, Electronic structure and luminescence of 1.1- and 1.4-nm silicon nanocrystals: oxide shell versus hydrogen passivation. *Nano Lett.* **3**, 2 (2003).
[23] J. Zhang, Interfacial charge carrier dynamics of colloidal semiconductor nanoparticles. *J. Phys. Chem. B* **104**, 7239 (2000).
[24] R. Kuse, M. Kundu, T. Yasuda, N. Miyata, and A. Toriumi, Effect of precursor concentration in atomic layer deposition of Al2O3. *J. Appl. Phys*. **94**, 6411 (2003).

[25] R. Ahrenkiel, N. Call, S. Johnston, and W. Metzger, Comparison of techniques for measuring carrier lifetime in thin-film and multicrystalline photovoltaic materials. *Sol. Energy Mat. Sol. Cell*, **94** (2010).
[26] R. Ahrenkiel, and S. Johnston, An advanced technique for measuring minority-carrier parameters and defect properties of semiconductors. *Mat. Sci. & Eng. B* **102** (2003).
[27] S. Johnston, R. Ahrenkeil, P. Dippo, M. Page, and W. Metzger, Comparison of silicon photoluminescence and photoconductive decay for material quality characterization. *Mater. Res. Soc. Symp. Proc.* **994** (2007).
[28] A. Feldman, and R. Ahrenkiel, Space charge limited current effect on photoconductive decay in silicon at high injection levels. *IEEE Xplore* **9**, 10 (2010).
[29] R. Ahrenkiel, and S. Johnston, An optical technique for measuring surface recombination velocity. *Sol. Energy Mat. Sol. Cell*, **93** (2009).
[30] R. Ahrenkiel, and D. Dunlay, Transmission modulated photoconductive decay system. US Patent No.: US 2010/0283496 A1, Nov. 11 (2010).
[31] H. Momida, T. Hamada, Y. Takagi, and T. Ohno, Theoretical study on dielectric response of amorphous alumina. *Phys. Rev. B* **73**, 054108 (2006).
[32] H. Momida, S. Nigo, G. Kido, T. Yamamoto, T. Uda, and T. Ohno, Effect of vacancy-type oxygen deficiency on electronic structure in amorphous alumina. *Appl. Phys. Lett.* **98**, 042102 (2011).

Ceramics for Next Generation Nuclear Energy

SiC/SiC FUEL CLADDING BY NITE PROCESS FOR INNOVATIVE LWR PRE-COMPOSITE RIBBON DESIGN AND FABRICATION

Yuuki Asakura[1], Daisuke Hayasaka[1], Joon-Soo Park[1], Hirotatsu Kishimoto[1,2], Akira Kohyama[1]

[1]OASIS, Muroran Institute of Technology, 27-1 Mizumoto-cho, Muroran, Hokkaido 050-8585, JAPAN

[2]Division of Mechanical Systems and Materials Engineering, Muroran Institute of Technology, 27-1 Mizumoto-cho, Muroran, Hokkaido 050-8585, JAPAN

ABSTRACT

Toward the establishment of industrial basis for producing high quality SiC/SiC fuel cladding by NITE process, a new concept of preform fabrication by pre-composite ribbon is provided and the current status of the pre-composite ribbon and the main features of the ribbon are also provided. By ribbon winding and lay-up procedure, the preform is semi-automatically fabricated. Thus the fiber architecture controllability and high density matrix formation capability have been drastically improved. The representing results utilizing carbon fibers with 1600 fibers/bundle or SiC fibers with 800 fibers/bundle and the preform was fabricated by semi-auto filament winding machine are described. Pre-composite ribbon is making many progresses in NITE process and these efforts are contributing improvement in quality and fabrication technology of fuel cladding for nuclear reactor.

INTRODUCTION

Since the accident at the Fukushima Daiichi Nuclear Power Station, the safety of nuclear reactor is strongly required. Although the potentiality of SiC/SiC for ensuring nuclear reactor safety has been widely recognized, SiC/SiC had not been required as one of near term nuclear reactor materials. Since 2012, Japanese energy policy has been drastically changed and SiC/SiC becomes highly prioritized candidate. OASIS (Organization of Advanced Sustainability Initiative for Energy System/Materials) has been developing SiC/SiC technology as a new site for more than 30 years R & D by the Kohyama research group. Under the new energy policy, "SCARLET (SiC Fuel Cladding/Assembly Research Launching Extra-safe Technology)" project was approved in September 2012 and funded by the Japanese Ministry of Education on "Innovative Nuclear Research and Development" program. This project aims to improve the safety of light water reactor by replacing the zircaloy from the core interior of the light water reactor, by a ceramic composite material[1]. The ceramic composite material of SiC/SiC for the project is fabricated by international patent; "Nano Infiltration and Transition Eutectoid (NITE)" method[2-3]. This method is an advanced concept of the liquid phase sintering method to fabricate SiC/SiC composite from nano-powders of high purity SiC and SiC fibers.

The "SCARLET" project consists of the following four tasks;

1. SiC/SiC fuel cladding production technology[4].
2. Assembly technology
3. Environment effect evaluation
4. Engineering / Safety design

OASIS carries out the fabrication of SiC/SiC fuel cladding mainly. Currently, prepreg sheets and green sheets are mainly used to preform fabrication of dry NITE method[5]. However, the use of prepreg sheet is difficult to fabricate preform for the fuel cladding shapes, because the diameter of the fuel cladding is small, there is a possibility that the fibers may be damaged during fabrication. This is not efficient, because cut and paste of SiC prepreg sheets are time consuming, so need a new intermediate material for easy to fabricate the preforms. In addition, the accuracy of the product is different by human fabricate because it is handmade. It is important to manufacture in automation product precision is required as fuel cladding. Therefore, we made a design and proof of "Pre-Composite Ribbon (PCR)" that is a new intermediate material suitable for new NITE process. The PCR is a wire-like intermediate material of SiC fibers and nano-powders, which the SiC fibers are spread and dipped in the SiC nano-powder slurry, and hardened flat. The PCR is very convenient for the mechanized winding of SiC fibers with appropriate orientation on a cladding. It also can be expected to obtain a uniform preform continuously by introducing a semi-auto filament winding machine for winding by the PCR. This paper describes the PCR of intermediate material, and fabrication of preform by semi-auto filament winding used it.

EXPERIMENTAL

Concept and Fabrication of the PCR

Figure 1 shows the conceptual images of prepreg sheet and PCR. The PCR is designed to be thinner than the prepreg sheet with finer fiber bundle structure and width smaller than a few millimeters. This is advantageous for fabrication of thin wall and small diameter tubes like fuel cladding for nuclear reactors. This makes it easy making preforms by filament winding machines. Figure 2 shows schematic image of how to fabricate a PCR. First, the fibers were de-bundled to make sufficient space for infiltrating NITE-slurry. Then NITE-slurry, which includes SiC nano-powders and sintering agents, is infiltrated. The infiltrated fibers were shaped into flat ribbon of about 100μm. Proof of

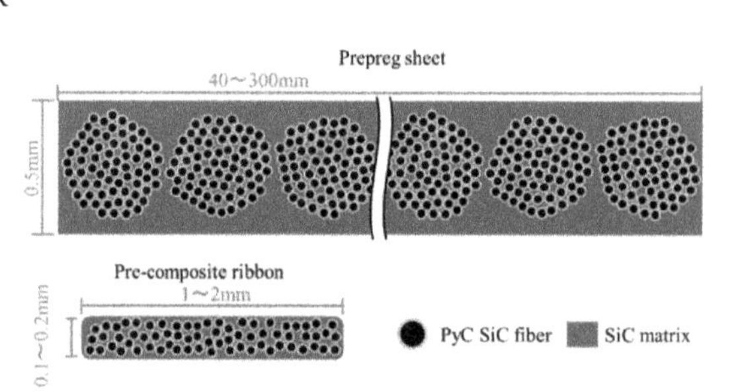

Fig.1. The conceptual images of prepreg sheet and PCR.

the concept for the first step was done by using carbon fibers with 1600 fibers/bundle, which is called as C/SiC PCR. As the second step SiC fibers with carbon coating was applied for making SiC/SiC PCR. The microstructure of PCR was investigated by the scanning electron microscopy (JEOL JSM6400).

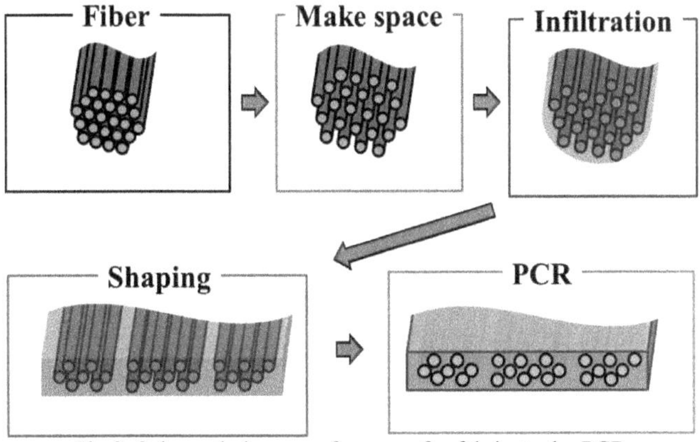

Fig.2. Schematic images of process for fabricate the PCR.

Fabrication of Preforms with the PCR

Figure 3 shows the schematic image of the preform fabrication method by semi-auto filament winding. The PCR was wound around the core rod under the rotation and travelling of the shaft. We were performed manually fine-tune the PCR to avoid overlapping. The final goal of the "SCARLET" project is 1 m length fuel cladding. The semi-auto filament winding machine for this size was designed and introduced. Up to now, preform of 500mm length was fabricated by the semi-auto winding machine. C/SiC PCR was used to make interconnected 2D fiber structure with 30 degrees to the hoop direction. As the second step SiC/SiC PCR was applied for making preforms. Surface of the preform was investigated by an optical microscope (KEYENCE VR-3000).

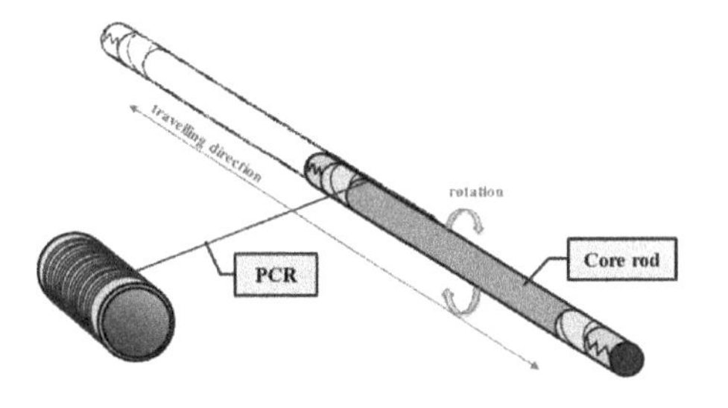

Fig.3. Schematic image of the auto filament winding method.

RESULTS AND DISCUSSION

Concept and Fabrication of the PCR

Figure 4 shows the external and cross-sectional images of the C/SiC PCR. Fabrication of C/SiC PCR has been successful in stabilizing shape by tension control of fiber, viscosity control of the SiC slurry and removal mechanism of the excess SiC slurry. As shown in the figure, width of C/SiC PCR is substantially uniform.

Fig.4. External and sectional images of the C/SiC PCR.

Figure 5 shows the external and cross-sectional images of the SiC/SiC PCR that was fabricated based on the technology for C/SiC PCR. The SiC matrix is formed in a highly dense condition as in the case of C/SiC PCR.

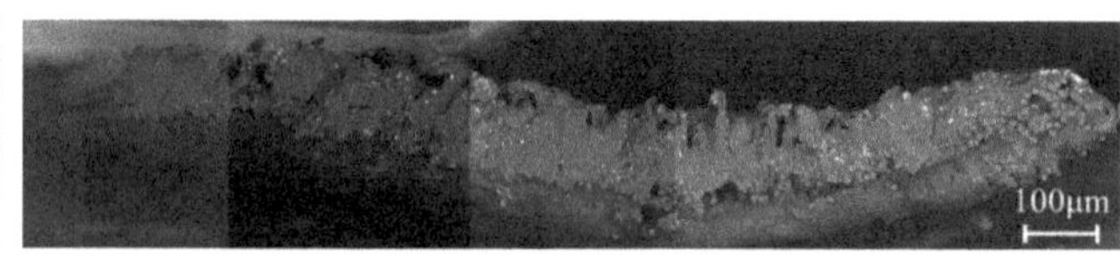

Fig.5. External and sectional images of the SiC/SiC PCR.

Fabrication of Preform using the PCR

Figure 6 shows the prototype of semi-auto filament winding machine. Figure 7 shows a preform fabricated using the semi-auto filament winding machine, where the numbers represent the fiber winding order of the 2D fiber structure. Odd layers are wound in the left direction at -30 degrees, even layers are wound in the right direction by 30 degrees with respect to the hoop direction, and they are presenting a highly controlled quality. As a result of the measurement, the irregularity of surface flatness with minimum-32.02μm and the maximum 169.14μm was confirmed. This improvement can be going further by changing the

Fig.6. Prototype of semi-auto filament winding machine.

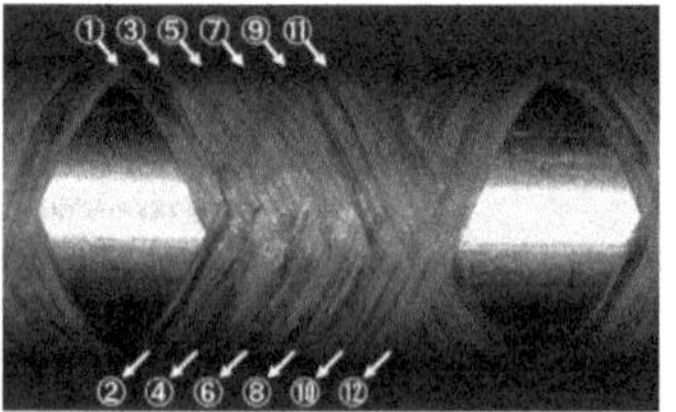

Fig.7. Preform using C/SiC PCR was fabricated by semi-auto filament winding machine.

winding interval control method and improvement of the accuracy of the thickness and width of the C/SiC PCR.

Figure 8 shows a preform fabricated by SiC/SiC PCR, where the surface irregularity caused by interconnecting SiC/SiC PCR can be controlled within about 100μm. The smaller irregularity for the case of SiC/SiC Preform is mainly due to the difference in fiber numbers per bundle.

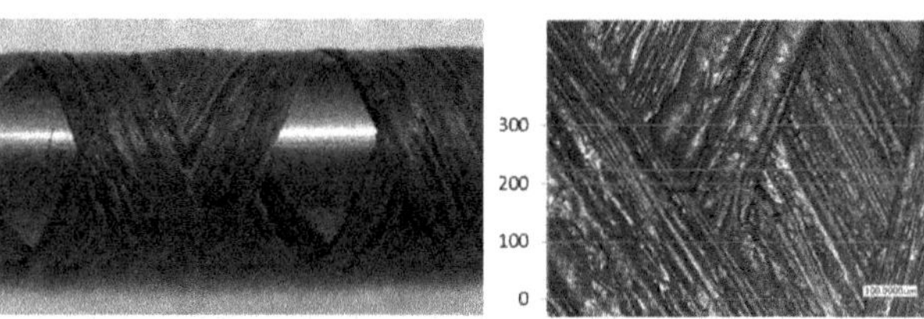

Fig.8. Preform using SiC/SiC PCR was fabricated by semi-auto filament winding machine.

CONCLUSION

The proof of the concept was done with carbon fibers as the first step. The C/SiC PCR satisfied the shape and quality of the PCR concept. Also preform fabricated with SiC/SiC PCR, as a new intermediate material for the fabrication of SiC/SiC fuel cladding showed advantages to SiC/SiC strips cut-out from UD SiC/SiC prepreg sheet. On-going fabrication of SiC/SiC PCR is making many progresses and these efforts are contributing improvement in quality and product ability of thin wall/ small diameter tubes like fuel cladding for nuclear reactors.

ACKNOWLEDGMENT

Present study is mainly supported by the Ministry of Education, Culture, Sports, Science and Technology of Japan (MEXT) on "Innovative Nuclear Research and Development" program. The authors would like to express their sincere appreciations to the members of OASIS, Muroran Institute of Technology, for their collaboration and support. This work is also partly supported by the president fund of Muroran Institute of Technology. The authors' appreciation is due to this support.

REFERENCES

[1]H. Kishimoto, T. Shibayama, Y. Asakura, D. Hayasaka, Y. Kohno and A. Kohyama, SiC/SiC fuel cladding by NITE process for innovative LWR-concept and process development of fuel pin assembly technologies, *Proceedings of PACRIM 10, June 2-7, 2013, San Diego, USA*, submitted.

[2]J.-S. Park, A. Kohyama, T. Hinoki, K. Shimoda, Y.-H. Park, Efforts on large scale production of NITE-SiC/SiC composites, Journal of Nuclear Materials, **367-370**, 719-724 (2007)

[3]K. Shimoda, T. Hinoki and A. Kohyama, Effect of additive content on transient liquid phase sintering in SiC nanopowder infiltrated SiC_f/SiC composites, Composites Science and Technology, **71**, 609-615 (2011).

[4]N. Nakazato, H. Kishimoto, Y. Kohno, A. Kohyama, SiC/SiC Fuel Cladding by NITE process for Innovative LWR -Cladding forming process development, *Proceedings of PACRIM 10, June 2-7, 2013,San Diego, USA*, submitted.
[5]A. Kohyama, J.-S. Park and H.-C. Jung, Advanced SiC fibers and SiC/SiC composites toward industrialization, Journal of Nuclear Materials, **417**, 340-343 (2011)

SiC/SiC FUEL CLADDING BY NITE PROCESS FOR INNOVATIVE LIGHT WATER REACTOR-COMPATIBILITY WITH HIGH TEMPERATURE PRESSURIZED WATER

C. Kanda[1], Y. Kanda[2], H. Kishimoto[1], A. Kohyama[1]

[1] OASIS, Muroran Institute of Technology, Muroran, Hokkaido 050-8585, Japan
[2] College of Environmental Technology, Graduate School of Engineering, Muroran Institute of Technology, Muroran, Hokkaido 050-8585, Japan

ABSTRACT

As an important part of the environmental resistance evaluation under light water reactor condition, NITE- SiC / SiC fuel cladding under R & D is subjected for the test. Ongoing "SCARLET" project includes compatibility evaluation task of NITE-SiC/SiC with high temperature pressurized water, mainly of PWR condition. The final goal of the project is to confirm the material soundness under reactor water environment in Halden reactor, Norway. The 5 years termed project plan is briefly introduced and the very preliminary results from the current high temperature pressurized water test equipment are provided. Immersion test to 250 C and 7 MPa pressurized static water for 30 hours exhibits surface damage on SiC and SiC/SiC materials tested. The reaction products at the surface have varieties in shape, size and density. Although qualitative and quantitative analyses are underway, preliminary results on microstructure and microchemistry are provided. For the case of SiC/SiC with carbon coating to SiC fibers, carbon layer was largely exfoliated and some of them were dropped-off. Time depending behavior of the carbon layer damage is discussed.

INTRODUCTION

Nuclear energy is currently major and essential power source for large scale electricity supply and the safety systems of nuclear power plants, especially of light water reactors, have been widely believed nearly perfect due to the multiple safety barriers. But the severe accident of Fukushima Daiichi Nuclear Power Station by the Great East Japan Earthquake has raised big and serious questions to the current safe system and technology. One of the detrimental events was caused by hydrogen generation from the reaction with water and Zircaloy. As an ultimate safety option to this reaction, the replacement of Zircaloy to SiC materials has been recognized for many years. After the Fukushima accident, Japanese energy policy has been changed and the new safety requirement light up SiC/SiC materials. The major attractiveness of SiC material comes from its

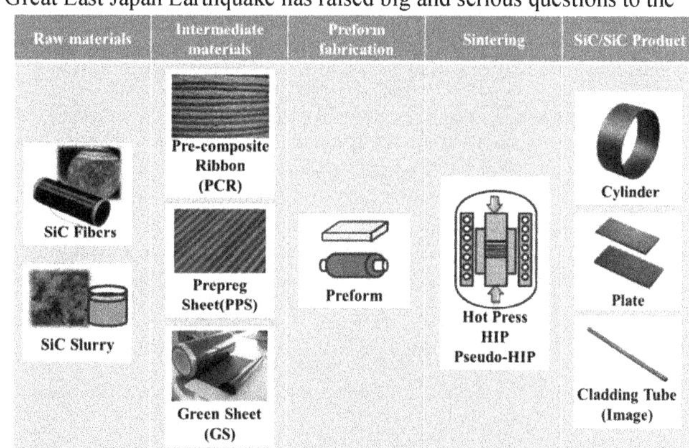

Figure 1. The production technique of SiC/SiC products by NITE process

high temperature stability on mechanical and chemical properties as well as excellent radiation damage tolerance. These features introduce ensuring safety of reactor core at severe accidents such as loss of coolant accident (LOCA) and station black out (SBO).

Organization of Advanced Sustainability Initiative for Energy System/Materials (OASIS), Muroran Institute of Technology has been devoted to the R & D of SiC / SiC products by our patented method; NITE process. The "SCARLET" project is trying to up-grade NITE technology to meet large scale industrialized products supply, shifted from wet-slurry to dry slurry method [1].

Figure 1 shows the manufacturing process of NITE-SiC / SiC products by dry system can solve the problem that is difficulty achievement by Reaction Sintering (RS) and Chemical Vapor Deposition (CVD) process. And NITE-SiC / SiC is only way of the commercial use. In SCARLET project, all of tests are assumed Pressurized Water Reactor (PWR) condition, and the compatibility test using high temperature and pressure water is planned for investigate of NITE-SiC corrosion stability.

The corrosion stability of SiC at high pressure and temperature water have been studied many researchers. Hiraide and Sugimoto reported that SiC shows superior to corrosion resistance in deionized water, H_2SO_4 and LiOH [2] compared with Si_3N_4, Al_2O_3 and Zi_2O_3. Kawakubo et al, reported that increase of corrosion weight loss of SiC is caused by increasing in pH value of the solution and dissolved oxygen concentration. In addition, type of corrosion of SiC is grain-boundary corrosion, and it cannot be find any deposits on the surface [3]. Tanaka et al. reported that using of β-SiC as precursors shows better corrosion resistance than α-SiC, also using of B and C as sintering additive show better than Al_2O_3 [4]. However, corrosion stability of NITE-SiC was not cleared up yet. In this paper, plan of water stability test in project SCARLET is explained and report shortly about result of pre-test.

THE PLAN OF COMPATIBILITY TEST IN PROJECT SCARLET

The compatibility test in SCARLET project (Figure 2) is classified as LOCA tests and SiC compatibility test with high temperature and pressurized water.

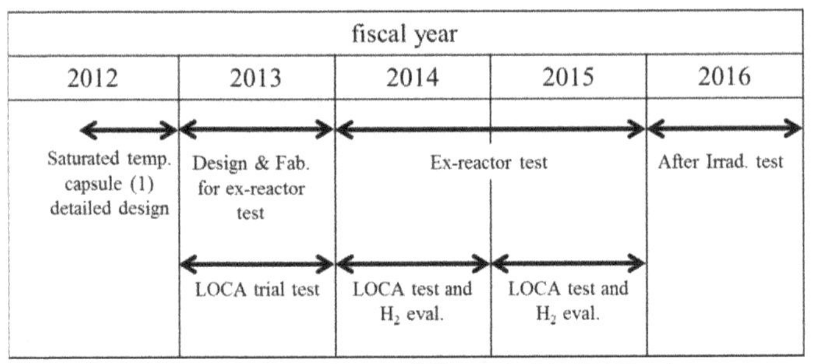

Figure 2. The plan of compatibility test in SCARLET project.

In compatibility test, it is going to evaluate interaction of fuel cladding tube and coolant with light water reactor environment under irradiation system (Figure 3). The test will be going on in Halden reactor in Norway. The rig (Figure 4) shows for operation in a Halden loop under PWR conditions, i.e. at 155 bar and 290°C inlet temperature. It contains 6 rods, and three Zircaloy cladding tubes filled up UO_2 fuel pellets are among them, and the other three rods without fuel inside are segmented manufactured by NITE-SiC / SiC. The test will be run under PWR conditions, i.e. 155 bar and 290°C inlet temperature. The irradiation test will occur in two Halden reactor cycles of 80 - 100 full power days and the irradiation contemplated in this proposal is currently foreseen to occur in two Japanese fiscal years (2014 and 2015). After irradiation, Inspection of all segments test will occur.

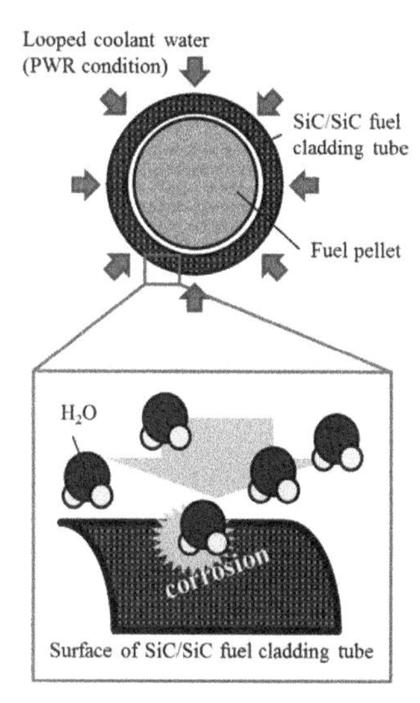

Figure 3. The image of compatibility test in Halden reactor.

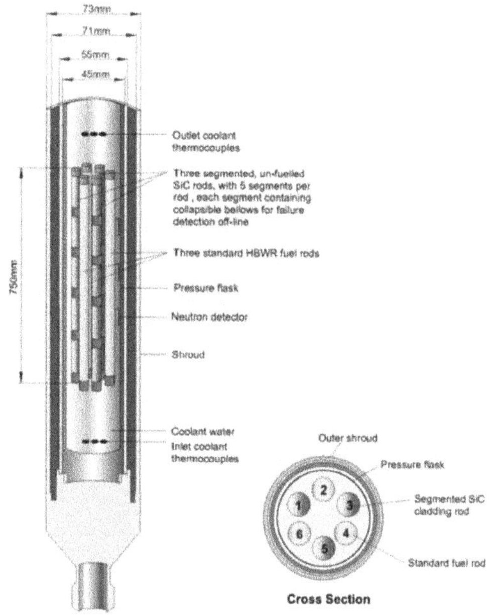

Figure 4. Outline of the test Rig for compatibility test.

Furthermore, safety in the application of SiC / SiC fuel cladding tube is going to be considered, and be planned to carry out safety system. One of severe accidents is LOCA. Figure 5 shows that's mechanism. The pressure vessel is filled with coolant water in normal run, but (a) some defects are broke out within the system, (b) coolant transmutes steam by decompression. That steam will escape from the hole ((c)), and (d) decrease levels of the coolant. Repeating (c) and (d) until lose all coolant. In this condition, surface of fuel cladding tube is exposed to high temperature and pressured steam. As the result, Fuel rods will be overheated and got serious

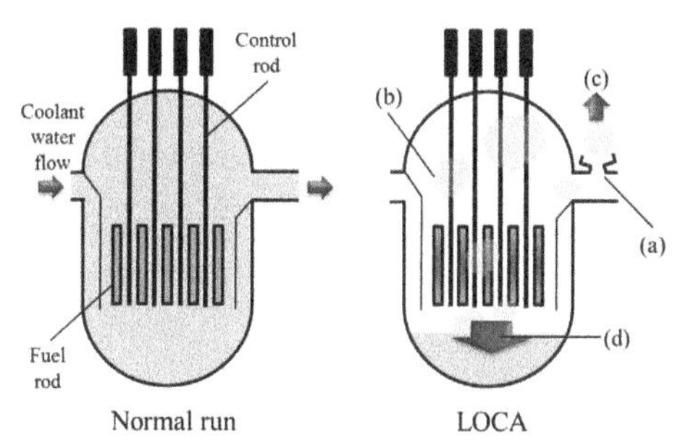

Figure 5. Normal run and LOCA mechanisms in PWR.

damage. The LOCA test in the SCARLET project is going to be carried out high temperature steam exposure test with consideration for those mechanisms.

THE COMPATIBILITY PRE-TEST

Coolant water transmutes high temperature and pressure in PWR reactor; therefor compatibility pre-test was carried out with near PWR condition. The task plan is briefly introduced and the very preliminary results from the current high temperature pressurized water test equipment are provided.

The samples are monolithic SiC and SiC/SiC plates prepared by PIP, RS and dry system NITE methods (Figure 6) and Hexoloy®. The extensive procedure and precursors of the samples were shown in Table 1. The samples were put into a metal meshed bag and were loaded to high temperature pressurized water test equipment.

Immersion test to 250°C and 7 MPa pressurized static water for 30 hours exhibits clear surface damage on SiC and SiC/SiC materials were tested. Figure 7 and Table 2 show the apparent condition and specifications of the apparatus for the corrosion tests. Scanning electron microscope (SEM) was used for sample surface observing and composition analysis (EDS). Surface roughness measured by 3D laser scanning microscope (3D-LSM).

Figure 6. Making methods of RS, PIP and NITE SiC (/ SiC).

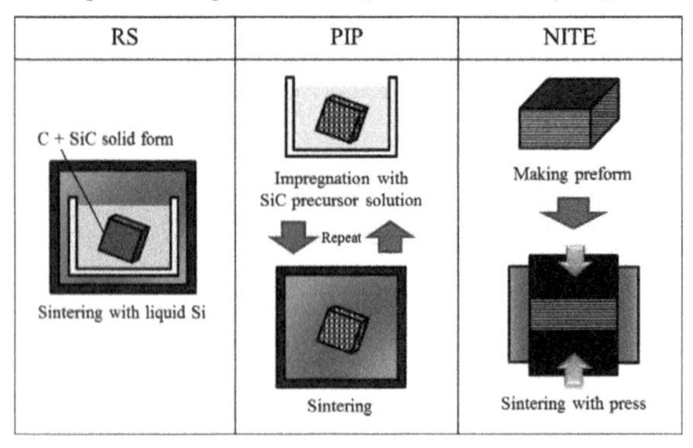

Table 1 Components of SiC and SiC / SiC materials.

Method	Fibers / fabrics	Matrix	End form
Hexoloy®	-	SiC	SiC
RS	-	SiC + C + Si	SiC
PIP	C coated SiC fabrics	SiC precursor solution	SiC/SiC
NITE	C coated SiC fibers (AG fibers)	Dried SiC slurry (Green Sheet)	SiC/SiC

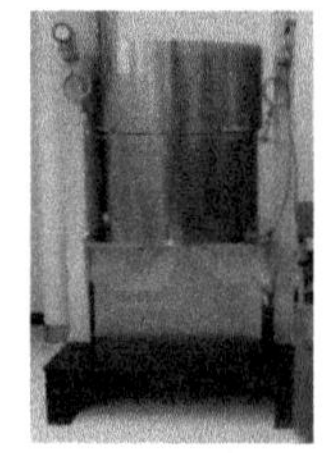

Figure 7. Overview of the current high temperature pressurized water test equipment.

Table 2 High temperature pressurized water test condition.

Article	contents
Sample condition	Stagnant
Medium	Pure water
Material of equipment	SUS304

Figure 8 shows close-up optical images of samples. Those expose parts of SiC fabric in (b) and AG fibers in monolithic layer in (d). Figure 9 shows SEM images before pre-test. It is found that the BEI image of (c) shows blighter parts than circumference exist on surface, therefore it is presumed that a Si-rich SiC phase[5]. The BEI image of (d) shows apparent blight parts, those are found that sintering additive in past our research [6].

Figure 8. Close-up optical images of SiC and SiC / SiC samples before test.
(a) Hexoloy®, (b) PIP, (c) RS and (d) NITE process.

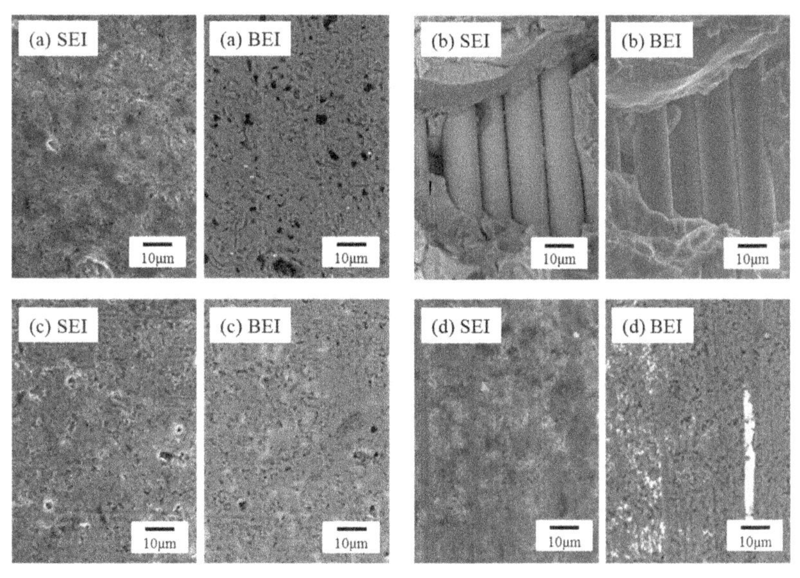

Figure 9. SEM images of various SiC and SiC / SiC samples before test.
(a) Hexoloy®, (b) PIP, (c) RS and (d) NITE process.

After pre-test samples were observed with SEI by SEM (Figure 10). The image at the top right is closed-up viewing that is indicated by arrows. Under this condition, no significant erosion or corrosion was observed. Furthermore, smooth surface fibers of (b) before test had been peeled (Figure 11). It is presumed that the carbon coating of fibers surface was dropped-out by physical or chemical factors.

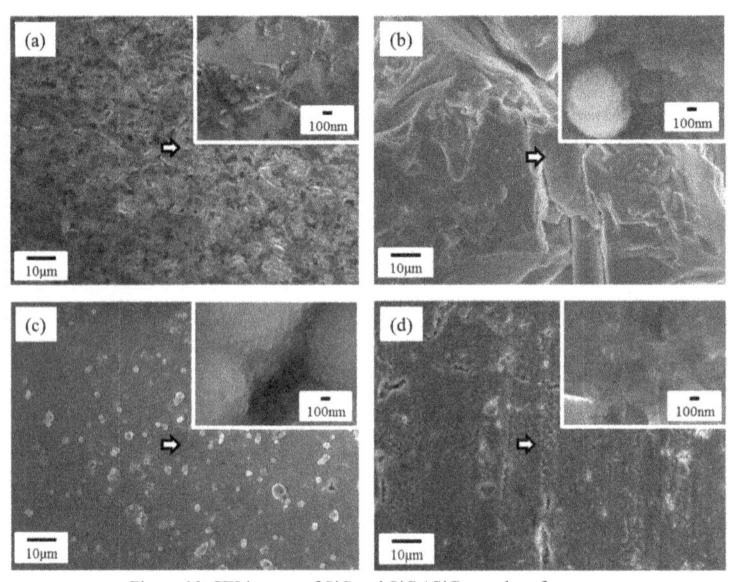

Figure 10. SEI images of SiC and SiC / SiC samples after test.
(a) Hexoloy®, (b) PIP, (c) RS and (d) NITE process.

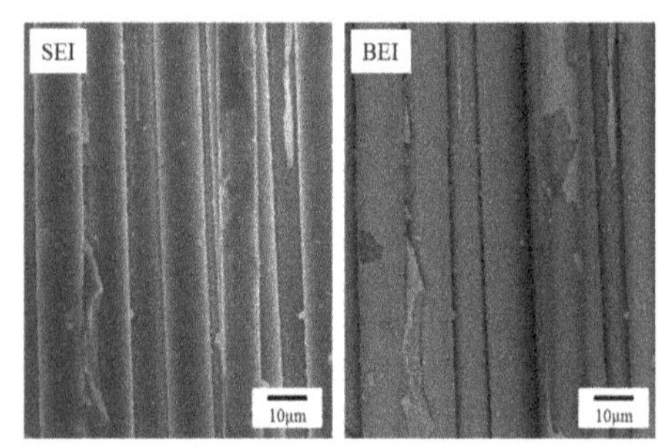

Figure 11. SEI images of the PIP sample (after test) focused on fibers.

Table 3 Weight change of SiC and SiC / SiC samples before and after test.

	Weight (g)	
	Before test	After test
Hexoloy®	0.263	0.263
PIP	0.286	0.279
RS	0.247	0.247
NITE	0.091	0.089

In Table 3, weigh loss was not detected each samples. But LSM image of NITE-SiC/ SiC (a) before and (b) after the compatibility pre-test (Figure 12), indicates that fiber of SiC / SiC be rough because of deposited green color materials on surface. EDS analysis presents Cr and Fe deposition on the surface forming silicate. Under the utilization of steels, these depositions would be enhanced. This should be precisely examined for the case.

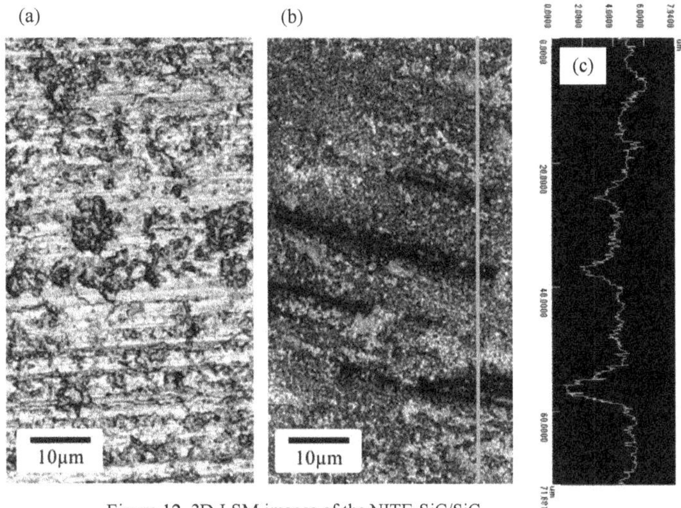

Figure 12. 3D-LSM images of the NITE-SiC/SiC.
(a) before test, (b) after test, (c) surface roughness after test.

CONCLUSION

As one of the task of the SCARLET Project, "Environmental resistance evaluation and Safety Design" is on-going. The major change of the SCARLET project in neutron irradiation experimentation site from JMTR to Halden Reactor and Belgium Reactor 2 is introduced. The brief introduction of the compatibility test under PWR water condition with LOCA simulation, which is the first step of the task, was performed. The preliminary test, feasibility test of materials response evaluation under high temperature and highly pressurized water condition performed using existing FEEMA facility, presented the feasibility and appropriateness of Halden reactor experiment under final planning. The slight erosion and corrosion, as was previously predicted, and also deposition of Fe and Cr on the surface of SiC were confirmed. These material responses without neutron irradiation will be related with those with neutron irradiation as the next step. Detailed analysis of these behaviors without neutron irradiation, together with the out of pile test in JAEA and OASIS, under preparation, may provide clear insights.

ACKNOWLEDGMENT

Present study is mainly supported by the Ministry of Education, Culture, Sports, Science and Technology of Japan (MEXT) on "Innovative Nuclear Research and Development" program, where SCARLET project; "SiC Fuel Cladding / Assembly Research, Launching Extra-Safe Technology", is one of 11 project. Authors acknowledge the support and encouragement by the members of OASIS.

REFERENCES

[1] N. Nakazato, H.Kishimoto, Y.Kohno, A.Kohyama, *Ceram. Eng. Sci. Proc.*, **32**, 103-108 (2011).
[2] N. Hiraide and K. Sugimoto, *CORROSION ENGINEERING*, **37**, 415 (1988).
[3] T. Kawakubo, H. Hirayama and T. Kaneko, *J. Soc. Mater. Sci., Japan*, **39**, 438-78 (1990).
[4] D. Tanaka, S. Souma and K. Sugimoto, *ZAIRYO-TO-KANKYO*, 42, 585 (1993).
[5] S.P Lee, J.O Jin, J.S Park, A Kohyama, Y Katoh, H.K Yoon, D.S Bae, and I.S Kim, *Proceedings of ICFRM-11*, **329-333**, 537 (2003).
[6] K. Shimoda, T. Hinoki and A. Kohyama, *Composites Science and Technology*, **71**, 609–615 (2011).

SiC/SiC FUEL CLADDING BY NITE PROCESS FOR INNOVATIVE LWR-CONCEPT AND PROCESS DEVELOPMENT OF FUEL PIN ASSEMBLY TECHNOLOGIES

Hirotatsu Kishimoto[1], Tamaki Shibayama[2], Yuuki Asakura[1], Daisuke Hayasaka[1], Yutaka Kohno[1], Akira Kohyama[1]

[1]OASIS, Muroran Institute of Technology, Muroran, Hokkaido 050-8585, Japan
[2]Hokkaido University, Sapporo, Hokkaido 060-8628, Japan

ABSTRUCT

SiC/SiC fuel cladding is a challenge to achieve a superior safety for light water reactors. A project aiming to produce just SiC/SiC cladding tube without zircaloy has been launched at Muroran Institute of Technology in Japan. The project named SCARLET is based on the NITE SiC/SiC technology. The NITE method has the potential to produce the cladding tube and the fuel pin assembly, but many technological issues, especially, relating to the mass production technologies need to be solved. The present paper will introduce the project SCARLET, the concepts of advances NITE SIC/SiC fabrication techniques, including new intermediate materials and SiC/SiC preform fabrication techniques. This paper also reports assembly techniques of SiC/SiC fuel pins.

INTRODUCTION

Following the Great East Japan Earthquake and the Fukushima 1st Nuclear Power Station accident, R and D for the enhancement of nuclear safety are being strongly pressed forward in Japan. SiC materials are very attractive for the objective because of the high temperature strength and excellent

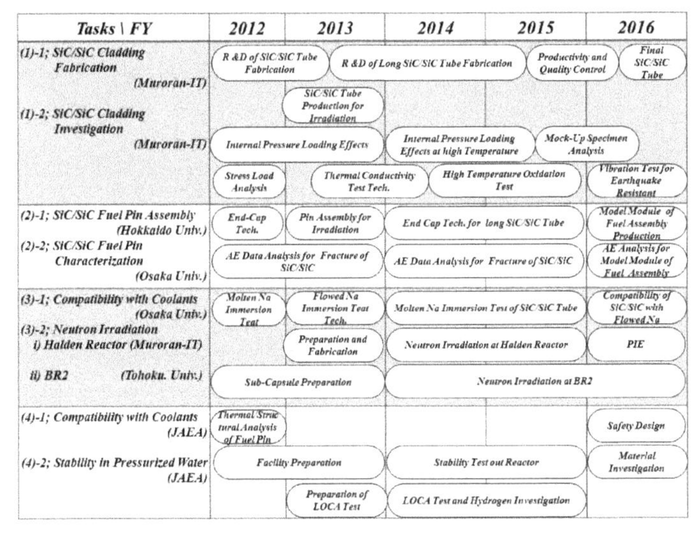

Figure 1 Schedule and Tasks of SCARLET project

chemical stability. Replacement of Zircaloy fuel claddings to SiC/SiC based materials is an considerable option. Ministry of Education, Culture, Sports, Science and Technology in Japan (MEXT) fund-based 5 years termed project has been launched at Muroran Institute of Technology. This project named *SCARLET is driven by five organizations of Hokkaido University, Tohoku University, Osaka University, JAEA and Muroran Institute of Technology. Following tasks of 1) Fabrication technology development of SiC/SiC cladding tube, 2) Assembly technology of SiC/SiC fuel pin, 3) Compatibility with coolants and Neutron irradiation effects, and 4) Nuclear safety and design were provided to the organizations and these are summarized in Figure 1. The schedule and tasks are also shown on the figure. The project is based on the continuous development of SiC/SiC composite named "Nano Infiltration and Transition Eutectoid (NITE)" method[1]. The NITE method is an applied liquid phase sintering method to fabricate SiC/SiC composite from nano-powders of high purity SiC and SiC fibers. SiC/SiC composites have been researched for fusion and fission reactor for a long time, and the irradiation resistance of high purity, crystalized SiC has been represented[2]. The NITE SiC/SiC is a dense material. This character results in the low gas permeability of NITE SiC/SiC composite[3]. The NITE SiC/SiC composite has a possible character for a cladding tube material. The present paper introduces the current status of SiC/SiC tube fabrication and pin assembly techniques in SCARLET project.

NITE SiC/SiC CLADDING FABRICATION

SCARET project aims to fabricate a mock-up SiC/SiC cladding of 1m length having about 10mm inner diameters with about 0.7mm wall thickness. The SiC/SiC has some issues on the mass-productive fabrication process as followings, 1) SiC fibers arrangement and orientation control techniques on the SiC/SiC preform fabrication process, 2) Reasonable fabrication technology for preforms, 3) Near-net shaping technology on the sintering process. For the mass production of the

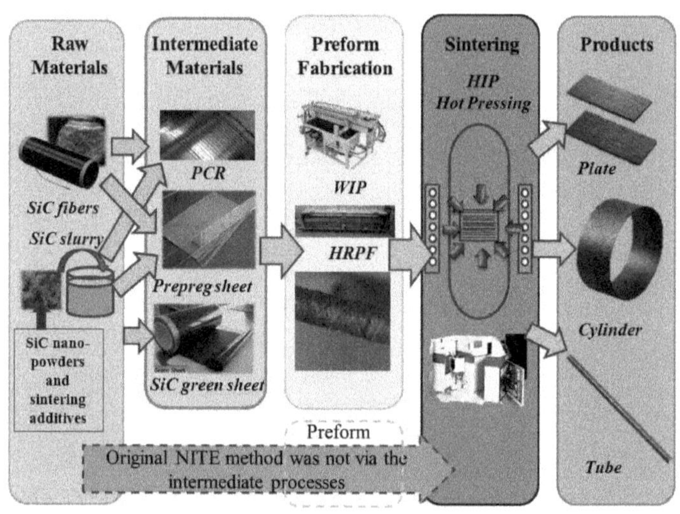

Figure 2 Advanced fabrication process of NITE SiC/SiC progressed in SCARLET project

SiC/SiC, R and D of intermediate SiC materials are important. Figure 2 shows the flow of advanced NITE SiC/SiC process. Originally, the NITE SiC/SiC fabrication process was that the SiC fiber was dipped into slurry of SiC powders, and arranged by hand to form a SiC/SiC preform. The original process did not use any intermediate materials and were not appropriate for the mass-production. To stabilize the quality of materials and to ensure the efficiency of production, some advanced technologies have been developed at OASIS in Muroran Institute of Technology in the past 5 years. A SiC green sheet is a sheet of SiC nano-powders consolidated by binders, it is flexible and easy to handle. A SiC prepreg sheet is a sheet of SiC powders with SiC fibers. The green sheets and the prepreg sheets are cut and stacked to become a SiC/SiC preform. These processes are effective for the quality control of SiC/SiC composites, and also contribute to mechanize the fabrication process. The NITE SiC/SiC production systems are advancing from the handcraft to the industry at OASIS[4]. In SCARLET project, some additional technologies are being developed to establish the fabrication of long SiC/SiC tubes.

Pre-composite ribbon

Normal SiC prepreg sheets need to be cut to be narrow for the winding around tubes. This is not efficient, and cut and paste of SiC prepreg sheets are time consuming. Thus a new intermediate material named Pre-Composite Ribbon (PCR) is being developed. The PCR is a wire-like intermediate material of SiC fibers and nano-powders, which the SiC fibers are spread and dipped in the SiC nano-powder slurry, and hardened flat. The PCR is very convenient for the mechanized winding of SiC fibers with appropriate orientation on a cladding. The PCR is able to wind the SiC fibers with SiC nano-powders at the same time resulting in forming a preform of a cladding tube efficiently[5].

Intermediate densification process

Because the NITE is an applied liquid sintering, the last step of the process is a sintering at elevated temperature. The volume of a SiC/SiC preform is reduced during the sintering resulted in the disorder of SiC fiber orientation on the product. For the suppression of the fiber disorder, densification processes for SiC/SiC preforms are very important. During the PCR winding on a SiC/SiC preform, the densification process needs to apply to the SiC/SiC preform repeatedly. The densification process suppresses macro pores between fiber bundles, and pushes SiC nano-powders in the SiC bundles. The process reduces the volume change during the sintering process, and results in the suppression of SiC fiber disorder. The densification process is brought about using a warm isostatic pressing (WIP) and a newly developed technique named as Hot-Roller Press Forming (HRPF). The WIP is to press the preforms by hydraulic pressure. Because the WIP is performed in a pressure chamber, the removal of pores is progressed efficiently, but its operation is relatively complex. The HRPF is a simpler mechanical method which uses three cylindrical rollers. The cylindrical three rollers press a SiC/SiC preform, and shape the SiC fibers along the circumference. The SiC/SiC preforms are densified using the WIP and the HRPF[6].

The densification of SiC/SiC preforms strongly affects the mechanical performance of the SiC/SiC cladding. Currently, a preform of 50 cm in length is successfully produced.

The sintering of the SiC/SiC cladding is performed by the hot-isostatic pressing (HIP). The projected length of the mock-up of SiC/SiC cladding in SCALET project is up to 1 m, thus, the HIP technique is also being developed. Current status of the HIP technique is that 5 cm SiC/SiC tube is successfully sintered. For the SiC/SiC cladding, some function must be applied to the SiC/SiC tube. The surface of SiC/SiC cladding is planned to be covered with monolithic SiC layer. It is a simple process for the NITE method to form the monolithic SiC layer on the tubes. Figure 3 shows a comparison of monolithic SiC coated and not coated NITE SiC/SiC tubes. The shape of fibers is shown on the surface of the not coated NITE SiC/SiC tube. Because the interfaces between SiC matrix and fibers are exposed to the environment, the tube is likely affected by the coolant. The monolithic SiC coated SiC/SiC does not show the shape of fibers, and this tube is expected to have higher resistance against the environments. The monolithic SiC formation technique is still unstable, and more developments are necessary. The compatibility investigation of SiC/SiC cladding and the coolants such as molten sodium and steam are ongoing[7]. Additionally, a chemical vapor deposition (CVD) device is being introduced to OASIS at Muroran Institute of Technology. SCARLET project will be able to choose the best method to ensure the safety and reliability of NITE SiC/SiC claddings against the coolants.

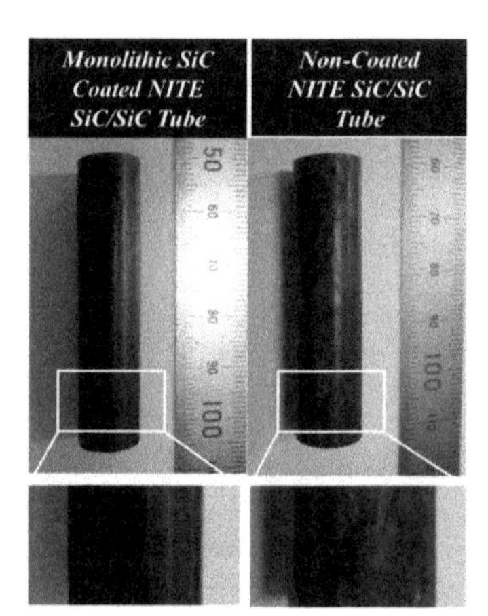

Figure 3 NITE SiC/SiC tubes fabricated in SCARLET project

FUEL PIN ASSEMBLY TECHNOLOGIES

SCARET project includes the task of assembly technology developments of fuel pins as shown in Figure 1. This task includes the end-cap connecting technique and non-destructive characterization technique for SiC/SiC tube. The end-cap is one of the difficult issues. SCARET project is planning to perform neutron irradiation experiments in fission reactors at Halden Reactor in Norway and at BR2 in Belgium. The first 2 years in SCARLET project, assembled pins for the neutron irradiation need to be produced. Based on the experience, end-cap technologies for the long SiC/SiC fuel pins will be developed in the later 3 years. An advantage of NITE SiC/SiC is that a screw-cutting is able to apply for the tube in spite of the ceramic composite. Figure 4 shows photos of basic technology development. Though the screw will be able to ensure some strength toward the axis of tubes, fixing and sealing need to be developed. Investigations of tungsten end-cap and SiC end-cap are ongoing, and a glass sealing method will be also tested. Table 1 describes possible fixing and sealing methods. The diffusion bonding of SiC and tungsten will be stable near or over 1500 °C. The diffusion bonding needs to introduce a diffusion bonding device in a radiation controlled area, or to develop a new technique using mechanical pressing and local heating methods. The SiC and tungsten has very closed coefficient of thermal expansion (CTE) to NITE SiC/SiC, the bonding itself

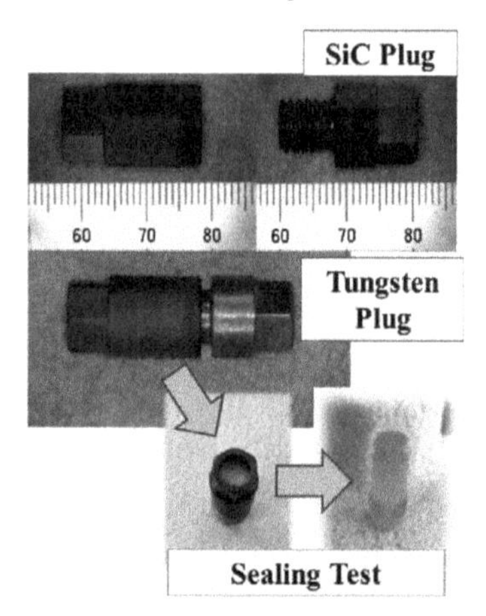

Figure 4 End-cap sealing test for the assembly of NITE SiC/SiC

Table 1 Possible technology of joining for the application of NITE SiC/SiC

Estimated available temp.(°C)	*Material*	*Methods*	*Sealing process condition*				*Previous works*
			Temp.(°C)	*Pressure (MPa)*	*Atmosphere*	*Facility*	
~1000	*Glass*	*Vitrification*	*~1400?*	-	*Vacuum*	*Glove box local heating device*	*Ref.8 Ref.9*
~1500	*W*	*Diffusion bonding*	*1500*	*~20*	*Inert gas*	*Diffusion bonding device*	*Ref.10*
~1900?	*SiC*	*CVD*	*~1200?*	-	*Reaction gas*	*CVD device*	-
		Diffusion bonding	*~1900*	*~20*	*Inert gas*	*Diffusion bonding device*	*Ref.11*

will be easy if the equipment is prepared. The vitrification by glass needs to find an appropriate grass having very closed CTE to SiC, and be investigated its stability under the neutron irradiation. The grass is expected not to have strength and high temperature stability, thus it is only for the sealing at normal operation condition. But the process is expected to be easy and needs some simple device in a glove box, it may be rapidly available for the neutron irradiation experiments in SCARLET project. The acoustic emission (AE) method will be developed as a non-destructive characterization technique for SiC/SiC fuel pins. The AE system developed in SCARLET project will be limited for the investigation of long SiC/SiC fuel pins in non-irradiated environments. The development need to be started from the accumulation of basic techniques for SiC/SiC plates at first, and the development for the long SiC/SiC tube is planned to be performed in later 3 years in the project.

CONCLUSION

Aiming to replace Zircaloy fuel claddings with SiC/SiC based materials is an option, a MEXT in Japan fund based 5 years termed project has been launched at Muroran Institute of Technology. This project includes 1m length SiC/SiC cladding mock-up fabrication, assembly techniques of fuel pins and characterization including material properties, compatibility with coolants, and neutron irradiation effects in fission reactors. R and D of the SiC/SiC cladding technique is ongoing, some new technologies such as SiC green sheet, SiC prepreg sheet, PCR, WIP and HRPF are applying to the fabrication of the claddings. Current status of the cladding fabrication is that 50 mm SiC/SiC cladding and 50 cm SiC/SiC preform are successfully fabricated. The R and D of fuel pin assembly technique has been also started. The end-caps are planned to be joined by mechanical joining, diffusion bonding, CVD and vitrification methods. The AE development for long SiC/SiC tubes is also included in the project.

ACKNOWLEDGEMENT

Present study is mainly supported by the Ministry of Education, Culture, Sports, Science and Technology of Japan (MEXT) on “Innovative Nuclear Research and Development” program.

FOOTNOTES

*SiC Fuel Cladding/Assembly Research Launching Extra-safe Technology

REFERENCES

[1]Y. Katoh, A. Kohyama, T. Nozawa, M. Sato, *J. Nucl. Mater.* **329–333** 587-591 (2004).
[2]Y. Katoh, H. Kishimoto, K. Jimbo, A. Kohyama, *J. Nucl. Mater.,* **307-311** 1221-1226 (2002).
[3]A. Kohyama, A. Moslang, S. Zinkle, H. Kishimoto, "R & D of Structural Materials for Fusion", *Proceedings of PBNC2008, October 13-18, 2008, Aomori, Japan,* Paper ID P16P127.
[4]A. Kohyama, J-S. Park, H-C. Jung, *J. Nucl. Mater.*, **417** 340–343 (2011).
[5]Y. Asakura, J-S. Park, H. Kishimoto, A. Kohyama, "SiC/SiC Fuel Cladding by NITE process for Innovative LWR -Pre-composite Ribbon Design and Fabrication", *Proceedings of PACRIM 10, June 2-7, 2013, San Diego, USA*, submitted.
[6]N. Nakazato, H. Kishimoto, Y. Kohno, A. Kohyama, "SiC/SiC Fuel Cladding by NITE process for Innovative LWR -Cladding forming process development", *Proceedings of PACRIM 10, June 2-7, 2013, San Diego, USA*, submitted.
[7]C. Kanda, Y. Kanda, H. Kishimoto, A. Kohyama, "SiC/SiC Fuel Cladding by NITE process for Innovative LWR - Resistance under high temperature water and steam" *Proceedings of PACRIM 10, June 2-7, 2013, San Diego*, USA, submitted.
[8]P.Colombo, B.Riccardi, A.Donato, G.Scarinci, J. Nucl. Mater., 278 (2000) 127-135
[9]M. Ferraris, M. Salvo, V. Casalegno, S. Han, Y. Katoh, H.C. Jung, T. Hinoki, A. Kohyama, *J. Nucl. Mater.*, **417** 379–382 (2011).
[10]H. Kishimoto, T. Shibayama, K. Shimoda, T. Kobayashi, A. Kohyama, *J. Nucl. Mater.*, **417** 387-390 (2011).
[11]H-C. Jung, Y-H.Park, J-S. Park, T. Hinoki, A. Kohyama, *J. Nucl. Mater.*,**386–388** 847–851(2009).

"INSPIRE" PROJECT FOR R&D OF SiC/SiC FUEL CLADDING BY NITE METHOD

Akira Kohyama
OASIS, Muroran Institute of Technology
Muroran, Hokkaido 050-8585, Japan

ABSTRACT

This paper introduces the new METI funded project, "INSPIRE", where SiC/SiC Fuel cladding R & D for ultra-safe nuclear reactors by new NITE method is ongoing since 2012. The goal of the "INSPIRE" project is to provide soundness of the SiC/SiC claddings by the current NITE method before and after reactor irradiation in BWR condition including nuclear fueled fuel-pin elements in water loop of Halden reactor under the contract with Halden Reactor Project. The "INSPIRE" project is to establish the technological basis to produce SiC/SiC cladding to replace Zircaloy cladding currently used in LWRs. The project emphasizes the importance of end cap joining technology integration and joint performance verification for SiC/SiC fuel pin and Zircaloy end cap. SiC/SiC claddings (10mm diameter, 1mm wall thickness and 200mm long) are satisfying dimensional specification standard of ASTM. Preliminary irradiation rig design and irradiation plan is presented.

INTRODUCTION

Japan has been suffering from "the crisis on March 11, 2011" which included severe accident of TEPCO Fukushima nuclear plants. After the crisis, Japanese energy policies had to be drastically changed from "The 3rd phase basic plan of science and technology of Japan" and "Framework for Nuclear Energy Policy" [1, 2]. The 4th phase basic plan of science and technology was issued on August 2011. In this plan, as the baseline understandings followed to the explanations about the present-status of Japan under "the after effects of the crisis", drastic changes/improvements of nuclear energy systems were pointed out.

Based on the "Basic Energy Policy"(June, 2010) and "The 4th Phase Basic Plan of Science and Technology"(Aug. 2011), The Revised Version of "Framework of Nuclear Energy Policy for 2012" had been prepared ,but was cancelled on October 2, 2012, where the followings were emphasized..

- ◆ Restructuring of nuclear energy policy towards safety assurance and recovery of public trust.
- ◆ Restructuring of nuclear energy policy under large drop of energy dependence on nuclear.
- ◆ Establishment of "The Highest Level of Nuclear Safety" is strictly required.

The Japanese Ministry of Economy, Trade and Industry (METI) supports "R & D towards Ensuring Nuclear Safety Enhancement", where "Innovative Silicon-carbide Fuel Pin Research"

(Project INSPIRE) was approved as five year termed project. Also, The Japanese Ministry of Education, Culture, Sports, Science and Technology (MEXT) Program on "Innovative Nuclear Research and Development" , which has been changed from 2012 to emphasizing "Basic Technology Development for Nuclear Safety Innovation", where "R & D of Basic Fabrication Process Technology of SiC/SiC Fuel Cladding for Extra-Safe Reactor Core" was approved as one of 12 proposals approved. This project is titled as "SiC Fuel Cladding/Assembly Research, Launching Extra-Safe Technology"; SCARLET Project. This paper is introducing the "INSPIRE" project; current status and future plan.

SiC and SiC/SiC related activities for nuclear application were initiated in LWR engineering and FBR engineering. Japanese inventions of PAN type carbon fibers and PCS type SiC fibers were strong basis for initiating C/C and SiC/SiC R & D activities for nuclear application in Japan. Those were basic research activities under JAERI (now, JAEA) and Universities/National Institutes. Figure 1 briefly summarizes SiC/SiC related activities. In this figure, shift from MMC (metal matrix ceramic composite; SiC/Al, C/Al) to CMC (ceramic matrix ceramic composite; C/C, SiC/SiC) can be seen together with the continuing efforts in nuclear and aero-space areas including domestic programs for Fission.

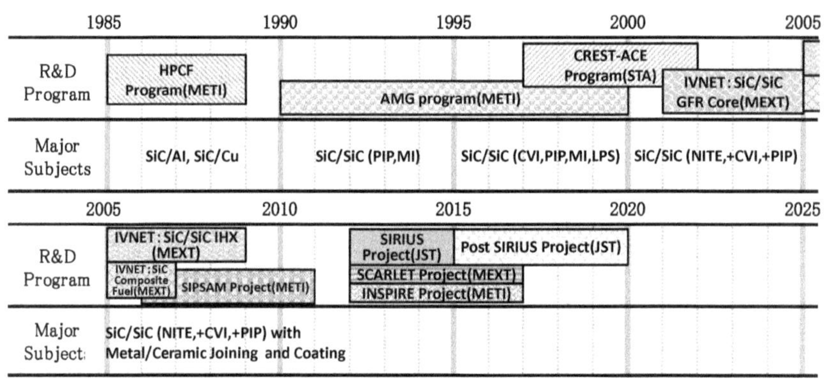

Fig. 1 The Past, Present and Future of Major SiC Composite R & D in Japan

INNOVATIVE SiC FUEL PIN RESEARCH; "INSPIRE" Project

Japanese Ministry of Economy, Trade and Industry (METI) supports "R & D towards Ensuring Nuclear Safety Enhancement" Program, where "Innovative Silicon-carbide Fuel Pin Research" (Project INSPIRE) was approved as five year termed project. Under the same program Toshiba has been approved joint project with IBIDE, Tohoku University and University of Tokyo, where R & D of innovative CVI technology for fabricating fuel pin for LWRs is on-going.

Objectives

The objectives of the "INSPIRE" project are to establish technological basis and to verify the feasibility of the NITE-method based SiC/SiC fuel pin elements and fuelled SiC/SiC fuel pin segment to be used as the replacement of Zircaloy fuel pin/assemblies for LWR core component. Important performances of NITE-based SiC/SiC fuel cladding to be confirmed are;

1) Compatibility with reactor water and nuclear fuel under PWR condition,
2) FP gas leak tightness,
3) Stability under Halden reactor irradiation.

Those verifications are using the NITE-based SiC/SiC fuel cladding produced by a large-scale production process at OASIS, Muroran Institute of Technology and Muroran Establishment of IEST Co., Ltd.

Task and Outputs

1) Ceramic Fuel Pin Fabrication Modification and Technology Integration

The goal of the fabrication modification is to establish large scale production process of NITE-based SiC/SiC fuel pin with the final goal of 4m length. The verification of the performance should be done with the 1 m length fuel pins. The important technology integration includes design and fabrication of end-cap by Zircaloy 2 and welding process establishment with SiC/SiC fuel pin.

2) Stability of Ceramic Fuel Pin Element under BWR Condition

To confirm the baseline stability of NITE-based SiC/SiC fuel pin, Halden reactor irradiation in BWR condition water loop with nuclear fuel capsuled fuel pin elements. The current irradiation plan is to finish reactor irradiation by the end of 2015. Based on the results, including PIEs, the modified NITE process and products based on the modified NITE-process will be provided.

3) Neutron Irradiation Effects of the NITE-SiC/SiC

To establish engineering database for the design and fabrication of NITE-based SiC/SiC fuel pin, neutron irradiation effect database is essential. This task will provide baseline neutron irradiation effect data from BR2 irradiation and Halden reactor irradiation. Currently, BR2 irradiation had been prepared and is suspended.

Accomplishments for Japanese Fiscal Year 2012

The "INSPIRE" project has been approved and been started on December 2012. Due to the time available for the budgeting procedure to Tohoku University, the task 3 for Tohoku University, mainly for BR2 irradiation sub-task was cancelled. Others are briefly summarized as;

1) Ceramic Fuel Pin Fabrication Modification and Technology Integration

200 mm NITE-based SiC/SiC fuel pins were successfully fabricated. By the three HIP treatments with 7 fuel-pins per treatment have been done during March, 2013. Final machining has been finished. The design and fabrication of end-caps by Zircaloy 2 and joint design and machining to 50mm length SiC/SiC fuel pins were done. The TIG welding condition confirmation and preliminary TIG welding has been finished.

2) Stability of Ceramic Fuel Pin Element under BWR Condition

The planning of the Halden reactor irradiation in BWR condition water loop with nuclear fuel capsuled fuel pin elements has been initiated and the conceptual design activity has been finished. Out of the reactor core experiments have been designed and some preliminary tests were done.

3) Neutron Irradiation Effects of the NITE-SiC/SiC

The preliminary discussion for the BR2 irradiation has been done and the conceptual definition of the capsule design and irradiation condition was finished. Irradiation coupon samples have been provided to Tohoku University to be included as a part of "MIKADO" project between Tohoku University and BR2.

The Current Project Plan

The "INSPIRE" project was approved as five year termed program. However the budget situation of METI only allows confirming for the Japanese fiscal year 2013 with the large budget reduction more than 30% from the original plan. The Halden reactor irradiation task is currently under approval only for JPY 2013. Also, under this condition, BR2 irradiation task should be suspended for JPY 2013. The current project plan is shown in Fig. 2.

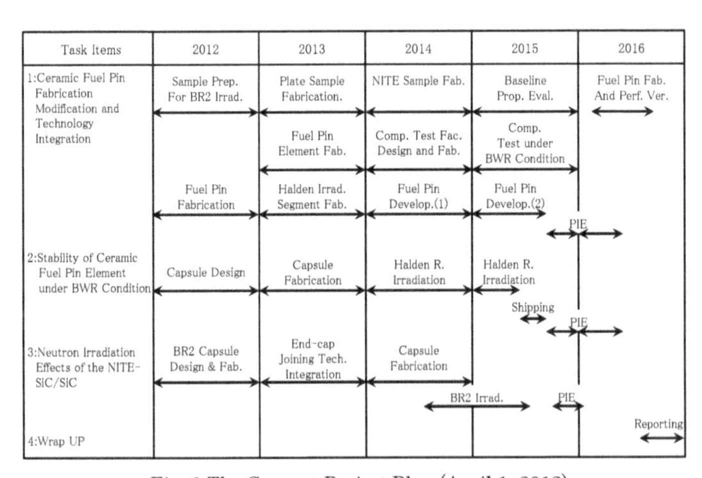

Fig. 2 The Current Project Plan (April 1, 2013)

JOINING OF ZIRCALOY END CAP TO SiC/SiC FUEL CLADDING

The ultimate goal of the replacement of a Zircaloy fuel pin to a SiC/SiC fuel pin is to use a SiC/SiC fuel cladding and SiC/SiC end caps. However, for the "INSPIRE" project, Zircaloy-2 end caps are used to give larger flexibility of the joining process and reliability of welded joint. The current candidate concept to make the end cap joining, the combination of mechanical joining (threading) and metallurgical joining (TIG welding) is anticipated. For the severe dimensional limitation, especially to wall thickness margin, very accurate machining of threading is required. Figure 3 is the accomplishment done in the "CREST-ACE" project [4], where 1.3 mm height threading was successfully and stably machined and has been proven the stability of the threads over 1000 times screw-on and screw-off. However, for the "INSPIRE" project, the wall thickness of the SiC/SiC is currently targeted to be 1 mm to 0.7 mm. Thus the height of the threading should be smaller than 0.5 to 0.3 mm. Although the accurate design based on strength requirement and fracture behavior analysis are still on-going, as a tentative testing and performance evaluation 10 mm (inner diameter) SiC/SiC fuel pins with 0.7 mm wall thickness were provided for machining to make 0.15 mm height threading. Figure 4 is presenting the both elements individually and as screw jointed. The high magnification 3D image of the threading surface is shown in the red column right hand side of the figure.

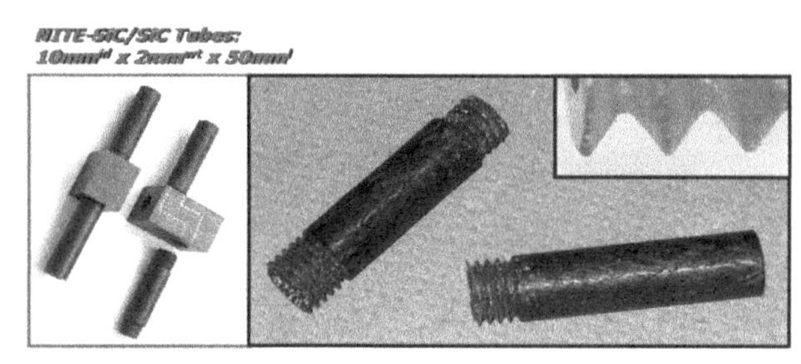

Stable even after 10 years handling by over one hundred persons

Fig. 3 Screw Type Joint (from "CREST-ACE" Project)

The as screwed joint sample was subjected to TIG welding with the arc striking depth corresponding to the thread height. This is providing appropriate weld zone formation which limits damage to the SiC/SiC thread as small as possible. The

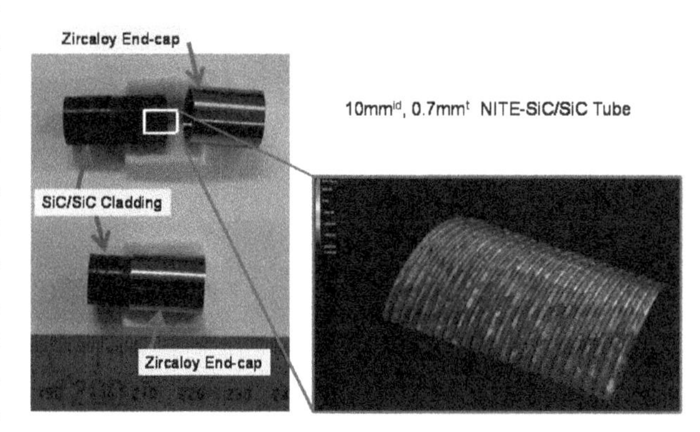

Fig. 4 Zircaloy-2 + SiC/SiC Joint Test Sample

final welding condition will be defined within this year for evaluation of welded joint performance. This is required before inserting test samples into out-of-core water-loop of Halden Reactor.

Preliminary Results of X-ray CT Analysis

For a potential non-destructive inspection method of SiC/SiC fuel pin and its joint with Zircaloy, X-ray CT analysis looks quite attractive and appropriate. The feasibility testing of this method has been started recently and the preliminary results became available. The X-ray CT imaging was performed at Industrial Research Institute, Hokkaido Research Organization. Micro Focus X-ray CT System (SMX-225CT) is providing 0.004 mm micro beam. Currently, imaging data was accumulated with the 1024 x 1024, as shown in Fig. 5. The maximum resolution is obtained with 2028 x 2028 and will provide more detailed information. Figure 5 (A) is the images of SiC/SiC tube, where diametric cróss section, axial image from right side, mid-wall cross-section of red dotted line position, and axial image from the bottom side are indicated crock-wise , respectively. Well controlled layer structure with 2D fiber orientation control can be identified. Figure 5 (B) is the images of joint with Zircaloy end cap, where fine threading and excellent mechanical joining with better accuracy than 0.1 mm can be clearly seen. The TIG welding applied to this joint is currently under analysis and the results will be provided soon.

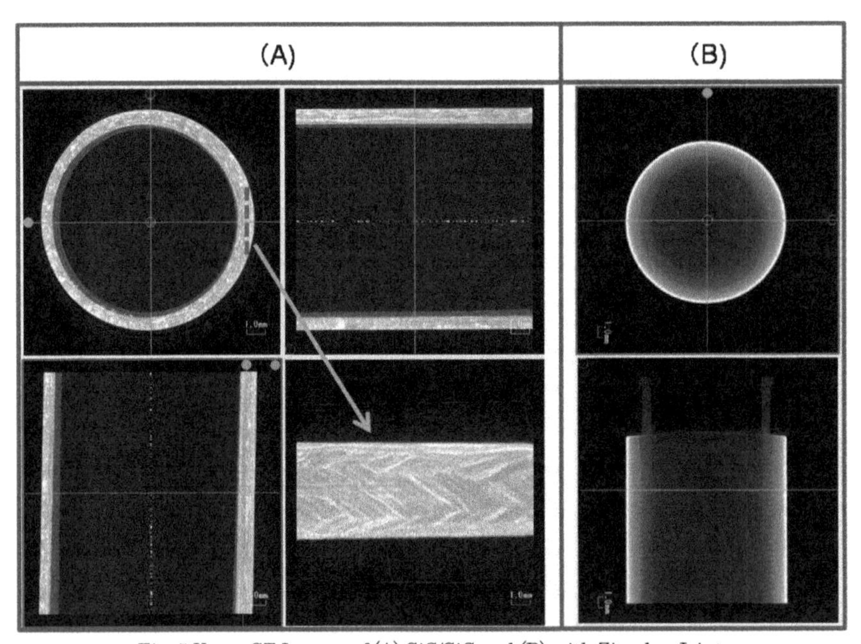

Fig. 5 X-ray CT Images of (A) SiC/SiC and (B) with Zircaloy Joint

THE CURRENT PLAN OF HALDEN REACTOR IRRADIATION

Due to the unclear budget situation of METI toward JFY 2014, Multi-year contract with Halden Reactor Project is suspended and the single year contract is now under final negotiation. The conceptual planning and design of the irradiation segment and rods has been finished at the end of JFY 2012. The following is the brief summary of the conceptual planning and design activity between "INSPIRE" project and Halden Reactor Project.

Test Rig to be Applied

- The irradiation is to be carried out in a test rig within a pressure flask, cooled by water at BWR thermal-hydraulics and chemical conditions. Some details are shown in Fig. 6.
- The rig will accommodate 6 test rods, typically 20 cm long, arranged in two clusters, as shown in Fig. 6. The rods will be equipped with cladding elongation detector for measuring on-line the amount of pellet-cladding mechanical interaction (PCMI) and cladding permanent growth due to PCMI and neutron irradiation.

Test Rods

- The six rods may contain different variants, such as different cladding or end-plug material.
- The fuel pellets will be fabricated at the IFE establishment at Kjeller, under agreed specifications. The cladding tube and the end plugs will be provided by the "INSPIRE" project, which is also to define, in consultation with Halden Reactor Project. The rod inner pressure is atmospheric pressure under current design. The fuel rod assembly, gas filling and end-plug welding will be realized by IFE following further discussions and under agreed specifications. To meet Halden reactor requirements for inserting those test rods into the reactor, preliminary discussions and test fabrication with TIG welding was done by the "INSPIRE" project.

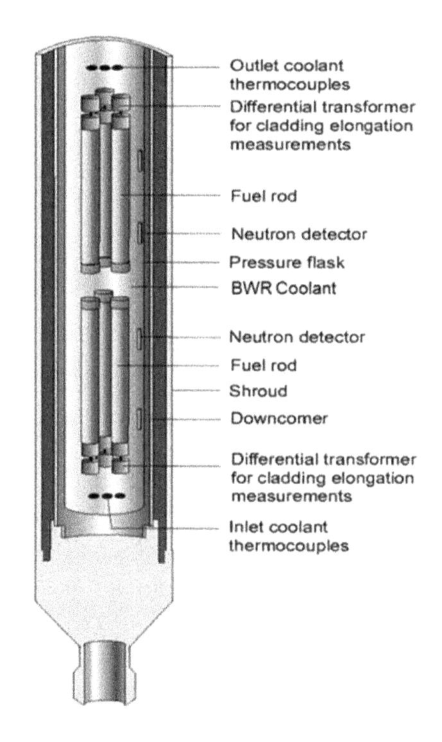

Fig. 6 Conceptual Drawing of the Test Rig

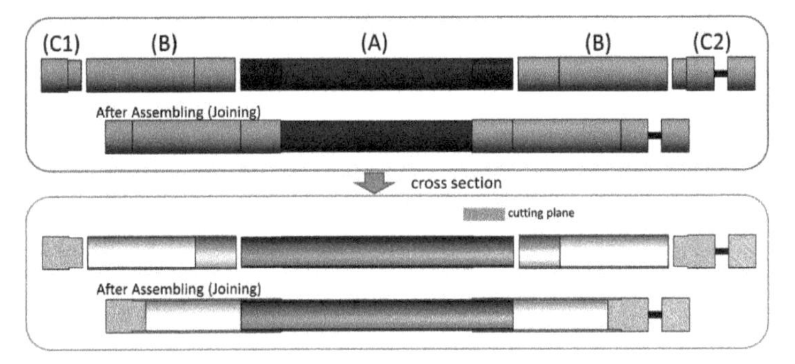

Fig. 7 Concept of the Test Rod

Based on very preliminary investigation, the first conceptual design of the test rod was basically agreed. Figure 7 indicates components to be jointed. For the operation at Halden reactor site for the final insertion of nuclear fuel and shield welding of end cap by EBW facility of Halden, three types of Zircaloy elements and a SiC/SiC fuel pin element with fine threading both ends will be fabricated. The element (A) is SiC/SiC fuel pin element with fine threading, element (B) is Zircaloy tube with inner threading and elements (C1) and (C2) are Zircaloy end cap. Two elements (B) are screwed up to the both ends of the element (A) and followed by TIG welding to make weld joint of Zircaloy and SiC/SiC. Then nuclear fuel is inserted into this component and elements (C1) and (C2) are electron beam welded to make perfect shieling of the rod. The elements (C1) and (C2) are connected by threading to make 6 rods into 3 segments to be inserted irradiation rig, as shown in Fig.6.

In-reactor Operation

- It is foreseen that the test will be run at power conditions typical for commercial fuel, i.e. 20-25 kW/m at beginning of irradiation and then gradually decreasing. The PCMI and permanent cladding strain will be assessed continuously during the irradiation.

THE FUTURE PLAN AND "SCARLET" PROJECT

The "INSPIRE" project planning toward JFY 2014 is on-going, searching for the better option to enforce the current plan including some options to combine with other projects, transfer to the other budget of METI and others. The important accomplishment is to provide realistic and economical technology available very soon based on the existing industrial organization/structure. For this final output, Halden reactor irradiation is inevitably important and essential.

The MEXT funded "SCARLET" project is an important supporting activity from further advanced fabrication technology with refined interconnecting fiber structure with advanced automation fabrication process. The "SCARLET" project introduction and some of the recent accomplishments are available in these proceedings.

CONCLUSIONS

The new METI funded project, "INSPIRE", where SiC/SiC Fuel cladding R & D for enhancing nuclear energy safety by new NITE method is introduced and the program outline, the current status with recent accomplishments and the future plan are briefly introduced.

The goal of the "INSPIRE" project is to provide soundness of the SiC/SiC claddings by the current NITE method before and after reactor irradiation in Halden Reactor. The current conceptual design activity is provided.

The "INSPIRE" project is to establish the technological basis to produce SiC/SiC claddings to replace Zircaloy cladding currently used in LWRs. The importance of end cap joining technology integration and joint performance verification for SiC/SiC fuel pins and Zircaloy end caps are emphasized and preliminary progress is provided.

The "INSPIRE" project may make a way to an early realization of SiC/SiC fuel pin/assembly into LWRs ensuring reactor safe technology in a realistic and economical way.

Acknowledgments

Present study is mainly supported by the Ministry of Economy, Trade and Industry (METI) on "Research and Development towards Ensuring Nuclear Safety Enhancement" program for the "INSPIRE" project. The supporting activities for making NITE-SiC/SiC materials and basic property evaluation done through OASIS with including utilization of FEEMA Facility are acknowledged. The author acknowledges Professors Y. Kohno and H. Kishimoto for their collaboration, support and encouragement. The members of OASIS, Muroran Institute of Technology are also acknowledged for their continuing efforts for this study. The special appreciation is due Mr. D. Hayasaka for his efforts to X-ray CT analysis.

References

[1] White Paper on Science and Technology 2009, MEXT, Japan.

[2] Framework for Nuclear Energy Policy, Atomic Energy Commission, Japan, 2005.

[3] Advanced SiC/SiC Ceramic Composites, Editors A. Kohyama, M. Singh, H.T. Lin and Y. Katoh, Ceramics Transactions vol.144(American Ceramic Society, 2002)

[4] A. Kohyama, Y. Kohno, H. Kishimoto, J. S. Park, and H. C. Jung, "Industrialization of Advanced SiC/SiC Composites and SiC Based Composites; Intensive activities at Muroran Institute of Technology under OASIS", Proceedings of ICC3 (2010)

[5] N. Nakazato, H. Kishimoto, Y. Kohno, A.Kohyama: Ceram. Eng. Sci. Proc. 32 (2011) 103-108

SiC/SiC FUEL CLADDING BY NITE PROCESS FOR INNOVATIVE LWR-CLADDING FORMING PROCESS DEVELOPMENT -

Naofumi Nakazato[1], Hirotatsu Kishimoto[2], Yutaka Kohno[2] and Akira Kohyama[1]

[1]Organization of Advanced Sustainability Initiative for Energy System/Materials (OASIS), Muroran Institute of Technology, Muroran, Hokkaido, 050-8585, Japan

[2]College of Design and Manufacturing Technology, Muroran Institute of Technology, Muroran, Hokkaido, 050-8585, Japan

ABSTRACT

After the Fukushima accident, the nuclear reactor safety improvement by utilizing SiC/SiC into reactor core became one of the top priority R&D (Research & Development) subjects. Among potential fabrication methods, nano-infiltration and transient eutectic-phase (NITE) process is the most realistic near term method to produce high quality SiC/SiC fuel cladding. In this study, the effectiveness of various forming techniques on SiC/SiC fuel cladding, such as the WIP (Worm Isostatic Press) method and the HRPF (Hot-Roller Press Forming) method, is examined. Pores in the inter-prepreg sheets were suppressed by the WIP process. By the HRPF process, in addition to preform densification, smoothing of preform surface was confirmed. As the result, wrap-up winding of prepreg sheet following the HRPF treatment becomes more easily, thus workability is improved. The combined process of the WIP and the HRPF methods, called "the maloti-stage preform forming process", greatly improved quality and workability of preform forming. Finally, the densified tubular preform was sintered by the HIP (Hot Isostatic Press) for making SiC/SiC tubes. The combined densification process drastically improved fiber architecture of the final SiC/SiC products. A SiC/SiC tube with monolithic SiC surface layers was fabricated under an excellent quality control.

INTRODUCTION

After the Fukushima accident, the innovative R&D on nuclear reactor safety improvement has been strongly required. The one of innovations for nuclear reactor safety improvement is replacement of cladding and reactor core structural materials from Zircaloy to silicon carbide composite (SiC/SiC composite). SiC/SiC composites have been recognized as attractive candidate structural materials for advanced nuclear fission and nuclear fusion due to their superior radiation stability coupled with the excellent baseline properties such as thermo-mechanical properties, chemical stability and etc. [1-3]. It is known that there are several processes for fabrication of SiC/SiC composites, such as chemical vapor infiltration (CVI), polymer impregnation and pyrolysis (PIP), reaction sintering (RS) and liquid phase sintering (LPS) [4-7]. Among the existing fabrication methods, nano-infiltration and transient eutectic-phase (NITE) process, which is an advanced liquid phase sintering method, is considered to be one of the most realistic near term method to produce high quality SiC/SiC fuel cladding. NITE process has been presenting the advantages in size flexibility, low porosity, and excellent neutron damage stability [8-10]. In order to fabricate high performance NITE-SiC/SiC fuel cladding, establishment of the preform forming technique with fiber-architecture control and the near-net shaping capability is essential. Especially, in the NITE process, since large volumetric shrinkage occurs during final forming, significant fiber-architecture and -strength damage in final product is concerned. Therefore, fiber-architecture control during fabrication of high dense preform is very important. Our previous study reported the effectiveness of preform densification on simple shaped (plate) and complex shaped NITE-SiC/SiC composites [11]. However, application cases on complex shaped composites still are a few. Also, development of the new preform densification method is necessary for improved quality of NITE-SiC/SiC fuel cladding. This study presents the effectiveness of various forming techniques on SiC/SiC fuel cladding. In particular, effects of two preform densification techniques, the worm isostatic press (WIP) and the hot-roller press forming (HRPF) methods on tubular preforms are evaluated. Finally, a method of dense preform fabrication by integration of these techniques, called "the multi-stage preform forming process", is presented.

EXPERIMENTAL PROCEDURE

SiC 'nano'-slurry for fabrication of green sheets consists of β-SiC nano-powder and sintering additives with Al_2O_3 and Y_2O_3. The green sheets are produced by facilities of Organization of Advanced Sustainability Initiative for Energy System/Materials (OASIS) / Facility of Energy and Environmental Material Assessment (FEEMA). Pyrocarbon (PyC) coated-Tyranno™ SA fibers were used as reinforcement for SiC/SiC cladding fabrications. The PyC coating was appropriately chosen at the thickness of 0.5 μm by chemical vapor deposition (CVD) process. Prepreg sheets were fabricated form PyC-coated Tyranno-SA fibers, green sheets and 'nano'-slurry. Prepreg sheets were wound in $\pm 30^{\circ}$ cross-plied (CP) for preparation of tube shaped preforms, which is followed by preform densification. The preform densification was performed by WIP method and/or HRPF method. The preforms prepared were sintered by the hot isostatic press (HIP) at 1800-2000 °C under a pressure. The surface and cross-section of samples were observed with an optical microscope.

RESULTS AND DISCUSSION

Effect of WIP treatment on tubular preform

The WIP is method, which a vacuum-packed preform is isostatically pressed in pressurized water under heating. In our previous study, preform densification was demonstrated to suppression of large volumetric shrinkage and fiber-architecture damage during hot-pressing [11]. Figure 1 shows optical microscope images of cross-section of the preform before and after WIP treatment. Before WIP treatment, many pores can be identified in the inter-prepreg sheets (Figure 1 (a)). After WIP treatment, decrease and/or disappearance of these pores are confirmed (Figure 1 (b)). Porosity as determined by an image analysis is reduced by WIP treatment from 3.9 % to 0 %. However, the wrinkle in the preform surface was slightly observed in after WIP treatment. Thus, in order to fabricate high quality preform without wrinkle in the surface, it is necessary to perform densification in every time of wrap-up winding prepreg sheets. The WIP operation is relatively complex process form reason of necessity of packing to the preform and, putting in and out of the preform to the pressure chamber. From these reasons, in order to progress industrialization and to improve quality of tubular preform, development of the new preform densification process and combined process of several densification processes are necessary.

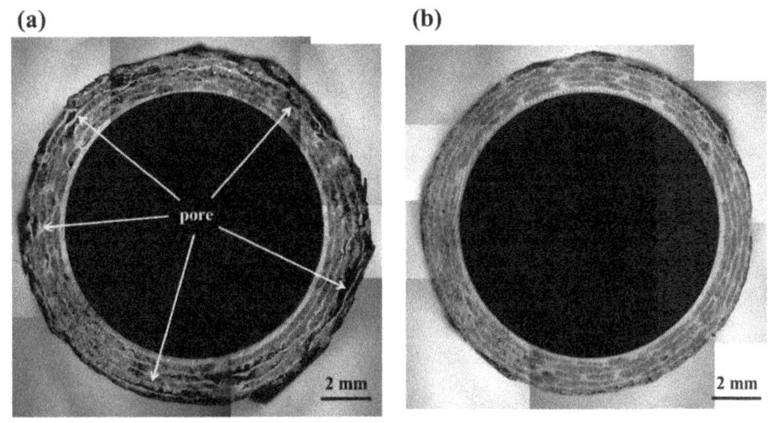

Figure 1: Optical microscope images of cross-section of the preform before and after WIP treatment.

Effect of HRPF treatment on tubular preform

The HRPF method is developed for "SiC Fuel Cladding/Assembly Research, Launching Extra-Safe Technology (SCARLET) Project". This method solved problems of WIP method and is able to correspond to fabrication of high quality preform and industrialization of NITE process. The HRPF method is designed as following basic concepts: (1) suppression of pores in the inter-prepreg sheets by heating roller, (2) maintain of circularity of preform and (3) improvement in workability. Figure 2 shows the schematic of HRPF method and appearance during HRPF treatment. The preform is set in the center of three hot-rollers and is pressed by these rotated rollers. Figure 3 shows optical microscope images of surface and cross-section of the preform before and after HRPF treatment. Before HRPF treatment, the preform has rough surface (Figure 3 (a)). In addition, many pores in the inter-prepreg sheets are observed from a cross-section image (Figure 3 (c)). Pores are mainly distributed in the inter-fiber bundles. Porosity as determined by an image analysis in before HRPF treatment is 3.9 %.

After HRPF treatment, the preform has smooth surface (Figure 3 (b)). Also, suppression of pores in the inter-fiber bundles are indicated (Figure 3 (d)). Porosity was decreased by HRPF from 3.9 % to 0.2 %. However, pores in the inter-fiber bundles are a few remained. By smoothing of preform surface, winding of prepreg sheet following the HRPF treatment became more easily, thus workability of preform forming improved.

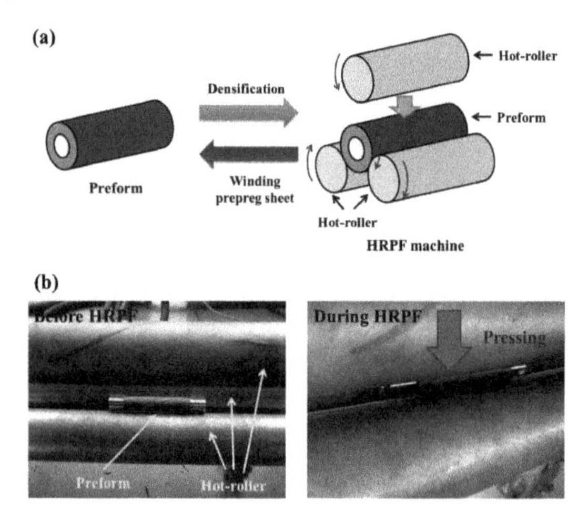

Figure 2: (a) Schematic of HRPF method and (b) appearance during HRPF treatment.

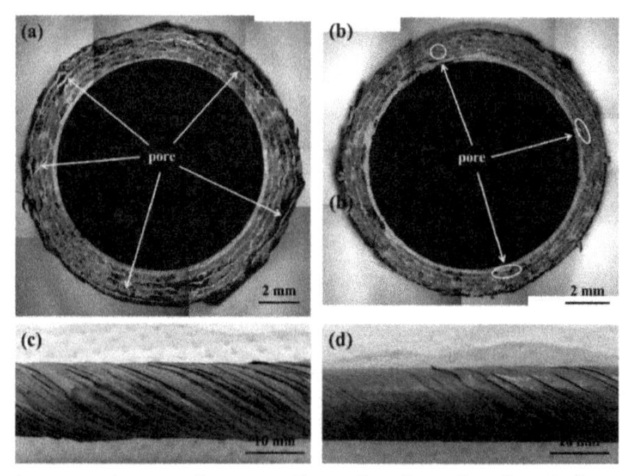

Figure 3: Optical microscope images of cross-section and surface of preform before and after HRPF treatment.

Effect of the multi-stage preform forming process on tubular preform

Toward establishment of the preform forming process for the tubular preform, combination of WIP and HRPF methods was performed. As the concept of the multi-stage preform forming process, HRPF method is used in wrap-up winding of each layer, and then WIP method is utilized in final preform densification. The appearance of preform after HRPF treatment in each layer and the optical microscope image of cross-section of preform after WIP treatment are shown Figure 4. Smooth surface was observed in all layers (Figure 4 (a)-(e)). After WIP treatment, wrinkle in the surface and pores in the inter-prepreg sheets are unobserved (Figure 4 (f) and (g)). The black area indicates fiber bundle region, which is not defect, as shown Figure 4 (g). In addition, aspect ratio of outer diameter of the preform after WIP treatment is 1.0, where the preform has maintained circularity. The multi-stage preform forming process greatly improved quality and workability of preform forming.

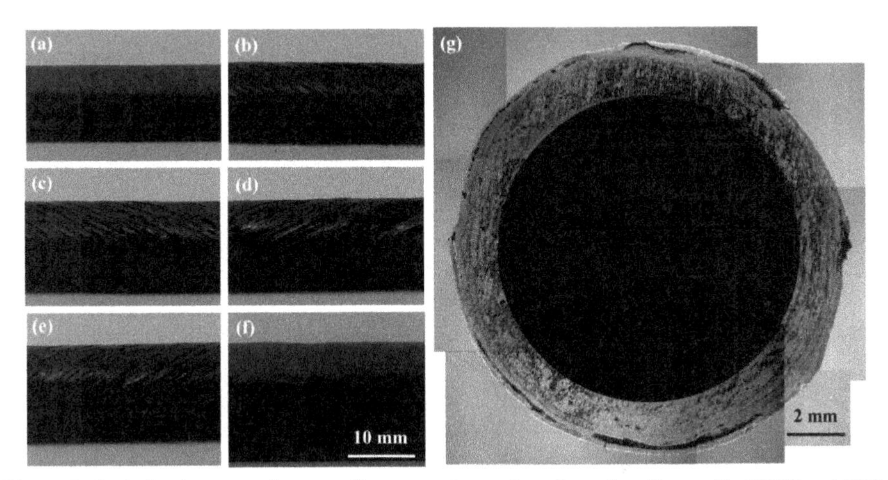

Figure 4: Optical microscope images of cross-section and surface of preform with HRPF and WIP treatments: (a) Inner monolithic SiC layer, (b) 1st SiC/SiC layer, (c) 2nd SiC/SiC layer, (d) 3rd SiC/SiC layer, (e) 4th SiC/SiC layer, (f) Outer monolithic SiC layer after WIP treatment, (g) cross-section of preform after WIP treatment.

SiC/SiC tube after HIP'ing and machining

The two types of densified tubular preforms were fabricated with/without surface SiC layers. Those preforms were HIP processed and surface machined. Figure 5 (a) shows appearance of the SiC/SiC tube with monolithic SiC surface layers. Dimension of the SiC/SiC tube for this study was 50 mm^L, 10 mm^{OD} and 1 mm^t. This SiC/SiC tube is an answer to environmental resistance. The thickness and quality of SiC layer is designed to meet the requirement. The combined densification can clearly contribute controllability of surface layer thickness and quality. In the surface of SiC/SiC tube without monolithic SiC layer, SiC fiber bundles are clearly observed under well controlled manner, as showen in Figure 5 (b). However, further research for improvement in fabrication techniques and their integration.

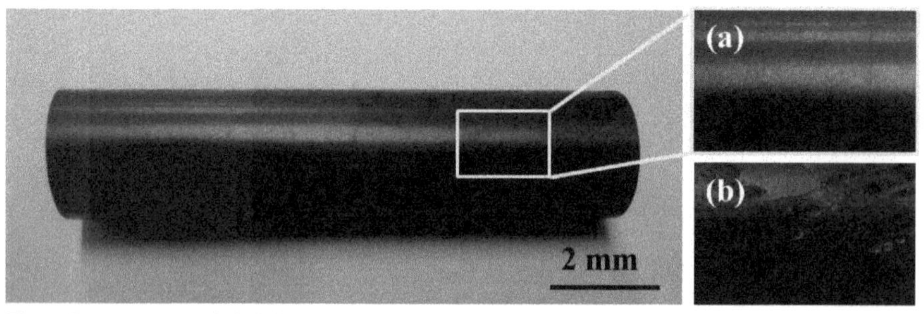

Figure 5: Appearance of SiC/SiC tube after HIP and machining (a) with monolithic SiC surface layer and (b) without monolithic SiC surface layer.

CONCLUSION

The effects of various forming techniques on SiC/SiC fuel cladding were evaluated. As to the effect of WIP treatment for the preform densification, pores in the inter-prepreg sheets were greatly suppressed. As to the effect of HRPF treatment as preform densification, smoothing of preform surface and suppression of pores in the inter-fiber bundles were confirmed. As a result, wrap-up winding of prepreg sheet following the HRPF treatment became more easily, thus workability of preform forming was improved. The multi-stage preform forming process greatly improved quality and workability of preform forming. Finally, densified tubular preforms were sintered by the HIP. The environmentally resistant SiC/SiC tube with monolithic SiC surface layers was fabricated under an excellent quality control by the multi-stage preform forming process.

ACKNOWLEDGEMENT

Present study is mainly supported by the Ministry of Education, Culture, Sports, Science and Technology of Japan (MEXT) on "Innovative Nuclear Research and Development" program.

REFERENCES

[1]R. Naslain, Design, Preparation and properties of non-oxide CMCs for application in engines and nuclear reactors: an overview, *Compos. Sci. Technol.*, **64**, 155-170 (2004)

[2]L.L. Snead, T. Nozawa, M. Ferraris, Y. Katoh, R. Shinavski and M. Sawan, Silicon carbide composites as fusion power reactor structural materials, *J. Nucl. Mater.*, **417**, 330-339 (2011)

[3]Y. Katoh, L. L. Snead, I. Szlufarska and W.J. Weber, Radiation effects in SiC for nuclear structural applications, *Curr. Opin. Solid State Mater. Sci.*, **16**, 143-152 (2012)

[4]H. Araki, W. Yand, H. Suzuki, Q. Hu, C. Busabok and T. Noda, Fabrication and flexural properties of Tyranno-SA/SiC composites, *J. Nucl. Mater.*, **329-333**, 567-571 (2004)

[5]R. Jones, A. Szweda and D. Petrak, Polymer derived ceramic matrix composites, *Composites: A*, **30**, 569-575 (1999)

[6]A. Sayano, C. Sutoh, S. Suyama, Y. Itoh and S. Nakagawa, Development of a reaction-sintered silicon carbide matrix composites, *J. Nucl. Mater.*, **271-272**, 467-471 (1999)

[7]K. Yoshida, Budiyanto, M. Imai and T. Yano, Processing and microstructure of silicon carbide fiber-reinforced silicon carbide composite by hot-pressing, *J. Nucl. Mater.*, **258-263**, 1960-1965 (1998)

[8]K. Shimoda, T. Hinoki, H. Kishimoto and A. Kohyama, Enhanced high-temperature performances of SiC/SiC composites by high densification and crystalline structure, *Compos. Sci. Technol.*, **71**, 326-332 (2011)

[9]H. Kishimoto, K. Ozawa, O. Hashitomi and A. Kohyama, Microstructural evolution analysis of NITE SiC/SiC composites using TEM examination and dual-ion irradiation, *J. Nucl. Mater.*, **367-370**, 748-752 (2007)

[10]A. Kohyama, Joon-Soo Park and H.C. Jung, Advanced SiC fibers and SiC/SiC composites toward industrialization, *Ceram. J. Nucl. Mater.*, **417**, 340-343 (2011)

[11]N. Nakazato, H. Kishimoto, K. Shimoda, J. S. Park, H. C. Jung, Y. Kohno and A. Kohyama, Effects of preform densification on near-net shaping of NITE-SiC/SiC composites, IOP Conf. Ser: Mater. Sci. Eng., 18, 202011 (2011)

Advances in Photocatalytic Materials for Energy and Environmental Applications

PREPARATION OF BROOKITE-TYPE TITANIUM OXIDE NANOCRYSTAL BY HYDROTHERMAL SYNTHESIS

S. Kitahara, T. Hashizume, A. Saiki
University of Toyama
Toyama, Toyama, Japan

ABSTRACT

Titanium oxide of the brookite structure was prepared by the hydrothermal treatment of a glycolic acid titanium complex solution. It was found that the crystal phase of the brookite could be obtained by the hydrothermal treatment from this precursor solution adjusted to basic pH condition, on the other hand the phase of rutile type can be obtained under from neutral pH to weakly basic pH condition. So it is considered that the brookite or the rutile phase could be synthesized selectively by adjusting the pH. Identification of the crystalline phase was perfomed by X-ray diffraction (XRD). The shape of the particle was observed by a field emission scanning electron microscope (FE-SEM). The shape of prepared brookite was a rod-like and nanometer particle (100×100×400 nm).

INTRODUCTION

In recent years, photocatalysts have been used as air cleaning, deodorization, antibacterial material, and purifing water. Main material for a photocatalyst is titanium oxide (TiO_2). TiO_2 has three crystal phases such as rutile, anatase, brookite [1]. Anatase shows photocatalytic activity in UV light, it is already widely used as a photocatalyst. Brookite shows photocatalytic activity even in a visible light. However, there is little industrial application. The reason is that purification is difficult because brookite is metastable phase thermodynamically in the TiO_2 species, and it transforms to rutile at high temperatures [2, 3].

For industrial, titanium oxide is synthesized using the chlorine method and sulfuric acid method, by dissolving mineral rutile and ilmenite ore as a raw material [4, 5]. On the other hand, in the laboratory, the hydrothermal method is widely used in the synthesis of the titanium oxide. Various solutions have been used as a raw material. For example, the strong acid aqueous solution of the titanium sulfate, strongly acidic aqueous solution of titanium tetrachloride, alcohol solution of titanium alcoxide, and titanium oxalate. However, these solutions have corrosive, volatile, flammable, and toxic, so handling has been difficult [1]. In addition, titanium oxide made from these solutions is almost rutile and anatase phase. Even if the brookite phase was obtained, the other phases are mixed as a by-product. So, the each solution, the synthesis of brookite single phase has been difficult. Therefore, the glycolic acid titanium complex have been attracting attention as a new precursor. This complex is stable in water and aqueous solution exhibits a weakly acidic with pH 5 ~ 6. There is not volatilization and corrosiveness. The glycolic acid is an organic acid present also in nature, so the risk of toxic by-products is low. Further, the complex has high stability even in the conditions in which the conventional precursors are hydrolyzed [1, 6]. Result of these characteristics, it is easier to handle much more than conventional precursor.

In this study, titanium oxide of the brookite structure was prepared by the hydrothermal treatment of a glycolic acid titanium complex solution. Identification of the crystalline phase was perfomed by X-ray diffraction (XRD). The shape of the particle was observed by a field emission scanning electron microscope (FE-SEM). The shape of prepared brookite had a rod-like and nanometer particle (100×100×400 nm).

EXPERIMENTAL PROCEDURE

Titanium metal powder was dissolved in a cooled aqueous solution containing 30 mass % H_2O_2 and 28 mass % NH_3 aqueous solution to obtain peroxo-titanium complex. Then glycolic acid was added to the solution. The solution was heated at 80 °C to remove excess H_2O_2 and NH_3. Peroxo glycolic acid titanium aqueous solution was prepared to a concentration and volume (0.1 mol / dm^3 or 0.04 mol / dm^3, 10 cm^3), added various concentrations of NH_3 (0.7 wt % ~ 28 wt %, 6 cm^3) and total volume of the solution was adjusted to 16 cm^3 in volume. The prepared solution was poured into a Teflon vessel of 20 cm^3, and heated at 170 °C for 24 h using a stainless-steel autoclave. After the reaction, the titanium oxide powder was filtered and dried.

The crystal structure of the obtained powder was analyzed by XRD (Shimadzu XRD-6100). The shape of the obtained particles was observed by FE-SEM (JEOL JSM−6700F).

RESULT AND DISCUSSION

Fig. 1 shows XRD patterns of the titanium oxide powder obtained by the hydrothermal treatment at 170 °C for 24 h of peroxo glycolic acid titanium aqueous solutions in different concentrations of NH_3 (a) 0.7 wt % ~ (f) 28 wt %. Fig. 2 shows relation between the profile intensity and NH_3 aqueous concentration. The brookite phase was obtained preferentially in accordance with increasing the ammonia concentration of the precursor solution. It is thought that the selective synthesis of the brookite phase is possible by adjusting the pH.

Fig. 3 shows SEM images of the titanium oxide powder obtained by the hydrothermal treatment at 170 °C for 24 of peroxo glycolic acid titanium aqueous solutions in different concentrations of NH_3, (a) 0.7 wt% (rutile phase), (c) 1.9 wt % (brookite and a small amount of rutile phase), (f) 28 wt % (brookite phase). The shape of prepared brookite was a rod-like and nanometer particle. Average particle size determined from FE-SEM micrographs was 100×100×400 nm.

Fig. 4 shows SEM images of the brookite obtained by the hydrothermal treatment at 170 °C for 24 h of peroxo glycolic acid titanium aqueous solutions (pH 10) in different precursor concentrations, 0.1 mol / dm^3 or 0.05 mol / dm^3. It was found that particle size of the brookite prepared by adjusting precursor concentration to 0.05 mol / dm^3 was smaller than one prepared by adjusting to 0.1 mol / dm^3. It is thought that it is possible to control the small particle size by adjusting the low concentration of the precursor.

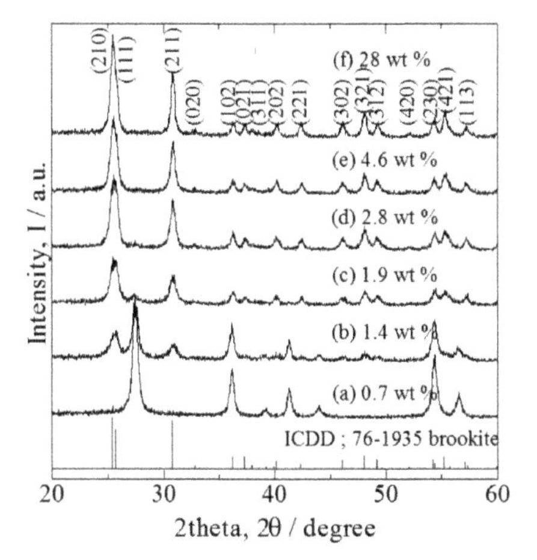

Fig. 1 XRD patterns of the titanium oxide powder obtained by the hydrothermal treatment at 170 °C for 24 h of peroxo glycolic acid titanium complex solutions in different concentrations of NH_3 aq, (a) 0.7 wt % ~ (f) 28 wt %.

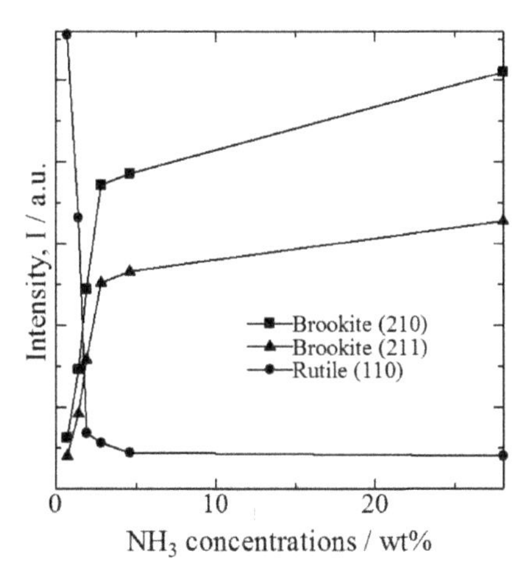

Fig. 2 Relation between the profile intensity and NH_3 aq concentration.

Fig. 3 SEM images of the titanium oxide powder obtained by the hydrothermal treatment at 170 °C for 24 of peroxo glycolic acid titanium aqueous solutions in different concentrations of NH_3 aq, (a) 0.7 wt % (rutile phase), (c) 1.9 wt % (brookite and a small amount of rutile phase), (f) 28 wt % (brookite phase).

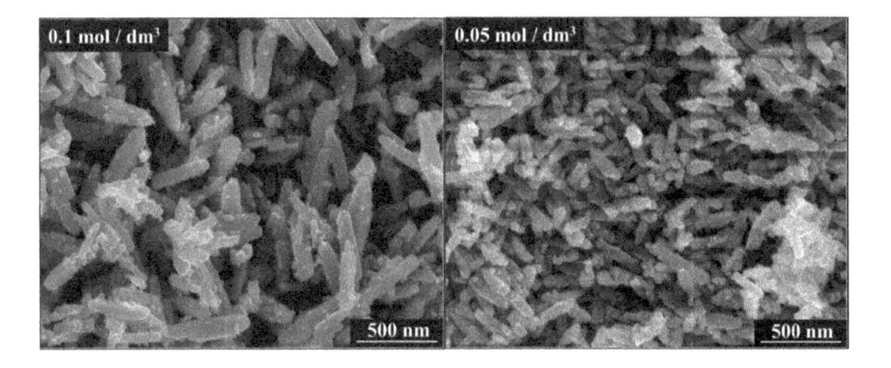

Fig. 4 SEM images of the brookite obtained by the hydrothermal treatment at 170 °C for 24 h of peroxo glycolic acid titanium aqueous solutions (pH 10) in different precursor concentrations, 0.1 mol / dm^3 or 0.05 mol / dm^3.

CONCLUSIONS

Titanium oxide of the brookite structure was prepared by the hydrothermal treatment of a peroxo glycolic acid titanium complex solution. Brookite phase was obtained preferentially in accordance with increasing the ammonia concentration of the precursor solution. It is seems that the selective synthesis of brookite is possible by adjusting the pH. It was found that particle size of the brookite prepared by adjusting precursor concentration to 0.05 mol / dm^3 was smaller than one prepared by adjusting to 0.1 mol / dm^3. It is seems that it is possible to control the small particle size by adjusting the low concentration of the precursor.

REFERENCES

[1] M. Kakihana, M. Kobayashi, V. Petriykin, and K. Tomita, J. Jpn. Soc. Powder Powder Metallurgy, 56, 188-193 (2008)

[2] V. Stengl, and D. Kralova, Mater. Chem. Phys., 129, 794-801 (2011)

[3] Y. Morishima, M. Kobayashi, V. Petriykin, S. Yin, T. Sato, M. Kakihana, and K. Tomota, J. Ceram. Soc. Japan, 117, 320-325 (2009)

[4] H. Cheng, J. Ma, Z. Zhao, and L. Qi, Chem. Mater., 7, 663-671 (1995)

[5] H. Sakai, J. Jpn. Soc. Powder Powder Metallurgy, 52, 891-895 (2005)

[6] K. Tomita, V. Petrykin, M. Kobayashi, M. Shiro, M. Yoshimura, and M. Kakihana, Angew. Chem. Int. Ed., 45, 2378-2381 (2006)

EFFECT OF ATMOSPHERE ON CRYSTALLISATION KINETICS AND PHASE RELATIONS IN ELECTROSPUN TiO_2 NANOFIBRES

H. Albetran[a,b], H. Haroosh[c], Y. Dong[d], B.H. O'Connor[a] and I.M. Low[a]
[a] Department of Imaging and Applied Physics, Curtin University, Perth, WA 6845, Australia
[b] Department of Physics, College of Education, University of Dammam, Dammam 31451, Saudi Arabia.
[c] Department of Chemical Engineering, Curtin University, Perth, WA 6845, Australia
[d] Department of Mechanical Engineering, Curtin University, Perth, WA 6845, Australia

ABSTRACT

Titanium dioxide has received much attention as an important photocatalytic material. In this paper, the effects of air and argon atmosphere on the crystallisation kinetics and phase relations in electrospun TiO_2 nanofibres have been investigated using high- temperature synchrotron radiation diffraction. Diffraction results showed that as-synthesized TiO_2 was amorphous initially, but when heated in air or argon atmosphere they crystallized to form anatase and rutile at different temperatures and in different rates. The crystallisation of anatase was delayed by 100 °C in argon when compared to in air. However, the transform rate of anatase to rutile was faster in argon than in air.

INTRODUCTION

Titanium dioxide (TiO_2) is one of the most widely studied materials due to its high photo-activity, photo-durability, mechanical robustness, low cost, and chemical and biological inertness.[1-6] Titanium dioxide TiO_2, has three polymorphs: anatase, brookite, and rutile.[7-10] Anatase and rutile are two main polymorphs that exhibit different properties and thus different photocatalytic performances. They have the same tetragonal crystal structure, but a different space group and atoms per unit cell, which are (*I4/amd*) and 4 for anatase, and ($P4_2/mnm$) and 2 for rutile.[9,11] Moreover, at room temperature, the lattice parameters of anatase are a = b = 3.785Å and c = 9.514Å, while for rutile are a = b = 4.594Å and c = 2.9589Å.[12] These differences in the spacing between the atom and crystal structures for anatase and rutile can cause a different electronic band structure and mass densities. In general, anatase is considered more photochemically active than rutile by virtue of its lower rates of recombination and higher surface absorptive capacity.[13] However, recent work has indicated that a mixed-phase photocatalyst tends to make a better photocatalyst than single- phase titania. For example, a mixed powder of 70% anatase and 30% rutile has been found as the best photocatalyst for the wastewater treatment.[14] Similarly, commercial TiO_2 photocatalyst (P-25) that contains 80 wt% anatase and 20 wt% rutile has been observed to exhibit excellent photocatalytic reactions.[12]

The optimal composition of the photocatalyst phase of anatase and rutile can be prepared by thermal annealing. In general, metastable anatase is formed at lower temperatures than stable rutile, but it is transformed dramatically to rutile at higher temperatures.[15,16] This transformation is affected by several parameters which include the synthesis method, amount of impurities, temperature, calcination time and atmospheres.[15,17] Reaction atmospheres have been observed to have a significant influence on the thermal transformation of titania phases because oxygen defect levels and interstitial titanium ions are influenced by the type of atmosphere.[15,18] For TiO_2 powders at a particular temperature, the effects of air, vacuum, argon, argon- chlorine mixture, hydrogen, steam and nitrogen on the anatase to rutile transformation have been studied.[18] Similarly, for titania films, the amorphous to anatase and anatase to rutile transformations were observed to occur at a lower temperature when annealed in hydrogen atmosphere than in air or vacuum.[19] The rate of anatase to rutile transformation was found to decrease in vacuum but increase in hydrogen atmosphere due to the formation of titanium interstitials that inhibited the transformation or oxygen vacancies which accelerated the transformation.[15,18]

Nanostructured titania with a noble morphology exhibit high surface area-to-volume ratio which is of great significance for increasing the decomposition rate of organic pollutants because photocatalytic reactions take place rapidly and drastically on the surface of the catalyst.[14,20] Hitherto, various techniques have been used to synthesise nanostructured TiO_2 with various morphologies (e.g. nanoparticles, nanowires, nanorods, nanotubes, and nanofibres) and these include microemulsion, sol-gel, hydrothermal and vapor deposition methods.[8,21] Elongated structures are of particular importance, as long, thin nanotubes and nanofibres can provide a high specific surface area.[9] Furthermore, a variety of techniques, such as self-assembly, evaporation, anodisation, and electro-spinning, have been developed for fabricating one-dimensional nano-TiO_2.[1,22] Among the fabrication methods, electro-spinning is a simple, and cost-effective technique applicable at industrial levels for fabricating one-dimensional nanofibres.[2] TiO_2 nanofibres can be synthesised by combining this technique with the sol-gel technique.[23] In this process, a solution of polymer (binder) and TiO_2 precursor is ejected through a needle in a high voltage electric field (kilovolts per centimetre) whereby composite nanofibres of polymer and amorphous TiO_2 nanofibres are formed.[24-28] The diameter, morphology and grain size of the TiO_2 nanofibres have optimal values for photocatalytic activity and are directly affected by the electro-spinning parameters. To the best of the authors knowledge, the effect of atmospheres on the in-situ crystallisation of anatase and rutile in electrospun TiO_2 nanofibres has not been reported.

In the present work, the effect of air and argon atmosphere on the transformation of anatase to rutile and the in-situ crystallisation kinetics of TiO_2 nanofibres was investigated using synchrotron radiation diffraction over the temperature range of 25-900 °C. TiO_2 nanofibres were synthesized using electro-spinning and their morphology, structure and composition were characterised by scanning electron microscopy and synchrotron radiation diffraction.

EXPERIMENTAL PROCEDURE

Material Synthesis

A homogenous sol-gel precursor solution was prepared by mixing titanium isopropxide (M_w = 284.22 g/mol, 97% purity), acetic acid (M_w = 60.05 g/mol, 99.7% purity), and ethanol (M_w = 46.07 g/mol, 99.5% purity) in a fixed volume ratio of 1:3:3, respectively. The sol- gel was stirred in a capped bottle for 5 min, then, 10-12 wt% of poly (vinyl pyrrolidone) (PVP) (M_w = 1300000 g/mol, 100% purity) was dissolved in the solution at 40 °C for 1h using a stirrer. In order to achieve complete dissolution and mixing, the sol-gel solution precursor was stirred ultrasonically for 5 min before it was loaded into a 10 ml plastic syringe with a 25-gauge stainless steel needle. During the electrospinning process, a high voltage power supply was used to provide 25 kV between the needle and a mesh collector covered by an aluminium foil at a distance of approximately 12 cm. A syringe pump was used to control the flow rate at 2 ml/h.

In Situ High-Temperature Synchrotron Radiation Diffraction

In this study, the in-situ crystallisation behaviour of electrospun TiO_2 nanofibres was evaluated using high-temperature synchrotron radiation diffraction (SRD) in air or in argon atmosphere. The specimens were mounted and heated using an Anton Parr HTK 16 hot platinum stage. All measurements were conducted at the powder diffraction beamline in conjunction with an Anton Paar HTK20 furnace, and the Mythen II microstrip detector. The SRD data were collected at an incident angle of 3° and wavelength of 1.125995 Å.

The phase transitions or structural changes were simultaneously recorded as SRD patterns. These patterns were acquired at high temperatures over an angular range of 84°. The SRD patterns were acquired in steps of 100 °C from 200°C to 900°C. The collected SRD data was analysed using Rietveld refinement methods and was carried out using commercial Rietica 2.1 software to compute the relative abundances, crystallite size, and lattice parameters of anatase and rutile.[29] Rietica 2.1 and CMPR programs were used to evaluate the full-width at the maximum intensity (FWHM) to calculate their crystallite sizes (L) using the Scherrer equation:[30]

$$L = \frac{K\lambda}{\beta \cos\theta} \quad (1)$$

$$FWHM^2 = U \tan^2\theta + V \tan\theta + W \quad (2)$$

where K is the shape factor, λ is the x-ray wavelength, θ is the Bragg angle, and β is the FWHM in radians, which can be calculated from equation (2). Rietica 2.1 was used to calculate the peak shape parameters U, V, and W.

Scanning Electron Microscopy

The morphology of electrospun TiO_2 nanofibres before and after the thermal annealing was examined using an EVO 40XVP scanning electron microscope with an accelerating voltage of 15 keV. Prior to microstructure observation, the samples were coated with platinum to avoid charging. The SEM images of both as–spun and annealed nanofibres were taken using secondary electrons and backscattered at working distances of 4 mm and 4.5 mm, respectively.

RESULTS AND DISCUSSION

Microstructures of Electrospun TiO_2 Nanofibres

Figure 1 illustrates the images of as-spun pure nanofibres using secondary and backscattered electrons before thermal annealing. The nanofibres have a smooth surface and individual lengths of one-dimensional structure with random orientation. The average diameters of 30 electrospun TiO_2/PVP composite nanofibres were determined to be 614 ± 190 nm and the nanosized TiO_2 were well dispersed and embedded within the PVP matrix as shown in Figure 1b.

As shown in Figure 1(c, d), after thermal annealing at 900 °C, the average diameters of the sintered TiO2 nanofibres shrank by 48.5 % in air and 80.5% in argon because the organic material evaporated completely during heating. The average diameters of 30 nanofibers were 298 ± 115 nm in atmospheric air and 120 ± 35 nm in argon. However, in air atmospheres, the fibres were broken and rougher while in argon they were still quite smooth, which indicated more brittleness nature of as-spun TiO_2 nanofibres in air. One possible reason is that an oxidation of the PVP polymer in air atmosphere which contains about 20% of oxygen causes damage and bond breaking in the chain. Moreover, it reduced the molecular weight of the polymer and also the mechanical properties like tensile stress to become weaker. However, argon is an ideal monatomic gas, that is, a gas containing just one atom per molecule and it is an inert atmosphere. However, it is also possible that the sample was damaged during preparation for SEM observation.

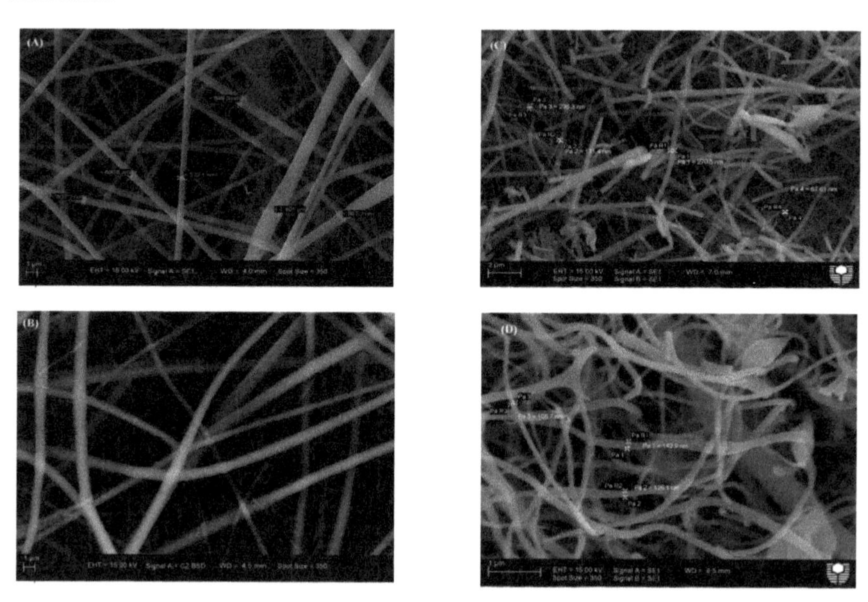

Fig. 1: SEM micrographs of as- spun pure TiO_2 nanofibres imaged with (a) secondary electrons (b) backscattered electron; and after thermal annealing in atmospheric (c) Air (d) Argon.

Effect of Atmosphere

Figure 2 shows the effect of atmosphere on the in-situ crystallisation kinetics of as-synthesized TiO_2 nanofibres over a temperature range of 25-900 °C, as revealed by synchrotron radiation diffraction. The TiO_2 nanofibres were initially amorphous but eventually crystallized to form anatase and rutile at elevated temperatures. When heated in air they crystallized to form anatase and rutile at 600 °C and 700 °C, respectively. However, in argon atmosphere, both anatase and rutile formed simultaneously at 700 °C. This suggests that the crystallisation of anatase was delayed by 100 °C relative to air when argon atmosphere was used. The reason for this delay in the argon atmosphere may be attributed to the presence of low oxygen partial pressure. It appears that oxygen atoms are essential to promote the transition and transformation from amorphous TiO_2 into anatase. But the lack of oxygen in argon atmosphere did not affect the transformation of metastable anatase into rutile. However, the lack of oxygen will inevitably lead to non-stoichiometry in the anatase and rutile formed with a composition of TiO_{2-x}, where x is oxygen vacancy. In contrast, anatase and rutile formed in the air will have a stoichiometric composition of TiO_2.

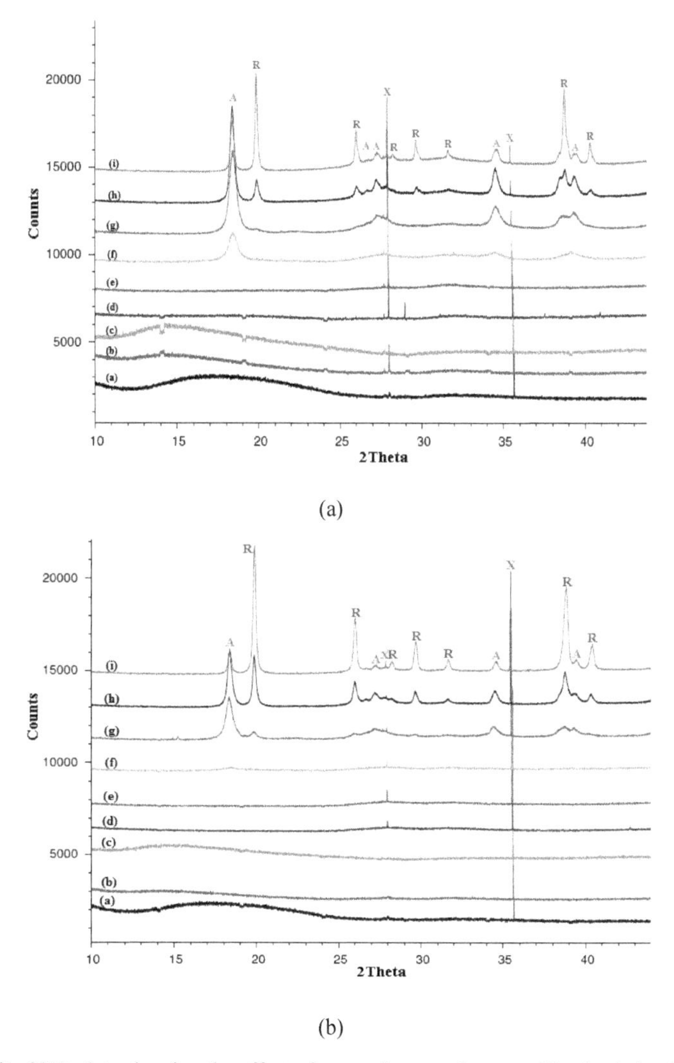

Fig. 2: In-situ SRD plots showing the effect of atmosphere on the crystallisation behaviour of as-synthesized TiO_2 nanofibres when heated in the temperature range 25-900 °C (a) in the air, and (b) in argon. [Legend: a=25°C, b=200°C, c=300°C, d=400°C, e=500°C, f=600°C, g=700°C, h= 800°C, i=900°C, anatase (A), rutile (R) and platinum (X)].

The Rietveld method was used to evaluate the phase abundance of anatase and rutile during the transformation in the air and argon atmospheres. Figure 3a shows the comparison of anatase phase content in both atmospheres as a function of temperature. At 600 °C, only anatase formed in air atmosphere. In argon atmosphere, 85% of anatase formed at 700 °C, which is 15% less than in air for the same temperature. Figure 3b shows the phase content of rutile increased dramatically as the temperature increased by virtue of the transformation of metastable anatase to rutile. Rutile formed at 700 °C in both atmospheres but with different abundances. It increased from about 15% at 700 °C to well over 85% at 900 °C in the argon atmosphere. However, in air, it increased from 4% to 65% for the same temperature range.

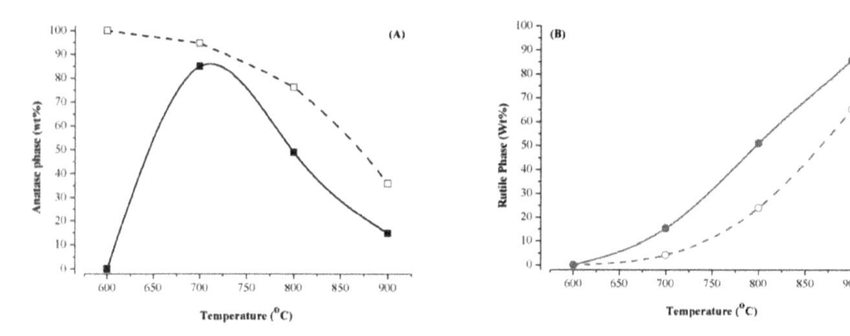

Fig. 3: Phase abundances of anatase and rutile in the temperature range 25-900 °C for as-synthesized TiO_2 nanofibres heated in (a) air, and (b) argon. [Legend: anatase in air (□); anatase in argon (■); rutile in air (○); rutile in argon (•)]

The crystallite sizes of the anatase and rutile were calculated from the Scherrer equation using Rietica 2.1 programs to evaluate the average full-width at the maximum intensity (FWHM). Figure 4 shows the effect of atmosphere on the crystal growth of anatase and rutile at various temperatures. The crystallite size of anatase was just under 10 nm at 600 °C in air and increased to just over 25 nm at 900 °C. However, atmosphere had no appreciable effect on the crystal growth of anatase. In contrast, the atmosphere had a noticeable effect on the crystal growth of rutile but only at 900 °C. The crystallite size of rutile was ~18 nm in both air and argon but increased to ~39 nm in air and 31 nm in argon.

The variations of unit-cell parameters and cell-volumes of anatase and rutile in different atmospheres with temperature and the corresponding linear and volumetric thermal expansion coefficients (TEC) are shown in Tables 1 and 2. The cell-volumes of both anatase and rutile inceased with increasing temperature by virtue of an increase of their lattice parameters. As can be seen, the cell-volumes for both anatase and rutile were affected slightly by the atmospheres which were higher in air than in argon. Thus, the linear and volumetric thermal expansion cofficients of both anatase and rutile were slightly greater in air than in argon. Moreover, the calculated unit-cell parameters, cell-volumes, and thermal expansion for anatase and rutile are comparable with results from Hummer and Heaney equations.[29]

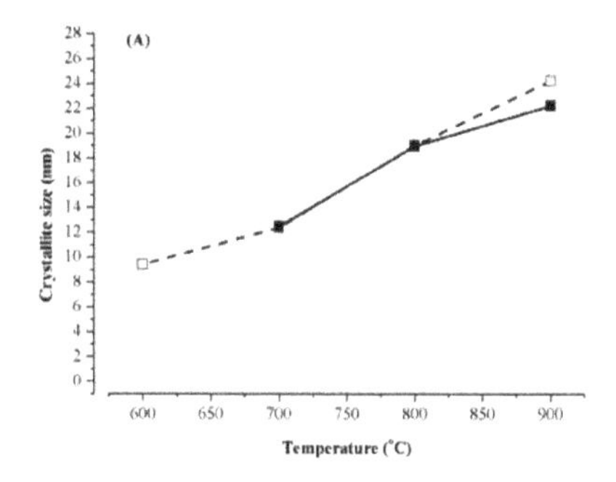

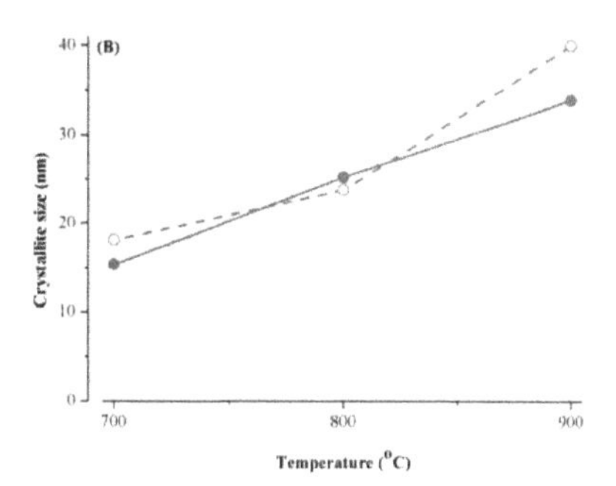

Fig. 4: Effect of atmosphere on crystallite size of (a) anatase and (b) rutile at various temperatures. [Legend: anatase in air (□); anatase in argon (■); rutile in air (○); rutile in argon (•)]

Table 1: Variation of the cell-volume of anatase and rutile in air and argon atmosphere as a function of temperature.

Temperature (°C)	**Air atmosphere** Anatase ($Å^3$)	Rutile ($Å^3$)	**Argon atmosphere** Anatase ($Å^3$)	Rutile ($Å^3$)
600	136.94(9)	-------	-------	-------
700	137.79(2)	63.12(4)	137.68(3)	63.41(4)
800	137.98(3)	63.67(2)	137.84(3)	63.60(1)
900	137.93(7)	64.01(1)	138.23(8)	63.95(1)

Table 2:The linear (α_a, α_c) and volumtric (β) thermal expansion coefficient at 800 °C.

Atmosphere	TEC $\times 10^{-5}$ (°C^{-1})	α_a	α_c	β
Air	Anatase	0.72	0.86	2.29
	Rutile	1.95	3.03	7.07
Argon	Anatase	0.52	0.84	2.03
	Rutile	1.08	2.01	4.24

CONCLUSIONS

The effect of air and argon atmosphere on the in-situ crystallisation behaviour of electrospun TiO_2 nanofibres was investigated using high-resolution synchrotron radiation diffraction in the temperature range 25-900°C. In both cases, as-spun TiO_2 nanofibres were initially amorphous but crystallized to anatase and rutile at elevated temperatures. Both atmospheres demonstrated a significant influence on the crystallisation kinetics of anatase and rutile. In air, anatase formed at 600°C and rutile at 700°C, but in argon they formed simultaneously at 700°C. The kinetics of transformation of anatase to rutile was also strongly affected by the atmospheric conditions. The rates of anatase to rutile transformation was faster in argon than in the air. The crystallite sizes of anatase and rutile increased with increasing temperature, and were only moderately affected by the atmospheres.

ACKNOWLEDGMENTS

We gratefully acknowledge the Australian Synchrotron (AS122/PDFI/5075) and the Australian Institute of Nuclear Science and Engineering (ALNGRA11135) for funding this work.

REFERENCES

[1]Low IM, Albetran H, Prida VM, Vega V, Manurung P, Ionescu M. A comparative study on crystallization behavior, phase stability, and binding energy in pure and Cr-doped TiO_2 nanotubes. *J Mater Res* 2013;**28**:304-12.
[2]Manurung P, Putri Y, Simanjuntak W, Low IM. Synthesis and characterisation of chemical bath deposited TiO_2 thin-films. *Ceram Int* 2013;**39**:255-9.
[3]Zhang P, Shao C, Li X, Zhang M, Zhang X, Sun Y, Liu Y. In situ assembly of well-dispersed au nanoparticles on TiO_2/ZnO nanofibers: A three-way synergistic heterostructure with enhanced photocatalytic activity. *J Hazard Mater* 2012;**237-238**,331-8.
[4]Yu C, Wei L, Li X, Chen J, Fan Q, Yu J. Synthesis and characterization of Ag/TiO_2– B nanosquares with high photocatalytic activity under visible light irradiation. *Mater Sci Eng B* 2013;**178**:344-8.
[5]Pan X, Zhao Y, Liu S, Korzeniewski CL, Wang S, Fan Z. Comparing graphene-TiO_2 nanowire and graphene-TiO_2 nanoparticle composite photocatalysts. *ACS Appl Mater Interfaces* 2012;**4**:3944-3950.
[6]Babu VJ, Nair AS, Peining Z, Ramakrishna S. Synthesis and characterization of rice grains like nitrogen-doped TiO_2 nanostructures by electrospinning-photocatalysis. *Mater Lett* 2011;**65**:3064-8.
[7]Liu G, Wang L, Yang HG, Cheng and HM, Lu GQ. Titania- based photocatalysts-crystal growth, doping and heterostructuring. *J Mater Chem* 2010;**20**:831-43.
[8]Chuangchote S, Jitputti J, Sagawa T, Yoshikawa S. Photocatalytic activity for hydrogen evolution of electrospun TiO_2 nanofibers. *ACS Appl Mater Interfa* 2009;**1**:1140-3.
[9]Kim DW, Enomoto N, Nakagawa ZE, Kawamura K. Molecular dynamic simulation in titanium dioxide polymorphs: Rutile, brookite, and anatase. *J Am Ceram Soc* 1996;**79**:1095-9.
[10]Liu R, Qiang LS, Yang WD, Liu HY. The effect of calcination conditions on the morphology, the architecture and the photo-electrical properties of TiO_2 nanotube arrays. *Mater Res Bull* 2013;**48**:1458-67.
[11]Li W, Ni C, Lin H, Huang CP, Shah SI. Size dependence of thermal stability of TiO_2 nanoparticles. *J Appl Phys* 2004;**96**:6663-8.
[12]Hanaor DA, Sorrell CC. Review of the anatase to rutile phase transformation. *J Mater Sci* 2011;**46**:855-74.
[13]Shang S, Jiao X, Chen D. Template-free fabrication of TiO_2 hollow spheres and their photocatalytic properties. *ACS Appl Mater Interfa* 2012;**4**:860-5.
[14]Li H, Zhang W, Pan W. Enhanced photocatalytic activity of electrospun TiO_2 nanofibers with optimal anatase/rutile ratio. *J Am Ceram Soc* 2011;**94**:3184-7.
[15]Shannon RD, Pask, JA. Kinetics of the anatase-rutile transformation. *J Am Ceram Soc* 1965;**48**:391-8.

[16]Eppler RA. Effect of antimony oxide on the anatase-rutile transformation in titanium dioxide. *J Am Ceram Soc* 1987;**70**:64-6.
[17]Iida Y, Ozaki S. Grain growth and phase transformation of titanium oxide during calcination. *J Am Ceram Soc* 1961;**44**:120-7.
[18]Gamboa JA, Pasquevich DM. Effect of chlorine atmosphere on the anatase-rutile transformation. *J Am Ceram Soc* 1992;**75**:2934-8.
[19]Huang JH, Wong MS. Structures and properties of titania thin films annealed under different atmosphere. *Thin Solid Films* 2011;**520**:1379-84.
[20]Lee JS, Ha TJ, Hong MH, Park HH. The effect of porosity on the CO sensing properties of TiO_2 xerogel thin films. *Thin Solid Films* 2013;**529**:98-102
[21]Low JM, Curtain B, Philipps M, Liu QZ, Ionescu M. High temperature diffraction study of in-situ crystallization of nanostructured TiO_2 photocatalysts. *J Aust Ceram Soc* 2012;**48**:198-204.
[22]Chuangchote S, Jitputti J, Sagawa T, Yoshikawa S. Photocatalytic activity for hydrogen evolution of electrospun TiO_2 nanofibers. *ACS Appl Mater Interfa* 2009;**1**:1140-3
[23]Huang ZM, Zhang YZ, Kotaki M, Ramakrishna S. A review on polymer nanofibers by electrospinning and their applications in nanocomposites. *Compos Sci Technol* 2003;**63**:2223-53.
[24]Li Q, Satur DJG, Kim H, Kim HG. Preparation of sol- gel modified electrospun TiO_2 nanofibers for improved photocatalytic decomposition of ethylene. *Mater Lett* 2012;**76**:169-172.
[25]Luo W, Hu X, Sun Y, Huang Y. Surface modification of electrospun TiO_2 nanofibers via layer-by-layer self-assembly for high-performance lithium-ion batteries. *J Mater Chem* 2012;**22**:4910-5.
[26]Wessel C, Ostermann R, Dersch R, Smarsly BM. Formation of inorganic nanofibers from preformed TiO_2 nanoparticles via electrospinning. *J Phys Chem* 2011;**C 115**, 362-72.
[27]Li H, Zhang, W, Pan W. Enhanced photocatalytic activity of electrospun TiO_2 nanofibers with optimal anatase/rutile ratio. *J Am Ceram Soc* 2011;**94**:3184-7.
[28]P. Zhu, Nair AS, Shengjie P, Shengyuan Y, Ramakrishna S. Facile fabrication of TiO_2-graphene composite with enhanced photovoltaic and photocatalytic properties by electrospinning. *ACS Appl Mater Interfa 2012*;**4**,581-5.
[29]Hummer DR, Heaney PJ. Thermal expansion of anatase and rutile between 300 and 575 K using synchrotron powder X-ray diffraction. *Powder Diffraction* 2007;**22**:352-7.
[30]Cullity BD, Stock SR. *Elements of X-ray Diffraction*. Prentice: Hall Inc; 2001.

ELECTRONIC AND OPTICAL PROPERTIES OF NITROGEN-DOPED LAYERED MANGANESE OXIDES

Giacomo Giorgi and Koichi Yamashita

Department of Chemical System Engineering, The University of Tokyo
7-3-1 Hongo, Bunkyo-ku, Tokyo 113-8656, JAPAN

ABSTRACT

Electronic and optical properties of nitrogen-doped layered MnO_2, the non-defective, the N-substituted defective, and the N-substituted in conjunction with O-vacancy forms, are analyzed theoretically based on Density Functional Theory. We have found that (1) N-subsitution in both the monolayer and the stacked system gives rise to an acceptor-like state close to the conduction band, (2) O vacancy is beneficial for the formation of a substitutional N_O defect, causing a stabilization of the doubly defective systems through a self-compensation mechanism, (3) the coexistence of both defects in the stacked system, resulting a quantitative energetic stabilization for the latter structure, and finally (4) the reduction of the oxidation state of Mn via N-doping reveals this process as extremely suitable to obtain catalysts with enhanced activity.

INTRODUCTION

Mn(IV) oxides are considered the most abundant and reactive polymorphs of MnO_2 available in the environment. Their activity ranges from metal sequestration to that of mediators in the redox reactions involving both inorganic and organic compounds. Mn(IV) oxides obtained by Mn(II) oxidation have the typical structure of phyllomanganates, whose class birnessite and δ-MnO_2 are part. Such oxides are the most powerful natural oxidants[1]: their wide versatility in various biological fields stems from this property. Few years ago it has been suggested this layered material as a plausible candidate for (photo)catalytic oriented devices and for H_2 production.

The optical properties of these layered MnO_2 materials have been experimentally investigated by Sherman[2]: its triclinic symmetry is reported to have a bandgap of 1.8 eV that in conjunction with a Conduction Band (CB) level at -0.16 V, indicates that the couple MnO_2/Mn^{2+} can be thermodynamically subjected to photoreductive dissolution in seawater. More uncertain is the optical activity of the hexagonal birnessite, even if it is well assessed that the presence of defects can give rise and act as a self-dopant process. In a series of paper Kwon et al.[3,4] have studied the impact of a typical defect in these materials, the Ruetschi defect. This defect consists in a Mn-vacancy whose charge is compensated by four H atoms that saturate the O-dangling bonds[5]. At DFT+U level, Sun et al.[6] has studied the formation of Oxygen vacancy via electron injection and found that in monolayers of birnessite-like structures Oxygen leaves the structure in a neutral atomic or molecular form. Notably, pure and stoichiometric layers of these materials have been not synthesized in laboratory yet, but many efforts have been spent in order to get them.

The very intriguing and yet unsolved role of MnO_2 in photoreductive dissolution of organic reductants and its nature of possible cromophore according to its semiconductor nature has motivated new campaigns of investigation on this material. A recent paper of Hashimoto et al.[7] where the activity of manganese oxides for water oxidation as function of pH is investigated. The very relevant characteristic of these systems is the fact that, at variance with all the others Transition Metal Oxides employed in photocatalysis and photovoltaics (i.e., TiO_2), both the

Valence Band Maximum and the Conduction Band Minimum lie on the *d* orbitals of the metallic atom (Mn), corresponding this gap to the crystal field splitting between the e_{2g} and t_{2g} orbitals.

In this paper we aim to computationally study and characterize layered materials constituted by MnO_2. In particular, we will study electronic and optical properties of both the mono-layered structure (ML) and the stacked structure. We intend to theoretically rationalize what experimentally found about electronic and optical properties and also to show how doping birnessite with Nitrogen gives rise to the possibility of tailoring new materials with bandgap suitable in the context of improved photocatalytic activity and sun-to-energy water splitting.

COMPUTATIONAL DETAILS

DFT spin-polarized calculations have been performed with the VASP code[8]. Calculations were performed using projected-augmented wave (PAW)[9]. The electron exchange-correlation part was considered under the GGA approximation using the PBE functional[10], while the cutoff for the plane-waves has been chosen to be 400. eV. The models we employed for our calculations stem from those reported by Kwon *et al.*[3]; indeed, since an ordered structure deriving from birnessite has not been synthesized yet in laboratory, even if the possibility of a triclinic symmetry of the system must be pointed out[2,3], here we have considered as parental structure a model with a $P_{63/mmc}$ symmetry (a=b=2.840 Å, c=14.031 Å), experimentally derived from $KMnO_4$ by Gaillot[11], and fully optimized both the stacked structure (birnessite-like) and the single sheet of MnO_2. Forces after the optimization were lower than 0.01 eV/Å, while the Brillouin sampling has been chosen (Γ-centred) according to the size of the system as reported in the following scheme. We studied the formation of O-vacancies (O_{vac}) and N substitutionals (N_O) in 2x2x1 and 3x3x1 supercells, for both the ML and for the stacked system. On the unit cell (1 x 1 x 1) we have also performed a GW[12,13] analysis in order to calculate the underestimation given by DFT on the bandgap of this novel material.

RESULTS AND DISCUSSIONS

We started from the geometry characterization of the two different systems, the ML and the stacked structure in Fig. 1. Differences between the geometry of the layers of these two systems are almost negligible, showing the correct choice of the sampling of the Brillouin zone. Lattice parameter and Mn-O bondlength do not vary after the full geometrical optimization of the two systems.

The correction we found at G_0W_0 for the ML is consistent with theoretical prediction and experimental bandgaps reported for monolayers. Sun et al.[6] clearly predict that the bandgap for perfect MnO_2 monolayer is smaller than 3.50 eV, due to the systematical underestimation of about 30-60% of the bandgap at DFT level. For the 1x1x1 we calculated the "one-shot" G_0W_0 correction, obtaining a value of the bandgap on Γ that is 3.30 eV, in perfect agreement with the prediction and in agreement with LDA+U calculations.

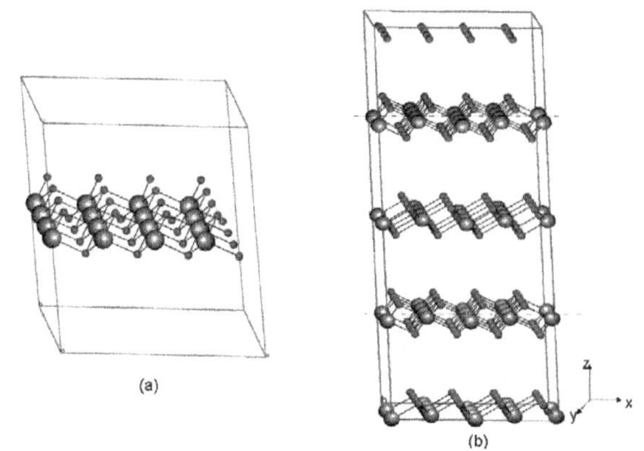

Fig. 1: (a) ML model of MnO_2 birnessite structure; (b) Stacked model of MnO_2 birnessite structure (a doubled cell along *a* and *b* is reported for the ML case and along *a* and *b* and *c* for the stacked case) [large atoms: Mn; small: O]

The 2 x 2 x 1 stacked supercell is characterized by an indirect up→dn transition occurring between some point along the Γ-to-K direction and Γ. This transition is characterized by an E_G= 0.86 eV (the direct transition on Γ is only few meV larger); at variance, still at Γ the up→up transition has an E_G= 2.73 eV. Quantum confinement effects make the ML bandgap slightly larger (up→dn indirect bandgap, E_G=1.04 eV. The direct up→up at Γ, E_G= 2.82 eV). Regardless the nature of the birnessite model, the most interesting result we observe is the "mixed" transition O_{2p}→Mn_{3d} in conjunction with the dominant *d-d* transition occurring in the Mn atoms, the latter corresponding to the $e_g \rightarrow t_{2g}$ crystal field splitting, Δ.

In the non-defective supercells, we replaced one oxygen with one N (N_O) in a concentration 0.125 (1N/8O) in the ML and 0.0625 (1N/16O) in the stacked structure, to study the impact of such doping process on the electronic properties of the two systems. In the former case, we both partly (ionic positions only) and fully optimized the cell, finding negligible difference between the two models in structural terms. In the case of the ML, the smaller transition is indirect, lying the VBM on the $\overline{K}$ point ($\overline{H}$ folded in $\overline{K}$). The indirect transition is thus $\overline{K} \rightarrow \overline{\Gamma}$; for this transition the E_G is 0.90 eV (up→dn), while the direct smallest one on $\overline{\Gamma}$ is 1.28 eV and on $\overline{H}$ ($\overline{K}$) is 1.17 eV. The Nitrogen atom gives rise to an acceptor-like empty state that lies 0.37 eV (0.22 eV) below the CBM on $\overline{\Gamma}$ ($\overline{H}$). Slightly different behaviour holds for the stacked structure. Here, similarly to the undoped case, the VBM lies on some point along the Γ→K direction. An indirect transition is observed between some point along the Γ→K direction and Γ, with the Nitrogen induced state that in this case is slightly deeper in the gap (0.47 eV below the CBM on Γ). We here mention the fact that the DFT+U calculation (U_d=4.0 eV) on both the N-doped systems does not provide a satisfactory correction for the induced state in the gap. For the case of the ML, indeed, while the bandgap of the MnO_2 host is "opened" up to 3.20 eV, the N-induced state lies 2.50 eV below the new CBM.

Table 1 reports the orbital population of the two systems. The electronic distribution of the N-induced state is almost identical for the two cases. The charge density localization slightly increases on Np_z orbitals in the case of the stacked structure with a subsequent reduced localization on the same p_z orbitals of Oxygen. The most noticeable difference between the charge density of the two cases is the fact that in the ML valence band maximum there is already presence of Np_z orbitals that, at variance, are totally absent in the case of the VBM of the stacked structure. Interestingly, still in the stacked structure both the N-induced state and the CBM reside on the defective layer, while the VBM is localized on the non-defective one. As stated, the acceptor state is extremely localized on Np_z orbitals, as witnessed by the comparison of charge density distribution of both the VBM and the acceptor state induced in the gap by the N-doping. In the case of the stacked system, this clear and net selectivity of the VBM on one hand, and of the CBM and the N-induced state on the other, ensures in principle a separation of the photogenerated charge carriers (electron/hole). In other words, in the stacked system the N-induced acceptor state behaves similarly as an Intermediate Band[14], able to facilitate the (photo)excitation process.

Table 1: Orbital population for the N_O birnessite-like structure MnO_2 at Γ point.

	↑ **VBM**	↑ **induced-acceptor state (N)**	↓ **CBM**
ML	**Mn** [24.4% d_z^2 + 14% d_{yz} + 13.5 d_{xz}] + **O** [17.3% p_z] + **N** [8% p_z]	**Mn** [32.% d_z^2 + 9.% d_{xz} +9.% d_{yz}] + **O**[9.% p_z] + **N** [22.% p_z]	**Mn** [19.% d_{x2-y2} + 18.0% d_{yz} + 47.4% d_{xz} + 7.% d_{xy}]
Stacked	**Mn** [58.5% d_z^2]+ **O** [25.3% p_z] (*non-defective layer)*	N [23.7% p_z] + **Mn** [9% d_{xz} + 9% d_{yz} +29.9 % d_z^2] + **O** [8.6 % p_z] *(defective layer)*	**Mn** [6.% d_{xy} + 16.7% d_{yz} + 50.3% d_{xz} + 17.9% d_{x2-y2}] *(defective layer)*

Since a few available theoretical literatures have focused on the effect of the presence of O-vacancies in similar systems, reporting the possible stabilization of MnO_2 nanosheets induced by the presence of O-vacancies[15], we have finally investigated the possible coexistence of O_{vac} and N_O. In those works, through a TEM analysis confirmed by DFT calculations, the possibility of an atomic reconstruction (2 x 1) and a MnO phase ruled out by the presence of O-vacancies has been found. To test the impact of the presence of O-vacancy in the N-doped system, we have considered both the case of the ML and the stacked structure where both a substitutional Nitrogen and an Oxygen vacancy are present at the same time. Moreover, for the stacked structure, both the structure with O-vacancy and N_O on the same layer and that with O-vacancy and N_O on different layers have been analyzed.

From the initial "one-shot" G_0W_0 calculation[12,13] on the unit cell of the ML we observed a sensitive correction of the bandgap, mainly associated to the raise of the CB minimum. Also the VB is corrected downward. In particular, the VB calculated at DFT level was 0.1 eV below the Fermi Energy. The G_0W_0 correction on the same VB lead to a correction of 0.26 eV with respect to the DFT calculated value i.e., 0.36 eV below the new Fermi Energy. Assuming a close correction for the O-vacancy ML, the states induced in the gap can step down to values below the Fermi Energy, without altering the semiconductor nature of the system, and making this defect a donor. Introducing then the second defect in the ML (N_O), once O-vacancy and N_O occur on the same ML, the system recovers a wide gap already at DFT level, confirming our hypothesis that a self-compensation mechanism takes place. It is also important to remind that experimental samples of birnessite always are characterized by the random presence of intralayer metallic atoms (Mn itself, but also H, Bi, K, Li) able to neutralize the partial negative charge of the layers.

Since our models do not take into account the possible presence of such charge compensating ions, it is easy to imagine that once the O-vacancy is formed (the thermodynamic bottleneck of the process), then to replace one Oxygen atom with a less electronegative N atom is a favoured process, i.e., a partial reduction of the negative charge on the layer. This is also a very interesting results in view of a possible usage of MnO_2 as (electro)photocatalyst. As indeed reported recently by Takashima et al.[7], the occurrence of Mn^{+3} in similar birnessite-like system, at pH>9, as result of the comproportionation between Mn^{2+} and Mn^{4+} is responsible for the enhanced activity of manganese oxides for water oxidation. Similarly, N-doping process seems to be another viable mechanism, pH independent, to get Mn with reduced oxidation state, being the reduction due to N-doping in the ML of about 0.5e per Mn atom with respect to the pure, non-defective ML. In the case of the stacked system we have both tested the possibility of having the coexistence of the two defects on the same layer and also one per layer. The comparison between the two cases reveals that still in this case the presence of both the defects on one layer is extremely favoured (ΔE=2.06 eV) over the case of the Nitrogen substitutional and the oxygen vacancy on different layers (ΔE=3.22 eV), confirming the result obtained in the case of the ML (self-compensating mechanism).

CONCLUSION

In this paper we have investigated the structural and electronic properties of both the monolayer and stacked MnO_2 birnessite-like systems, 1) in its non-defective, 2) N-substituted, and 3) N-substituted plus O-vacancy defective form. We observed that the formation of an N subsitutional gives rise to an acceptor-like state close to the Conduction Band in both the monolayers and in stacked systems. Such acceptor-like state in the stacked system seems to operate as an Intermediate Band system, making the excitation between the valence and the conduction band a more facile process. The reduction of the oxidation state of Mn via N-doping reveals the paramount importance of this process in obtaining photocatalysts with enhanced activity regardless the pH conditions.

REFERENCES

[1]B. M. Tebo, J. R. Bargar, B. G. Clement, G. J. Dick, K. J. Murray, D. Parker, R. Verity, S. M. Webb, Annu.Rev.Earth Planet Sci., (2004) 32 287

[2]D. M. Sherman, Geochim. Cosmochim. Acta, (2005) 69, 3249.

[3]K. D. Kwon, K. Refson, G. Sposito, Phys Rev. Lett., (2008), 100, 146601.

[4]K. D. Kwon, K. Refson, G. Sposito, Geochim. Cosmochim. Acta (2009) 73 4142.

[5]P. Ruetschi, J. Electrochem. Soc. (1984) 131, 2737.

[6]C. Sun, Y. Wang, J. Zou, S. C. Smith, Phys. Chem. Chem. Phys., (2011),13,11325.

[7]T. Takashima, K. Hashimoto, R. Nakamura, J. Am. Chem. Soc. (2012), 134, 1519.

[8]G. Kresse, J. Furthmüller, Comput. Mater. Sci. (1996) 6, 15; (b) G. Kresse and J. Furthmüller, Phys. Rev. B (1996) 54, 11169.

[9](a) P.E. Blöchl. *Phys. Rev. B*, (1994) 50 17953; (b) G. Kresse, J. Joubert. *ibid.*, (1999) 59 1758.

[10]J.P. Perdew, K. Burke, M. Ernzerhof, Phys. Rev. Lett., (1996) 77 3865.

[11]A.-C. Gaillot, D. Flot, V. A. Drits, A. Manceau, M. Burghammer, B. Lanson, *Chem. Mater.*, (2003) 15 4666.

[12]M. Shishkin, G. Kresse, Phys. Rev. B (2006) 74, 035101.

[13]M. Palummo, G. Giorgi, L. Chiodo,A. Rubio, K. Yamashita, J. Phys. Chem. C, (2012), 116, 18495.

[14]A. Luque, A. Marti', Phys. Rev. Lett. (1997) 78, 5014.
[15]Y. Wang, C. Sun, L. Wang, S. Smith, G. Q. Liu, D. J. H. Cockayne, Phys Rev. B, (2010), 81, 081401(R).

Ceramics Enabling Environmental Protection: Clean Air and Water

UNDERSTANDING THE EFFECT OF DYNAMIC FEED CONDITIONS ON WATER RECOVERY FROM IC ENGINE EXHAUST BY CAPILLARY CONDENSATION WITH INORGANIC MEMBRANES

Melanie Moses DeBusk*, Brian Bischoff, James Hunter, James Klett, Eric Nafziger and Stuart Daw
Oak Ridge National Laboratory
Oak Ridge, TN 37831

ABSTRACT

An inorganic membrane water recovery concept is evaluated as a method to recover water from the exhaust of an internal combustion engine. Integrating the system on board a vehicle would create a self-sustaining water supply that would make engine water injection technologies "consumer transparent." In laboratory experiments, water recovery from humidified air was determined to evaluate how different operating parameters affect the membrane system's efficiency. The observed impact of transmembrane pressure and gas flow rate suggest that gas residence time is more important than water flux through the membrane. Heat transfer modeling suggests that increasing membrane length can be used to improve efficiency and allow higher exhaust flow through individual membranes, important parameters for practical applications where space is limited. The membrane water recovery concept was also experimentally validated by extracting water from the exhaust of a diesel stationary generator. The insight afforded by these studies provides a basis for developing improved membrane designs that balance both efficiency and cost.

INTRODUCTION

Consumer and government demands to simultaneously reduce the United States' dependence on foreign oil and reduce the production of greenhouse gases have resulted in the automotive industry continually updating and implementing new vehicle technologies that improve vehicle fuel economy and lower harmful emissions. Increases in fuel economy standards for cars and light-trucks require US fleets to attain a Corporate Average Fuel Economy (CAFE) of 54.5 mpg by 2025 which is large increase from the 2016 standard of 35.5 mpg.[1] A 2012 report by the EPA and the NHTSA states that this level of improvement will require an array of technologies.[2] Even under the most aggressive electric vehicle integration scenarios, gasoline and diesel engines would still be found in 74% or more of the sales fleet. When evaluating technologies for improving engine efficiency, their effect on vehicle emissions must be taken into account so that EPA's stringent emission standards are still achievable.

It has been known for decades that introducing water along with the fuel into the combustion chamber of an engine can both improve fuel efficiency and reduce NO_x emissions.[3] A major engineering barrier to water injection is the need to carry a large quantity of water on the vehicle. The ability to efficiently produce the required water on-board the vehicle as it is needed would make the technology "consumer transparent" (i.e. no consumer involvement) and eliminate the need for a large storage tank.

Internal combustion engines (ICEs), such as those used on vehicles, run on hydrocarbon (HC) fuels which are combusted in the presence of oxygen. The HC components of the fuel produce carbon dioxide (CO_2) and water (H_2O)upon combustion regardless of the type of fuel burned (diesel, JP-8, gasoline, natural gas). Dodecane is a good single component representative of standard diesel.[4] Therefore, from the stoichiometric relationship (eq. 1), one mole or 170.34g of dodecane, will produce 13 moles or 234.2g of water (normalizing with density, 1 gallon of diesel will furnish about 1 gallon of water).[5] This water, present in the vehicle's exhaust, could be used as the on-board water source if recovered efficiently.

$$C_{12}H_{26} + 18.5\ O_2 \rightarrow 13\ H_2O + 12\ CO_2 \quad (1)$$

Use of exhaust water vapor for engine water injection technology requires both separation and a phase change to the liquid state. Water recovery by condensation provides separation and phase change in a single step making it a more favorable route than gas phase separation of the water by porous membranes. Traditional condensation requires low temperatures, necessitating the employment of refrigerant-based compressive cooling which is energy intensive and bulky, making it undesirable for a vehicle application. Our technology provides water recovery from the vehicle's exhaust using nanoporous, inorganic membranes to efficiently capture the water via capillary condensation.[6] Because, capillary condensation can liquefy the water at higher temperatures, the need for compressive cooling is eliminated offering the potential for a lighter system with a smaller footprint. Capillary condensation can also recover more water while at a higher purity than a traditional heat exchanger approach. Instead of using compressive cooling, membrane temperature is regulated through several approaches. First, a novel light-weight, highly conductive graphite foam is used; furthermore, the combination of current vehicle emission control systems and reduced contact time of the exhaust gases with the condensed water help to minimize dissolved contaminants in the water. Here, the results of demonstrating this concept in the laboratory and evaluation of the effect of different membrane and system parameters are described.

EXPERIMENTAL

Membranes

All membranes evaluated in this study were fabricated by applying a thin porous ceramic layer to a 9-inch long porous tubular 434 stainless steel support tube. The ~11.3 mm OD supports had an average pore size of ~4.3 µm with a wall thickness of approximately 0.55 mm. A ~5 µm thick ceramic separative layer made of aluminum oxide with 30% porosity was applied to the inside of the support tube resulting in ~10.2 mm ID tubular membranes. The flow-weighted pore size distributions of the ceramic separative layers were determined by first filling the pores of the membrane with a condensable fluid and then measuring the gas flux as the pores are emptied from largest to smallest diameter.[7] The median pore diameters of the ceramic membranes investigated in this study were 5.8, 8.2 and 9.5 nm. Each membrane was bubble tested for leaks with isopropanol before use.

Lab-Scale Experiments

A humidified feed gas was produced by injecting metered water (deionized) into a metered flow of heated dry compressed air. An MKS 1179 mass flow controller was used to meter the air flow and an ISMATEC IPC high precision multichannel dispenser was used to meter the injected water. The feed gas stream contained 10% water vapor and was heated to 250°C before entering the membrane. A small Welch-Ilmvac (MP 101Z) vacuum pump, attached to a regulator, was used to develop a variable transmembrane pressure (ΔP). The transmembrane pressure creates a slight vacuum on the support side of the membrane to create a driving force to continuously pull the water condensed in the membrane (i.e. permeate) from the pores of the separative layer and support layer into a collection vessel. The vacuum on the support side of the membrane does not affect the gas pressure on the feed side of the membrane. The permeate collection vessel was placed in an ice bath to prevent the recovered water from vaporizing. The gas stream containing the water vapor that did not condense in the membrane pores (i.e.

retentate) was directed to a second collection vessel located in the ice bath so that all water vapor not recovered by capillary condensation in the membrane pores was collected for a material balance. Water from a recirculated chiller, set at 25°C, continuously flowed through a water jacket surrounding the membrane holder to control the temperature. Relative humidity and gas stream temperature were monitored using a Vaisala HMT337 HUMICAP® humidity and temperature detector. All data reported are based on 3 hour experimental runs and efficiency is based on total water collected (permeate and retentate). Figure 1 shows a schematic of the lab-scale test set-up.

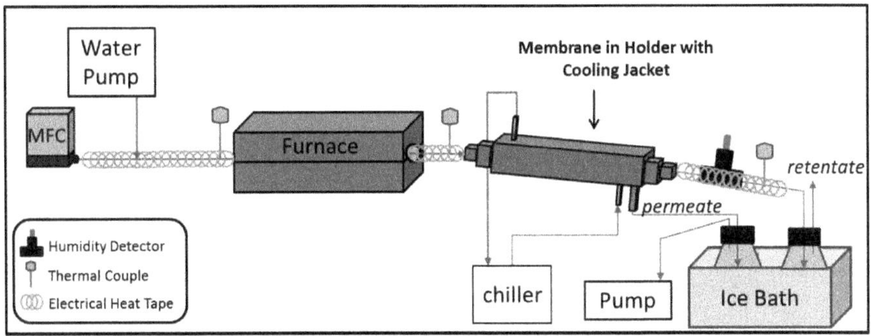

Figure 1. A schematic of the set-up used for lab scale testing.

Diesel Generator Experiments

In order to evaluate the membrane performance using diesel engine exhaust, we measured the water recovery for both single and 9-tube (3x3) membrane prototypes in the setup illustrated schematically in Figure 2. Exhaust from a PRAMAC Model P6000s 5kW stationary diesel generator (engine manufactured by Yanmar) was routed first through a diesel oxidation catalyst (DOC) and a catalyzed diesel particulate filter (DPF) to remove unburned hydrocarbons and particulates. The DOC and DPF were commercially available catalysts from a 2008 Ford F-250 light-duty truck and were installed unaltered. A slipstream from the conditioned exhaust was then passed through the prototype membrane and into a cold trap to collect any remaining water. A backpressure valve near the downstream end of the exhaust train was used to control the slipstream flow. Thermocouples were installed at several locations: engine out; DOC in; DPF in; membrane in; membrane out; chiller in; and inside the cold trap. A wide band lambda sensor was used to monitor air/fuel ratio and oxygen content in the exhaust.

Experiments were conducted using certified ultra-low sulfur diesel fuel (ULSD). Typically, the generator was started and the exhaust temperatures and exhaust flow allowed to stabilize at 80% load (at constant 3600 rpm speed). The electrical output was continuously dumped to a large resistor bank. After the operating conditions were stabilized, water collection by the membrane and cold traps was monitored for 3 hours. An auxiliary fuel injector was installed upstream from the DOC to provide extra heating to regenerate the DPF in case high back pressures were detected. However, auxiliary fuel injection was never needed because the DOC and DPF combination was very effective at passively oxidizing any accumulated soot. Heat losses from the exposed exhaust lines were controlled with insulation and heat tape such that the slipstream temperature entering the membrane was maintained at 250°C. Both membrane prototypes were cooled with cooling fluid from a chiller set at 25°C.

A second generation holder design was used for both single and multi-tube prototypes that included graphite foam (Koppers, Inc.) around the membrane to facilitate heat transfer. The graphite foam is 1/4th weight of aluminum with a thermal conductivity of 240 W/m•K resulting in ~30% higher thermal conductivity and 10 times the thermal diffusivity. Thus the foam's low thermal mass and high conductivity facilitates a very rapid thermal equilibrium with minimal loss in effectiveness during startup.

The single-membrane measurements used a slipstream flow set at 2 L/min (STP) while 18 L/min (STP) was used for the 9-tube prototype. The membranes in both prototypes had median pore diameters of 8.2 nm and tube lengths of 9 in (22.8 cm). The graphite foam/membrane assemblies were housed in cylindrical aluminum shells fabricated from 1.5" and 3" schedule 40 pipes, respectively. The foam was in intimate contact with both the membrane support and the wall of the holder allowing efficient heat transfer from the membrane to the holder. A helical copper coil on the outside of the aluminum carried the circulating cooling fluid.

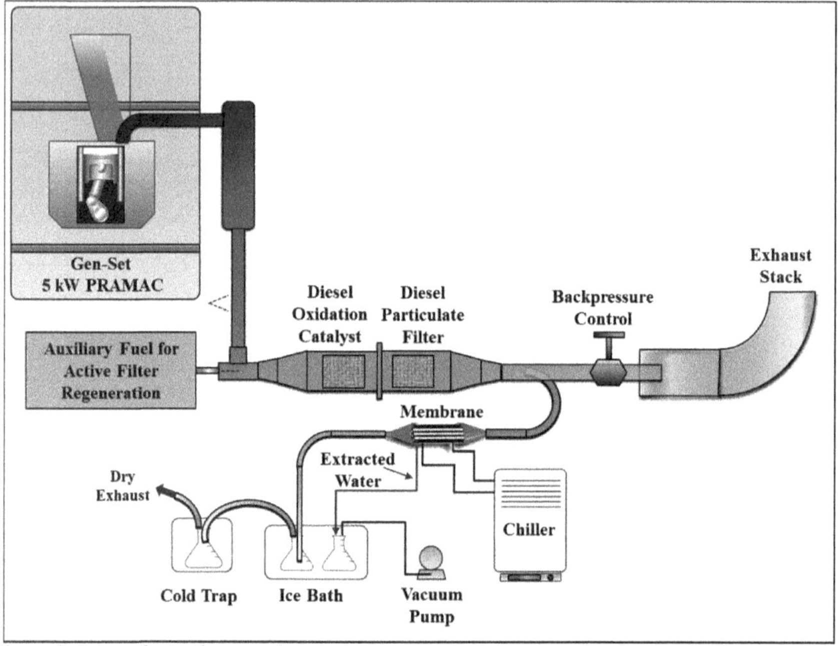

Figure 2. Set-up for testing membrane water recovery systems on diesel exhaust slipstream from a stationary generator (GenSet).

As in the lab-scale experiments, the transmembrane pressure in the diesel portion of the study was controlled at 150 Torr (20.0 kPa). Due to the high porosity of the graphite foam used in the 2nd generation holder, the emergence of the recovered water was delayed as the pores of the graphite foam slowly filled with water. Therefore, the operating procedure was modified such that the water not collected in the retentate cold trap was assumed to be recovered by the membranes. However, to ensure that essentially all of the water vapor in the retentate stream was recovered, a dry ice/acetone cold trap was added after the ice bath, see Figure 2.

RESULTS AND DISCUSSION

Lab-Scale Measurements of Membrane Water Recovery

The major design element of our water recovery system is the series of nanoporous tubular inorganic membranes. The tubular design allows the exhaust gas to flow down the center of the tubes unrestricted, while the water vapor condenses in the porous wall of the membrane by capillary condensation, see Figure 3. Limited research has been done on applications employing capillary condensation as a membrane separation mechanism. Some work has been done related to water vapor and latent heat recovery from the flue gas of power plant boilers.[8] While other studies have focused more on the basic behavior of water in pores of various ceramic materials.[9-11] Our membranes have a thin nanoporous alumina membrane supported on a porous stainless steel support. The stainless steel support provides a more robust membrane for real-world application and a conductive structure for the transport of the heat of condensation from the membrane layer to the graphite foam.

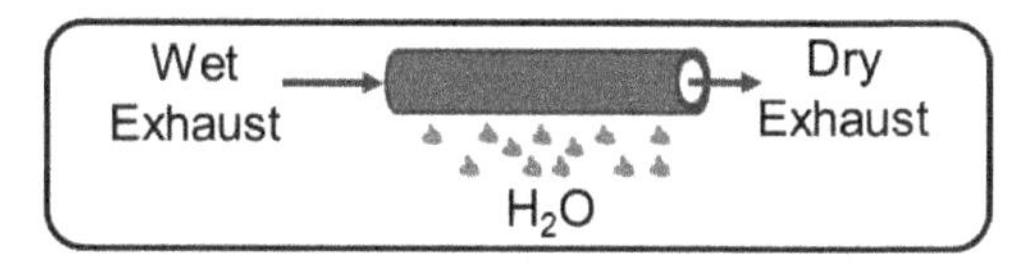

Figure 3. A schematic illustration of the gas and water flow through the hollow tubular membranes investigated in this study. Wet exhaust enters the upstream end and water vapor condenses in the pores on the walls. Liquid water permeates through the wall and the remaining dryer exhaust exits the downstream end of the tube.

During capillary condensation, van der Waals forces cause water vapor to undergo multilayer adsorption in the membrane capillaries (pores) until the pores fill with the liquid.[12] The shape of the resulting menisci depend on the surface tension and size of the pores. The menisci that form in hydrophilic pores are concave, reducing the water's equilibrium vapor pressure below its normal saturation value. The difference between the equilibrium vapor pressure (P) and the saturation vapor pressure (P_0) is related in the Kelvin equation (eq. 2) as a function of the surface tension (γ), molar volume (V_m), Kelvin radius (r_m), and the temperature (T). The Kelvin radius is the radius of the open capillary with an adsorbed layer of water molecules when capillary condensation occurs (i.e. the pore radius minus the thickness of the adsorbed layer). The lower equilibrium vapor pressure results in more water vapor condensing out of the exhaust than is possible by traditional condensation methods.

$$\ln (P/P_0) = (-2\gamma V_m) / (r_m RT) \tag{2}$$

The Kelvin equation is derived assuming thermodynamic equilibrium, which does not account for transport effects, so it is important to understand how such effects reduce the potential water recovery from the ideal Kelvin limit. Membranes with median pore sizes between 5.8 and 9.5 nm were used in this investigation. In the lab experiments, a feed gas of humidified air flowing at a rate of 2 L/min containing 10% water vapor was heated to 250°C prior to entering the membrane. The latent heat of the condensation process was transferred from the membrane support to 25°C chilled water that was circulated through the holder's water jacket.

Under these conditions, the 9”- membranes were cable of recovering 54-60% of the water from the feed, see Table I.

Table I. Effect of Transmembrane Pressure on Efficiency of Water Recovery at 2 L/min

	Transmembrane Pressure		
Pore Size (nm)	100 Torr	150 Torr	350 Torr
5.8	53.5%	56.8%	60.2%
8.2	58.1%	57.4%	59.1%
9.5	54.3%		

According to the Kelvin equation, the recovery efficiency is inversely affected by the pore size, within the range studied. However, the water recovery efficiencies of the membranes studied did not support this trend (Table I). This inconsistency might be related to the dynamic testing conditions of our tests, which are not expected to provide the equilibrium environment described from by the Kelvin equation. One such variable present in the system that is not accounted for by the Kelvin equation is the transmembrane pressure, which the results presented in Table I suggest has a noticeable impact on the efficiency of the membranes. The water recovery efficiency of the membranes increased as the transmembrane pressure (ΔP) was increased. When 100 Torr of vacuum was applied, both the 5.8 and 9.5 nm membranes recovered about the same amount of water suggesting the systems may be flux limited. Increasing the vacuum to 150 and 350 Torr, moderately improved the recovery. The inconsistency of the 8.2 nm membrane at 100 Torr may be related to its wider pore size distribution, but more testing is needed. The 9.5 nm membrane could not maintain condensed water in all of the pores as the ΔP was increased, eliminating the potential for a direct correlation between the water recovered and efficiency of capillary condensation.

Flow rate is another dynamic parameter not accounted for by the Kelvin equation. The rate at which a feed gas passes through the membrane affects the water vapor-pore contact time, and thus it can influence water recovery when transport processes limit the approach to equilibrium. Table II summarizes the observed membrane efficiency as a function of feed flow rate. Water recovery increased as the flow rate decreased, regardless of pore size, indicating that residence time has a significant impact on efficiency. As residence time increased, the efficiency of the 9.5 nm membrane measured with a ΔP of 100 Torr increased similarly to the 8.2 nm membrane with more than twice the ΔP. This suggests that the membrane's ability to recover water from the feed gas is more greatly impacted by gas residence time than ΔP. The increase in efficiency observed for the 5.8 nm membrane when the flow rate was reduced by half was less significant than that observed for the 9.5 nm membrane even though a greater ΔP was applied. This indicates that transmembrane pressure may be a more significant variable at smaller pore sizes. This is not surprising considering the fact that the capillary forces, which are used to retain water in nanopores, are greater at smaller pore sizes. The greater transmembrane pressure used for the 8.2 nm membrane measurement has a greater impact on efficiency at lower flow rates when membrane efficiencies are compared. As the gas residence time inside the membrane decreased, the impact of the transmembrane pressures diminished, underscoring the importance of residence time on dynamic efficiency.

Modeling the Impact of Membrane Length

With help from CD-Adapco, a simulation of the system was developed for STAR-CCM+. Importing CAD data from SolidWorks, the program used the finite volume method to solve equations within each cell of a volumetric mesh created from the boundaries of the CAD model. The mesh consisted of polyhedral cells within each body and a layer of prism cells between the condensing vapor and the solid holder. The prism layer, Figure 4, represented the porous membrane and was necessary for the activation of the Kelvin equation (eq. 2) in the simulation. The Kelvin equation was converted into custom field functions applied to the prism layer in the condensing vapor region of the model. Vapor entering the prism layer cells was subject to these field functions. Any liquid condensed in these cells was removed from the system entirely using a mass sink function.

Table II. Effect of Feed Gas Flow Rate on Efficacy of Water Recovery

	Flow Rate of Feed Gas		
PORE SIZE	1 L/min	2 L/min	3 L/min
5.8 nm(1)	63.1%	56.8%	
8.2 nm(2)	70.9%	59.1%	49.6%
9.5 nm(3)	65.5%	54.3%	48.3%

(1)150 Torr displacement vacuum; (2)350 Torr displacement vacuum; (3)100 Torr displacement.

The membrane's inner diameter, the temperature of the inlet gas, the mass flow rate of that gas and the percent of water vapor present were all initial conditions. The temperature of the membrane was regulated by a boundary condition placed on the outside of the holder's geometry. Two other conditions, pore size of the membrane and transmembrane pressure, were defined parametrically in the simulation's field functions. For the purposes of this model, the transmembrane pressure was considered a volumetric mass sink and determined by an equation with a user-defined coefficient.

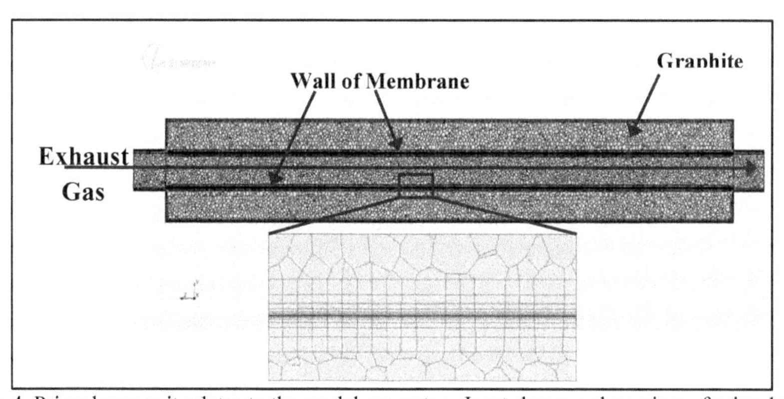

Figure 4. Prism layer as it relates to the model geometry. Inset shows a close view of prism layer cells.

In order to determine the proper mass sink coefficient, the simulation was run with conditions of 8.2 nm pore size, 250°C inlet temperature, 2 L/min flow rate at 10% water vapor, 25°C chiller temperature and 9 inches of membrane. Lab measurements for these conditions had already been obtained with a transmembrane pressure of 350 Torr. For an arbitrary mass sink coefficient, the simulation's efficiency was compared to the lab-scale efficiency. The coefficient was adjusted as needed until the two efficiencies matched. Additional simulations have been run using different initial and boundary conditions to provide trends to base future physical tests.

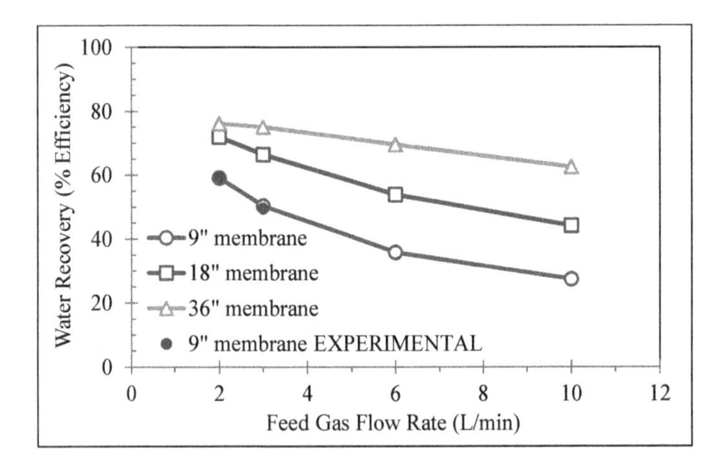

Figure 5. STAR CCM+ modeling of water recovery efficiency as function of the feed gas flow rate for different length membranes. The closed circles (•) are experimental data.

The model was used to evaluate how changes to membrane dimensions can be used to increase the gas contact time with the membrane pores. If the limiting factor is the gas residence time, then increasing the length of the membrane should increase efficiency until the system approaches equilibrium. Once the optimal residence time is obtained, further increases in the length should have essentially no further measurable benefit. Figure 5 shows the results of modeling the system's water recovery efficiency while employing 9, 18 and 36 inch long, 8.2 nm membranes as the flow rate is increased. As the length of a membrane is increased, its efficiency also increases but is not directly proportional. This can be seen by comparing efficiencies from different examples where the ratio of membrane length to flow rate is maintained, such as the 9"-membrane with a flow rate of 3L/min (50.4%) and the 18"-membrane at 6 L/min (53.9%). Interestingly, the model predicts that by quadrupling the membrane length from 9" to 36", a greater efficiency can be attained even at 5 times the flow rate.

Membrane Water Recovery from the Diesel Exhaust

The measured membrane water recovery for the single tube membrane was 50.7% (±0.1%) when the diesel exhaust contained ~10.0% (±0.1%) water. This was in close agreement to the 59.1% efficiency seen in the laboratory using the humidified air feed gas. For the 9-membrane prototype with the exhaust containing ~8.58% (±0.05%) water, the measured recovery efficiency averaged 47.4% (±1.4%). The similar efficiencies obtained in the single and 9-membrane (50.7% and 47.4%, respectively) show the favorable potential for scaling-up the size of the system. We speculate that the reduced efficiency for the 9-membrane system can be attributed to the reduced water content in exhaust caused by the humidity in the intake air, but it could have also resulted from slightly unbalanced gas flows and/or reduced heat transfer efficiency.

After measuring the membrane performance with the diesel exhaust, the membrane was re-measured in the lab. The 8.2 nm membrane showed no change in efficiency, at a transmembrane pressure of 350 Torr, indicating no deterioration of the membrane's performance after roughly 50 hours of exposure to diesel exhaust. Since the lab-scale retesting of the membrane gave the same 59.1% efficiency, the differences were attributed to the properties of the feed gas and the holder design, since different holder designs were used for the engine and laboratory tests. If the holder design was the cause for the drop in efficiency, understanding how different design details influence water recovery could facilitate the development of an improved holder design.

CONCLUSION

Our results demonstrate that a tubular membrane water recovery system can be used to recover at least half of the exhaust water typically present in diesel engine exhaust when appropriate flow rates/membrane are maintained. Laboratory measurements have shown that applying a sufficient transmembrane pressure is necessary to optimize the water recovery from the system. Our measurements also showed that increasing the gas residence time improves water recovery efficiency. This suggests that vapor-liquid-pore mass and heat transfer processes that depend on gas-pore contact time can be critical factors in maximizing overall efficiency. The laboratory results have been confirmed with exhaust from a diesel stationary generator. Computational simulations indicate that increasing the length of the membrane tubes may be a good way to improve recovery efficiency without sacrificing feed flow rate. Evaluating how membrane temperature, increased flow rate per membrane, and membrane length affect water recovery efficiency are other key parameters in designing a practical, full-scale system that are currently being investigated in our laboratory

ACKNOWLEDGEMENT

This research was sponsored by the Laboratory Directed Research and Development Program of Oak Ridge National Laboratory, managed by UT-Battelle, LLC, for the U. S. Department of Energy.

REFERENCES

1. United States. Dept. of Transportation. Natl. Highway Traffic Safety Administration, "Corporate Average Fuel Economy Standards Passenger Cars and Light Trucks Model Years 2017-2025 - Final Environmental Impact Statement." July 2012.
2. U.S. Environmental Protection Agency and National Highway Traffic Safety Administration, "Joint Technical Support Document: Final Rulemaking for 2017-2025 Light-Duty Vehicle," EPA-420-R-12-901, August 2012.
3. Hountalas, D. T.; Mavropoulos; G. C.; Zannis, T. C. "Comparative Evaluation of EGR, Intake Water Injection and Fuel/Water Emulsion as NOx Reduction Techniques for Heavy Duty Diesel Engines." *SAE Technical Paper Series,* no. 2007-01-0120, 2007.
4. Kang, I.; Bae, J.; Bae, G. *J. Power Source,* 163 (2006) 538-546.
5. Heck, R. M.; Farrauto, R. J. "Catalytic Air Pollution Control," p 148, Van Nostrand Reihold, New York, 1995.
6. a). Judkins, R.; Bischoff, B.; DeBusk, M.M.; Narula, C.K. "Reclamation of Potable Water from Diesel Exhaust." US Patent #8,511,072, 2013.
   b). Judkins, R.; Bischoff, B.; DeBusk, M.M.; Narula, C.K. US Patent Appl. #13/915,182, 2013.
   c). Judkins, R.; Bischoff, B.; DeBusk, M.M.; Narula, C.K. International Patent PCT Appl. #PCT/US/1230558, 2012.
7. Fain, D.E. "A dynamic flow-weighted pore size distribution," in *Proc. 1st Intl. Conf. Inorganic Membranes,* 1-5 July 1989, 199-205, Montpellier.
8. Wang, D.; Bao, A.; Kunc, W.; Lis, W. "Coal power plant flue gas waste heat and water recovery." *Appl. Energy,* 91 (2012) 341-348.

9. Solveyra, E. G.; de la Llave, E.; Molinero, V.; Soler-Illia, G. J. A. A.; Scherlis, D.A. "Structure, Dynamics, and Phase Behavior of Water in TiO_2 Nanopores." *J. Phys. Chem. C,* 117 (2013) 3330–3342.
10. Seshadri, A. K. and Lin, Y.S. "Synthesis and water vapor separation properties of pure silica and aluminosilicate MCM-48 membranes." *Sep. Purif. Technol.,* 76 (2011) 261-267.
11. Ng, E.-P. and Mintova, S. "Nanoporous Materials with Enhanced Hydrophilicity and High Water Sorption Capacity." *Microporous Mesoporous Mater.,* 114 (2008) 1–26.
12. Gregg, S. J. and Sing, K.S.W. Adsorption, Surface Area and Porosity, 2nd Ed. Academic Press, London, 1982.

RELIABILITY OF CERAMIC MEMBRANES OF BSCF FOR OXYGEN SEPARATION IN A PILOT MEMBRANE REACTOR

Pfaff, E.M., Oezel, M., Eser, A., Bezold, A.
RWTH Aachen University
Aachen, Germany

ABSTRACT

In the framework of CCS Technology an oxyfuel combustion process was designed for coal fired power plants. Heart of the process is an oxygen separation module which allows enriching recirculated flue gas with oxygen for combustion in a special developed burner. This German project OXYCOAL-AC has choosen BSCF (Ba0.5Sr0.5Co0.8Fe0.2O3-x) as membrane material. BSCF is well known as the MIEC ceramic with highest oxygen flux but information about thermo mechanical data are relatively poor. The membrane components in the reactor are manufactured as one side closed tubes from BSCF powder by isostatic pressing and firing in air at 1100 °C. The dimensions of the sintered tubes are 500 mm length and 15 mm outer diameter. The wall thickness is about 0.85 mm. The tubes are integrated in a special designed module which can hold up to 600 membrane tubes so that a membrane are of 14 m^2 can produce about 0.5 tons oxygen per day. The module operates at 850 °C with air pressure of 15-20 bar on the feed side and a vacuum of 0.5 – 0.8 bar on the permeate side. For calculation a reliable long term operation in a power plant respectively other applications firstly the long term behavior of the ceramic material and the chosen joint patch has to be investigated. Neglecting the chemical reactivity of BSCF this paper concentrates on the mechanical strength at operating temperature.

1. INTRODUCTION

In the course of actions to reduce CO_2-emissions of fossil burning power plants so called Oxyfuel-Processes are under development. The aim is a "CO_2-free combustion" of coal using recirculated exhaust gas enriched with approximately 20 % oxygen. In this case the burning exhausts consist of mainly CO_2 and H_2O which can be easily extracted. Finally nearly pure CO_2 remains which subsequently can be liquefied for a long-term underground storage in geological formations. Such processes are called Carbon Capturing and Storage (CCS). Supplying oxygen by cryogenic air fractionation is an industrial standard but for a power plant process a loss of efficiency in the order of 5 – 10 % points is estimated. Oxygen generation by oxygen transport membranes OTM is the subject of intensive research activities because this technology promises a loss of efficiency in the order of 2–5% points only[1]. OTMs operate at high temperatures which make them ideally suited for direct integration in coal combustion power plants[2].

OTMs are built up with so called mixed ionic-electronic conductor materials (MIEC). Perovskites like $Ba_{0.5}Sr_{0.5}Co_{0.8}Fe_{0.2}O_{3-\delta}$ (BSCF) show the highest oxygen permeabilitiies[3]. The transport mechanism of oxygen through the dense ceramic material is by oxygen diffusion via oxygen vacancies in the crystal lattice as known also from solid oxide fuel cells. Using its inherent electron conductivity the membrane material has an inner circuit and needs no electrical connections. Oxygen will be transported selectively at temperatures of 800- 900 °C. The driving force for diffusion is the difference in oxygen partial pressure between both membrane sides.

A membrane module for producing oxygen using this type of membranes was built up in the German OXYCOAL-AC research project[4]. It consists of 600 membrane tubes with 500 mm length, an outer diameter of 15 mm and a wall thickness of 850 μm. This module is working at 850 °C with pressurized air on the feed side and a low pressure of about 0.5 bar using a vacuum pump on the permeate site. It can produce 0.5 tons oxygen per day[5].

The reliability of the module depends on the strength of the ceramic material and its degradation process during operation such as subcritical crack growth, creep, chemical decomposition of the complex structure and corrosion related to the environment (i.e. Cr from high temperature alloys). The strength depends on the microstructure and therefore on the manufacturing process. To increase reliability of the ceramic tubes several efforts were made to optimize the manufacturing process and the quality assurance.

BSCF material shows an untypical behavior of the temperature dependent strength with a minimum strength at temperatures between 200 and 400 °C. First experimental observations on ring-on-ring tests were discussed by[6]. An alternative test arrangement with a biaxial stress condition namely the ball-on-three ball method[7], was used in this study in which particular importance were taken into account to achieve same sample properties as the manufactured tubes for the membrane module.

2. MANUFACTURING AND JOINING OF MEMBRANE TUBES

There are different possibilities to design membrane components as tubular or planar structures. Independent of this the main objective is to realize a robust component with a minimum wall thickness while the wall thickness is inversely proportional to maximum oxygen flux. The highest strength is found in monolithic components and asymmetric membranes have the smallest membrane thickness[8]. The module was designed with a tubular concept using single closed membrane tubes. Wet plastic extrusion is an alternative shaping process which showed problems relating to the water instability of BSCF powder. Thermoplastic extrusion seems to be a good method for producing tubes with small diameter and small wall thickness[9].This process is still under development. In this study isostatic pressing was found to be an efficient shaping method.

One of the most important concerns in order to build up a pilot production for the objective of testing the module operation is to have ceramic powder of same properties and quality. Therefore 600 kg BSCF powder was delivered by Treibacher Industrie AG, Austria. The powder was granulated in an industrial spray granulator at DORST, Germany. Special attention was paid to the filling of the press mould using a vibration table since there were difficulties in realizing a homogenous wall thickness. Pressing at 180 MPa in an isostatic press (EPSI, Belgium) delivered the highest green density which allowed good handling. The tubes were pressed with one end closed and were finished by green shaping. The BSCF tubes were put in the space between two alumina tubes and sintered at 1100 °C for 5 hours in an electric furnace.

In order to integrate the tubes in the membrane module first metal mounting tubes of the high temperature steel X15CrNiSi25-20 were first joined to the BSCF tubes. Since the joining zone is located in a water cooled flange during operation a glue stable up to 150 °C could be used. Due to the abnormal temperature dependent strength behavior of the BSCF material and the risk of decomposition at temperatures between 700 and 840 °C [10], so called reactive air brazing RAB with a silver-CuO solder were tested to move the joining zone from the insulated flange to the hot air in the module vessel[11,12]. On the other hand brazing BSCF to steel introduces joining stresses which reduce the strength of the tubes in the brazing zone. These stresses are originated from different thermal expansion of the joining partners and the difference has its highest value at room temperature. Therefore temperature cycles (running up and shutting down the module) should be more critical than long term operating at 850 °C though long term operation could result in creep of the silver based solder and has to be considered as well.

3. EXPERIMENTAL INVESTIGATIONS

Quality assurance

During operation the tubes are loaded by an outer air pressure of 15-20 bar resulting in compressive stresses only. The only critical zone where tensile stresses appear is the outer face of the tube where it is fixe. Ovalization and variation of wall thickness were seen critically since it was expected that they lead to higher tensile stresses. Therefore all fired tubes were first tested regarding wall thickness at three different locations along the tube as presented in Fig. 1. At this three locations the wall thickness was measured continuously around and the maximum and minimum values each are stored. The wall thickness was measured using a Minitest 7200 FH (ElectroPhysik) equipment which works by the magnetic attraction principle shown in fig.2. The probe's Hall effect sensor measures the distance between the probe tip and ball with an accuracy of 1%. These investigations were used to optimize the manufacturing process. The influence of variations in wall thickness on the stress distribution in the ceramic tube was calculated by finite element method (FEM) using ABAQUS software.

Before the tubes were integrated into the module each joined tube was tested in outer air pressure of 30 bar. This pressure is twice that of the operating pressure and was used because the strength at operating temperature (850°C) is up to 57 % lower than that at proof test temperature (22°C), see table 1.

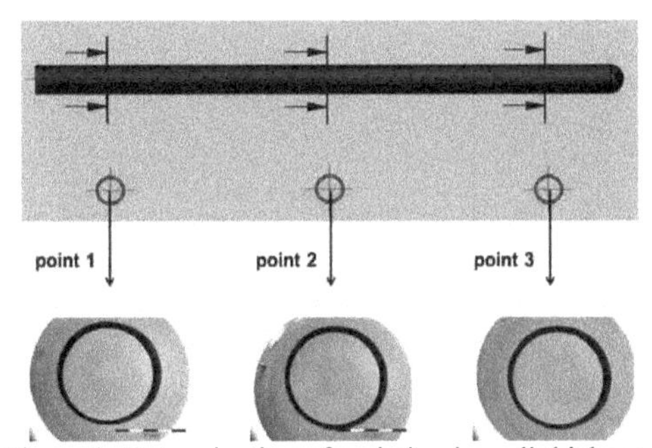

Figure 1. Determination of variation in wall thickness

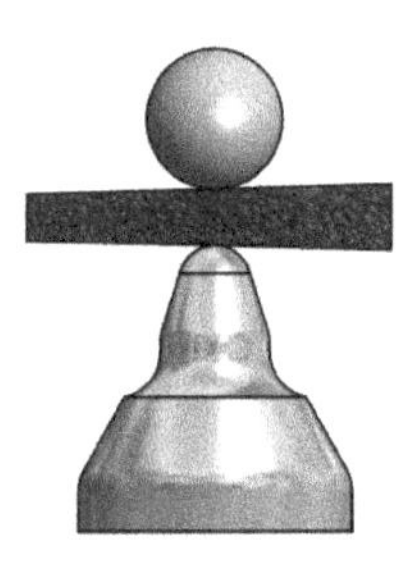

Figure 2. Measuring principle for determination of the wall thickness

Strength measurements

The strength of the BSCF tube material was tested by the ball-on-three-ball method. Samples of 25.8 mm diameter and 3.87 mm thickness were manufactured by die-pressing using the same granulate powder and densification conditions as used for the tubes. Because the tubes will be used without surface finishing the samples were also tested in a first step in the as-fired condition as well. Later these values should be compared to grinded and polished samples. The samples were put on 3 Si_3N_4 balls with a diameter of 19,05 mm and centrally loaded through a fourth ball on the upper side (Figs. 3 and 4). This arrangement was initially fixed by a ring which was removed before testing to ensure free movement and no implementation of outer loads. The diameter of the bearing is 22 mm.

The sample was then heated up in air to the test temperature with a heating rate of 200 K/h and held for a dwelling time of 15 min to ensure oxygen equilibrium in the sample. The loading was increased motion controlled with 0.5 mm/min up to fracture. Selected fracture surfaces were investigated afterwards by SEM to determine the fracture origin.

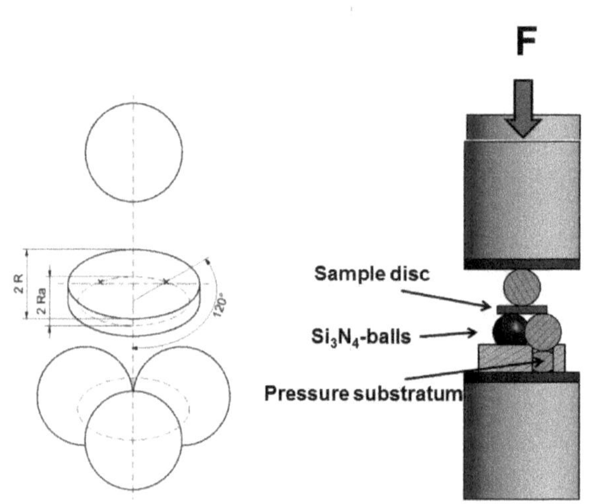

Figure 3. Sample and test arrangement for ball-on-three-ball tests

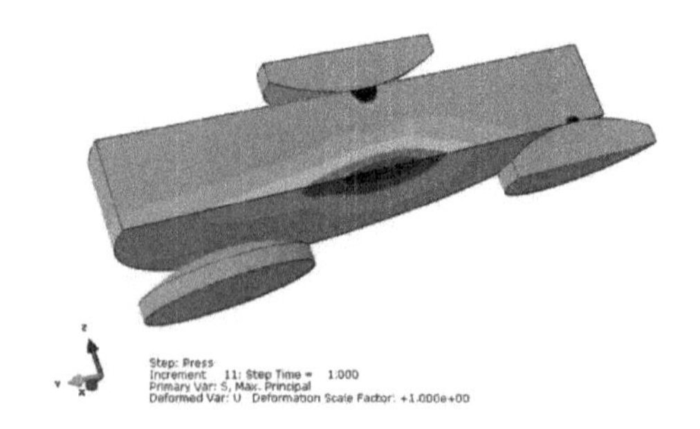

Figure 4. Stress distribution in a sample of a ball-on-three-ball test

The creep behavior has been experimentally investigated some years ago on o-rings. These results can only be used for a first orientation since the material is not the same as the presently used material. BSCF samples for creep testing were produced by wet extrusion and manufactured at a different laboratory. New tests are running but results could not be presented at this time.

4. PILOT MODULE

The membrane module consists of a pressure vessel with inner thermal insulation and heating elements. The feed gas is pre-heated compressed air streaming in from one side of the vessel vertically towards the tubes. An electric heating element inside the vessel assures an air temperature of 850 °C. The membrane tubes are mounted in a plate (mounting flange) which is cooled by water to temperatures below 100 °C. In the holes of this flange the metal jackets glued to the end of the ceramic tubes are fixed and sealed by conventional polymer o-rings. The flange is insulated from the vessel by ceramic fiber material. A schematic drawing and a picture of the module are given in fig. 5 and 6. The module operates at 850 °C with an air pressure of 15-20 bar. The oxygen on the feed side of the membranes is removed by a vacuum pump working in the low pressure range of 0.3 – 0.8 bar. In this membrane module 300 tubes are mounted hanging in the lower part of the vessel and 300 tubes are upright in the upper half. The lower tubes are loaded additionally to outer pressure by a small tensile stress related to the weight of the tubes. The upper ones are endangered if they would be deformed like a banana by creep.

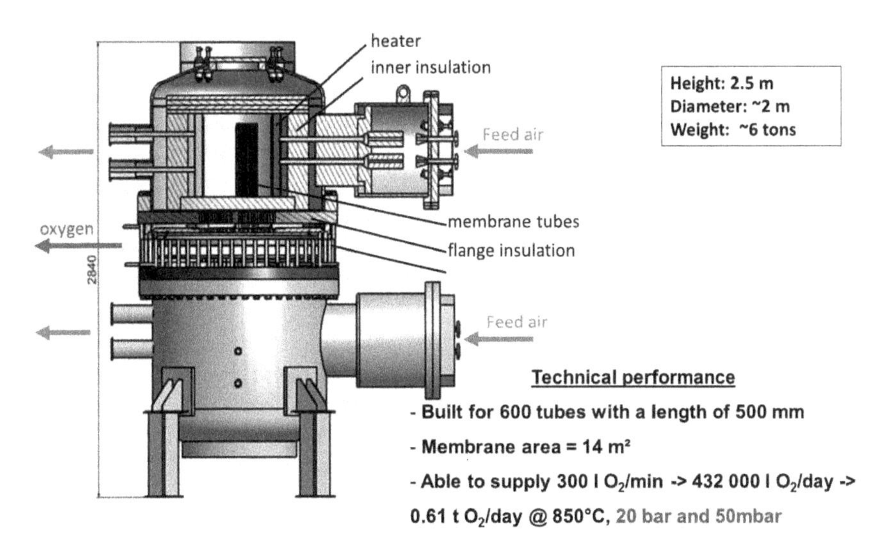

Figure 5. Schematic drawing of the oxygen pilot module

Figure 6. View to the pilot module

The module was built in order to perform long term tests but during operation it was observed that the mounting of the tubes is much more critical than expected. A slight tilt during mounting could result in contact of the tubes and lead to flexural stresses. In order to avoid such contacts it is necessary to have a certain amount of space between the tubes, thus reducing the maximum possible packing density.

5. RESULTS

The sintered isostatically pressed tubes and the die-pressed samples show the same microstructure after sintering characterized by a closed porosity of about 7 % and grain sizes of 5 – 30 μm (Fig. 7). Differences can be seen in the surfaces. Since the outer mould of the isostatic press consists of weak synthetic material the granules are not fully destroyed resulting in a rough surface. It is well known that surface flaws are critical for failure through tensile loads and can determine the ultimate strength of a sample but for the special application of the membrane tubes the outer surface will not be loaded by tensile stress but only by compressive stress. Therefore no necessity was seen to finish the surface. On the contrary the roughness of the outer surface increases the specific surface area needed to decompose oxygen molecules to ions and should therefore improve the oxygen flux.

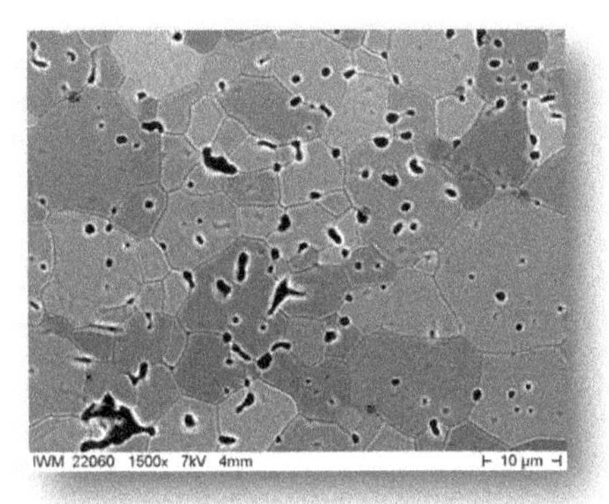

Figure 7. Microstructure of BSCF membrane tubes and samples

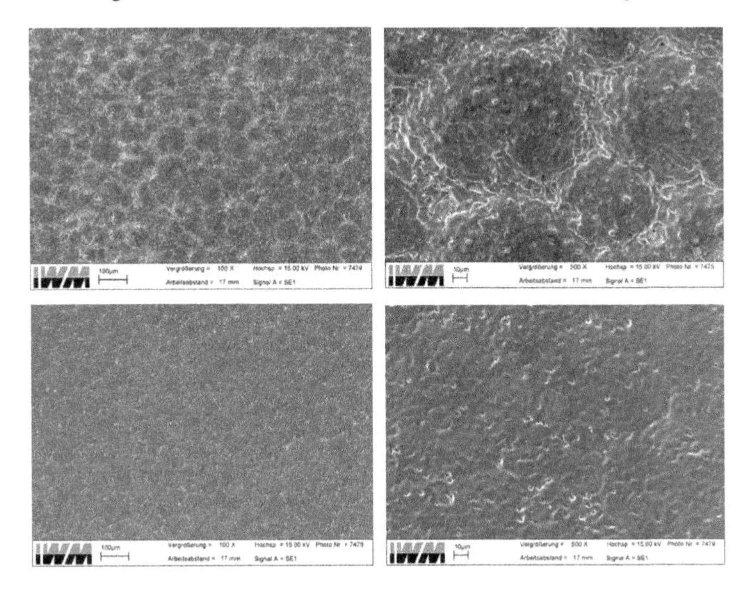

Figure 8. Comparison of outer surface (upper pictures) and inner surface (lower pictures) of isostically pressed membrane tubes

All membrane tubes were tested after gluing to the metal joining socket with 30 bar air pressure at room temperature before mounting in the module. About 7 % of the joint tubes failed during this test.

The results of the ball-on three-ball tests (30 samples each at different temperature) were evaluated by Weibull statistic. The loaded volume was calculated by FEM analysis. As an example the Weibull plot for the room temperature values is presented in fig. 9. All results are summarized in tab. 1 and fig.10.

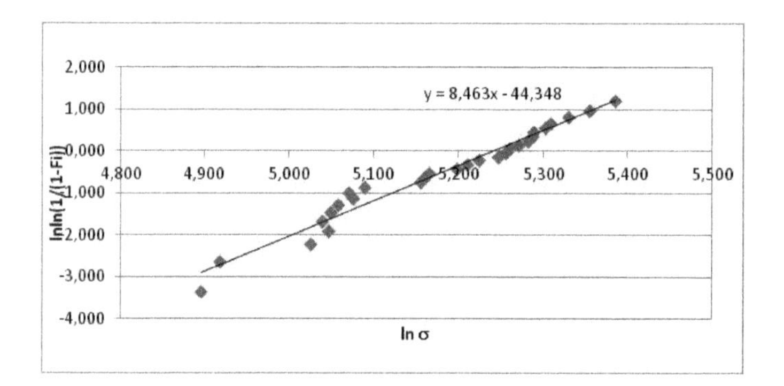

Figure 9. Weibull evaluation of the ball-on-three ball test at room temperature

Table 1. Results of ball-on-three-ball tests at different temperatures

Temperature [°C]	T	20	100	200	300	400	850
Characteristic strength [MPa]	σ_0	188	169	75	69	71	108
Weibull modulus [-]	m	8.5	5.4	4.8	4.4	4.8	7.4
Effective volume [mm^3]	V_{eff}	1.1	7.1	9.4	11.6	9.5	3.7
Weibull scale parameter [MPa]	σ_{0v}	189	241	119	120	113	128

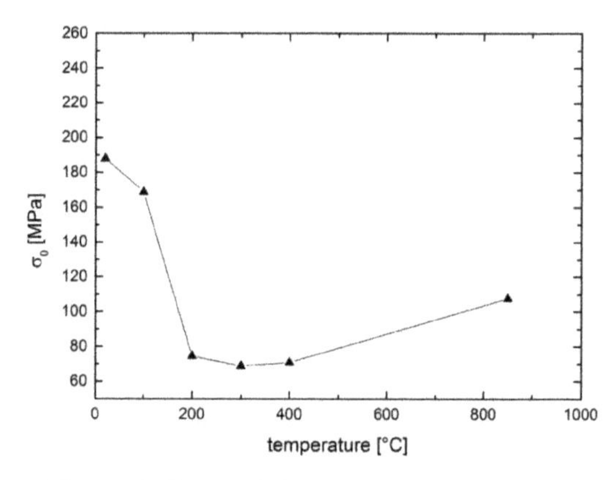

Figure 10. Characteristic strength of BSCF as results of ball-on-three-ball tests at different temperatures

The results show at room temperature a characteristic strength value of 188 MPa. To get a better feeling this value can be compared to four-point-bending strength by a transformation considering the different effective volumes of the test arrangements using Weibull theory. For a 4-point-bending test the characteristic strength should be 135 MPa assuming that the Weibull parameter m is constant. Figure 10 shows a minimum in strength between 200 and 400 °C and an increase at higher temperatures corresponding to biaxial ring-on-ring tests described by Huang et.al.[6]. The minimum strength at 300 °C is only 37% of the room temperature strength and increased again to 57% at 850 °C. Comparing to the results of Huang the minimum strength was observed at 400 °C with a strength of about 50% and at 800 °C of about 80%. In our study the lowest strength values at 300 °C are accompanied by the highest scattering compared to the results of Huang where the lowest strength values at 400 °C show also the smallest scattering.

Both studies give the same advice not to operate the membranes at intermediate temperatures. During operation at 850 °C and 15 bar air on the outer side of the membrane tubes the ceramic is loaded uncritically by compressive stress only but at the joining zone tensile stress can occur caused by hindered expansion. FEM calculations showed maximum principle stresses at the end glued to the metal socket of about 33 MPa for a tube wall thickness of 850 µm. If the wall thickness is smaller the stress increases. During wall thickness measurements large deviations from the intended value of 850 µm were observed. Thus, a location was analyzed with a maximum thickness of 1130 µm and a minimum value of 450 µm. For this minimum wall thickness the tensile stress increased to 54 MPa which becomes quite critical.

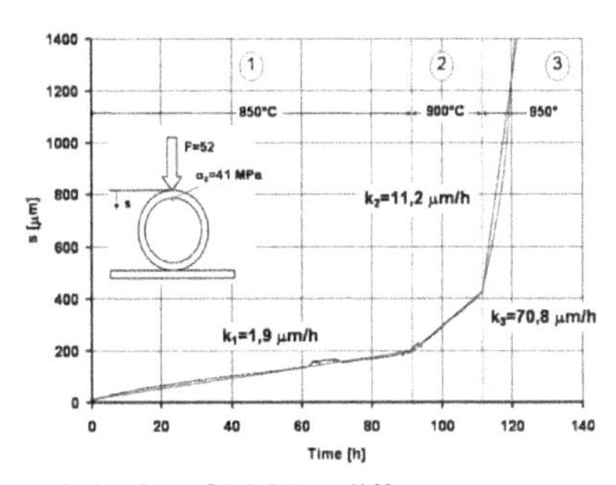

Figure 11. Creep behavior of BSCF at different temperatures measured on rings with Ø=10mm, l=10mm, wall thickness s=1.1mm [13]

For long term reliability also creep of the ceramic material could be of importance as well. In the first investigations creep test were conducted on BSCF rings cut off from extruded tubes (HITK, Germany) at temperatures of 850, 900 and 950 °C. Fig. 11 shows the strong increase of creep with temperature.

Much more critical could be chemical changes in the BSCF material. In the membrane pilot module the water cooled flange holding the membrane tubes is insulated with respect to the gas chamber with ceramic fiber wool. In the transition zone where the temperature is still high corrosion of the BSCF tube was found as a blue colored area. This area was analyzed by SEM/EDS and indicates decomposition with enrichment of Strontium and Silicon Oxides.

BSCF is sensitive to CO_2 in the environment and will form $BaCO_3$ at the surface accompanied by a decomposition of the perovskite, resulting in decreasing oxygen flux through the membrane. The effects on strength are unknown. During operation in compressed air no formation of carbonates and no decrease of oxygen flux were observed. Another chemical effect which could become critical is a reaction with chromium oxide evaporated from high temperature resistant steels of parts of the membrane module such as pipelines. When analyzing the above described blue corrosion zone small amounts of chromium were also found but these small amounts did not affect the oxygen flux so far. This will have to be observed during the next operational interval.

6. CONCLUSION

The first test phase of the pilot membrane module was affected by handling problems. Most of the ceramic tubes failed after assembly. The start-up and shut down phases are particularly critical because of the thermal shock sensitive material. Emergency cutouts caused by problems in the water cooling system or the power supply resulted in failures. A further problem was the corrosion of the membranes by the SiO_2 containing fiber insulation. Thus the fiber material will be changed for further operation.

The BSCF material has the highest oxygen flux but is very brittle and has low strength. The abnormal strength behavior at intermediate temperatures gives recommendations for operation. Loads at temperatures between 200 and 400 °C should be avoided. Because of the tube design tensile stress will occur only near the joining zone and has to be considered carefully. Temperatures between 800 and 840 °C should also be avoided because of a slowly progressing phase transformation of the cubic perovskite BSCF to the hexagonal structure with low oxygen flux. Considering also the creep behavior BSCF can only be used in a very small temperature window of 840 – 900 °C.

The first operation phase show that, if damage during assembly can be avoided, thermal shock by emergency cutouts could be dampened and chemical reactions can be prevented a reliable operation of a membrane module is possible.

ACKNOWLEDGEMENTS

The authors gratefully acknowledge the financial support from the German Federal Ministry of Economics and Technology as well as from the companies RWE Power AG, Vattenfall Europe Generation AG and Linde AG within the OXYCOAL-AC project

REFERENCES

[1] M. Czyperek, P. Zapp, H.J.M. Bouwmeester, M. Modigell, K. Ebert, I. Voegt, W.A. Meulenberg, L. Singheiser, D. Stöver, Gas separation membranes for zero-emisission fossil power plants: MEM-BRAIN, *J.Membr.Sci.* **359** 149-159 (2010)
[2] R. Kneer, D. Toporov, M. Foerster, D. Christ, C. Broeckmann, E. Pfaff, M. Zwick, S. Engels, M. Modigell, OXYCOAL-AC: Towards an integrated coal-fired power plant process with ion transport membrane-based oxygen supply, *Energy & Enviromental Science,* **3**, 198-207 (2010)
[3] J. Sunarso, S. Baumann, J.M. Serra, W.A. Meulenberg, S. Liu, Y.S. Lin, J.C. Diniz da Costa, Mixed Ionic-Electronic Conducting (MIEC) ceramic-based membranes for oxygen separation. *J.Mem.Sci.* **320,** 13-41 (2008)
[4] E. Pfaff, A. Kaletsch, C. Broeckmann, Design for a MIEC oxygen transport membrane pilot module, Chemical Engineering & Technology, **35,** 455-463 (2012)
[5] E.M. Pfaff, A. Kaletsch, M. Zwick, C. Broeckmann, N. Nauels, M. Modigell, Generation of 0.5 t oxygen per day with ceramic oxygen separation membranes in a pilot module, Int. Symposium on Ceramic Materials and Components for Energy and Environmental Applications, CMCEE Dresden 20.-23.05.2012
[6] B.X. Huang, J. Malzbender, R.W. Steinbrech, L. Singheiser, Discusion oft the complex thermo-mechanical behaviour of $Ba_{0.5}Sr_{0.5}Co_{0.8}Fe_{0.2}O_{3-\delta}$, *J.Mem.Sci.* **359,** 80-85 (2010)

[7] A. Börger, P. Supancic, R. Danzer, The ball on three ball test for strength testing of brittle discs: stress distribution in the disc, *J. Eur.Cer.Soc.* **22,** 1425-1436 (2002)
[8] S. Baumann, J.M. Serra, M.P. Lobera, S. Escolástico, F. Schulze-Küppers, W.A. Meulenberg, Ultrahigh oxygen permeation flux through supported $Ba_{0.5}Sr_{0.5}Co_{0.8}Fe_{0.2}O_{3-\delta}$ membranes, *J.Mem.Sci.* **377,** 198– 205 (2011)
[9] M. Salehi, F. Clemens, E.M. Pfaff, S. Diethelm, C. Leach, T. Graule, B. Grobéty, A case study of the effect of grain size on the oxygen permeation flux of BSCF disk-shaped membrane fabricated by thermoplastic processing, *J.Mem.Sci.,* **382,** 186-193 (2011)
[10] C. Niedrig, S. Taufall, M. Burriel, W. Menesklou, S.F. Wagner, S. Baumann, E. Ivers-Tiffée, Thermal stability of the cubic phase in $Ba_{0.5}Sr_{0.5}Co_{0.8}Fe_{0.2}O_{3-\delta}$, *Solid State Ionics* **197,** 25-31 (2011)
[11] A. Kaletsch, E.M. Pfaff, C. Broeckmann, Joining Oxygen Transport Membranes by Reactive Air Brazing, International Brazing and Soldering Conference IBSC2012, Las Vegas, NV USA, R. Gourley and C. Walkers, Eds. ISBN 978-1-61503-377-7 437-443 (2012)
[12] A. Kaletsch, A. Bezold, E.M. Pfaff, C. Broeckmann, Effects of copper oxide content in the AgCuO solder on microstructure and mechanical properties of reactive air brazed Ba0.5Sr0.5Co0.8Fe0.2O3-δ (BSCF), *J. Ceramic Science and Technology*, **3**, 95-104 (2012)
[13] E.M. Pfaff, M. Zwick, Oxyfuel combustion using perovskite membranes, *Mechanical Properties and Performance of Engineering Ceramics and Composites III, Ceramic Engineering and Science Proceedings,* **28**, 23-31 (2007)

Advanced Materials and Technologies for Electrochemical Energy Storage Systems

IN SITU EXPERIMENTATION WITH BATTERIES USING NEUTRON AND SYNCHROTRON X-RAY DIFFRACTION

Neeraj Sharma

School of Chemistry, The University of New South Wales,
Sydney NSW 2052, Australia

ABSTRACT

There has been a recent growth in using neutron diffraction to study *in situ* the structural and electrochemical evolution of lithium-ion battery materials. The information extracted from these *in situ* neutron diffraction studies complement the laboratory-based or synchrotron-based X-ray diffraction studies. Essentially these studies harness the advantages provided by each technique to probe the evolution of long-range ordered or crystalline materials in batteries. This paper highlights some of the recent progress on *in situ* neutron and synchrotron X-ray powder diffraction studies and helps to explain which type of elements should be considered by researchers for a successful experiment. To facilitate these explanations new results are presented building on previous *in situ* neutron powder diffraction studies. It is envisioned that future studies will use the combination of *in situ* synchrotron X-ray and neutron diffraction to probe electrode evolution.

INTRODUCTION

Lithium-ion batteries are found in many everyday portable electronic devices, such as mobile phones and laptop computers, and the motivation to improve these batteries is driven by consumer demands for longer-lasting devices and emerging higher power applications such as electric vehicles [1,2]. The chemistry of most lithium-ion batteries revolves around removing lithium ions from the cathode and inserting them into the anode via an electrolyte during charge and the reverse during discharge [3]. This in turn produces an electric current during discharge or utilises power from an external circuit during charge.

In order to produce a good lithium-ion battery, the electrodes need to withstand thousands or more charge/discharge cycles. A number of chemical reactions can occur at the electrodes, including lithium insertion/extraction from a crystalline structure [3], conversion reactions [4] and alloying reactions [5]. These processes need to occur reversibly and the majority of electrode materials are based on the lithium insertion/extraction processes, where lithium is removed from a crystalline substance (or at least some lithium is removed from a crystalline component) and the structure changes to accommodate this. The ideal crystal structure for an electrode would show minimal structural perturbations resulting in a high propensity for reversibility with the maximum amount of lithium being inserted/removed resulting in a high energy density.

These facets of structural transformations, especially long-range ordered or crystalline, and lithium distribution are probed by diffraction methods. Of particular note, is that these processes can be probed in real-time as a function of battery charging and discharging with *in situ* powder diffraction [6-14]. *In situ* X-ray powder diffraction (XRPD) methods are a relatively mature method for probing battery materials, with experiments undertaken at both laboratory and synchrotron X-ray sources using modified coin-type cells or custom designed cells [15]. These experiments provide information on structural phase evolution, such as lattice expansion and intermediate phase formation, but are limited in sensitivity for atomic parameters of the lighter elements such as lithium and oxygen. This is because the scattering power is derived from the atomic number of the elements and the scattering from the lower atomic number elements produce relatively weak signal compared to the higher atomic number transition metals.

Although high-energy synchrotron X-ray diffraction can be used to probe larger batteries such as cylindrical 18650-type or prismatic cells, conventional laboratory and the majority of the synchrotron X-ray sources are used to probe smaller batteries, coin-type, Swagelok-type or custom batteries with small electrode surfaces [15]. *In situ* neutron powder diffraction (NPD) requires larger sample sizes due to the weaker interaction of neutrons with matter, making the larger 18650-type and prism/pouch cells more suited to this technique. Relatively fewer *in situ* NPD studies have been undertaken compared to *in situ* XRD mainly because of the availability of fewer neutron diffractometers, fewer researchers working with larger batteries, and often the need to custom-make batteries [9]. However, there has been recent growth in *in situ* NPD which can be attributed to factors such as development of new instruments, better custom-made batteries and ability to resolve perturbations in both the lattice and atomic parameters, especially with respect to lithium [7,9,10,12,13,16-29]. The key disadvantage with *in situ* NPD is the large contribution to the patterns made by hydrogen-containing and liquid components such as the conventionally-used carbonate-based electrolyte. These components contain elements (with low atomic numbers) that are effectively transparent to X-rays but can dramatically influence the signal-to-noise ratios of NPD patterns. One element, hydrogen is particularly problematic due to its large incoherent neutron scattering cross section that contributes to a large background in NPD patterns [12]. Methods have been developed to overcome these detrimental influences which include minimising hydrogen-content (replacing hydrogen with its isotope deuterium that gives a significantly weaker background contribution [7]) and maximising the electrode quantity in a battery, e.g. double coating current collectors [27] with thicker layers. Progress is continuously being made on both *in situ* XRD and NPD experiments to improve signal and elucidate pertinent information that can be used to improve battery performance.

In this work, two factors are considered for both *in situ* XRD and NPD; the time-resolution and the type of detectors used for these *in situ* studies in order to direct researchers to use the most appropriate instrument for their scientific question.

DETECTORS

Conventional laboratory XRD's often feature 'line' detectors which move through a certain angular range (e.g. 2θ angle) to collect a XRD pattern. The need to transverse the angular range determines the minimum time required for data collection (time resolution). In other words, the minimum time for data collection is a factor of the time required at each detector point for sufficient statistics to be collected and the time required for the detector to transverse the required angular range. The alternative to a single 'line' detector is an array of line detectors that simultaneously collect distinct angular ranges, e.g. 0-1.25, 2.50-3.75, 5-6.25°, and so on. High resolution NPD instruments located at research nuclear reactor sources [30] and some powder diffraction beamlines at synchrotron sources [31,32] feature these types of detectors. The detector regions are stepped to collect angular regions that were not collected in the first positions of the detectors, e.g. 1.25-2.50, 3.75-5°, and so on. Although, the example shown here uses only one step to cover the missing regions, there are examples where many steps can be taken to cover the missing regions. The number of steps depends on the angular region covered by each detector and the overall angular range required for the analysis. In some cases, a smaller individual detector angular region (e.g. 1.2-1.3 compared to 1.2-2) can result in a higher resolution pattern. Therefore, a compromise must be made considering the resolution required, number of detectors that can be used and the number of steps that are ideal for a collection.

One major issue of using a step-scan type of *in situ* measurement is whether the material of interest, e.g. electrode, changes during one collection over a full angular range. In other words, during a step-scan the peak position shifts by a certain amount which can either be averaged out in the data collection or produce unconventional peakshapes which are difficult to model. One way to overcome this is to equilibrate the battery and then collect XRD or NPD data [18,20]. Another method, which is used for some *ex situ* measurements on high-resolution instruments, is to add a number of steps that feature overlapping detector regions, such that certain angular regions are collected more than once in a stepping procedure. This can then be post-processed (which can be difficult in itself) to an appropriate pattern. However, these methods result in the loss of time-resolved and industrial pertinent information about battery materials. The resulting patterns are effectively equilibrium or quasi-equilibrium experiments. The advantage is that a relatively high quality diffraction pattern (even on high resolution instruments) can be collected without the need to open up batteries, extract the electrodes in argon glove boxes and re-seal them for *ex situ* diffraction experiments. The extraction process can result in unwanted side reactions and deterioration of the electrode materials. Thus keeping an electrode under study in a battery and performing an electrochemical step (equilibrating) and then collecting high resolution NPD or XRD data can remove unwanted experimental errors with extraction processes and sample handling.

Another alternative detector available on conventional XRD's and at synchrotron sources are image-plate type detectors. The simplest description of these detectors are photographic film, which when exposed to X-rays (or neutrons in some cases) darken and the amount of darkening is related to the intensity of the diffracted X-ray beam. Modern image-plate type detectors are based on CCD or similar technology and data can be continuously collected in short time intervals, where the time between datasets is determined by the time the electronics require to refresh the CCD screen. The angular range probed is determined by the size (or area) of the screen and the distance from the diffracting object. Researchers need to take into account the resolution they wish to have and the speed of data acquisition when choosing between 'line' detectors with step scanning or image-plate detectors.

For *in situ* NPD studies another alternative is available at reactor sources, these are area detectors that cover an effective continuous angular range, e.g. 120° in 2θ. The angular coverage and detector height are optimised for different instruments and their applications. For *in situ* NPD of batteries the large angular range and fast read-out times afforded at instruments such as Wombat, the high-intensity powder diffractometer at the Australian Nuclear Science and Technology Organisation (ANSTO) [33], D20, at the Institute Laue-Langevin (ILL) [34], and SPODI the high resolution neutron diffractometer at the Heinz Maier-Liebnitz Zentrum (MLZ) [35] make such experiments accessible to researchers. The neutron flux (number of neutrons impinging on the sample) and the detectors on Wombat and D20 allow for fast data collections, on the order of milliseconds for some experiments, but generally for *in situ* NPD studies of battery materials the collection time is on the order of minutes in order to obtain good signal for analysis. Examples of material investigated using these instruments, include $LiCoO_2$ [10,12,23], $LiFePO_4$ [13,24,25,28] and $LiMn_2O_4$-based [7,27] cathodes, and $Li_4Ti_5O_{12}$ [17], MoS_2 [16] and graphite [10,12,23,24,29] anodes. These data contain time-resolved information revealing details about the rates of reactions which can be correlated to the electrochemistry.

Other *in situ* NPD experiments have been conducted on spallation neutron sources [20-22,26] where the intensity of neutrons is generally higher compared to reactor sources. Larger detector coverage over *d* (or 2θ) can be achieved by increasing the number of detector banks placed around the battery. Data handling (and complexity), and the quality of data in certain detector banks have limited the widespread use of spallation sources, however, specially designed instruments for battery research such as SPICA at J-PARC are expected to show some exciting results.

Different detectors are available at synchrotron and neutron diffraction instruments and researchers should choose the instrument and detector based on the problem they are wishing to solve.

TIME-RESOLUTION

Time-resolved *in situ* experiments on lithium-ion batteries are becoming more and more common. The collection time is determined in part by the detector used and the angular range covered. For detectors that are stepped through angular ranges to collect a 'full' diffraction pattern the time resolution depends on the time required at each collection angle and the time to step through the range. As stated above, detectors that cover a larger angular range (and that do not require stepping) can be used to collect shorter (faster) datasets. This assumes that the signal-to-noise ratios and information obtained from the shorter datasets are of sufficient quality.

There are multiple methods to ensure sufficient signal from the material of interest, the first and possibly the simplest approach is to adjust the current rates applied, applying a lower current rate will increase the time for charge/discharge and thereby allow more diffraction patterns of a set time to be collected relative to a faster charge/discharge. Unfortunately, this limits the maximum current rates applied and the extreme case of this would be a slow charge/discharge that is representative of a quasi-equilibrium lithiation/delithiation process rather than the more non-equilibrium processes that occur during conventional battery use. Nonetheless, this can be applied to both *in situ* XRD and NPD experiments. Another method to increase signal is to adjust the quantity of electrode material of interest, however, this may also adversely influence battery performance. Although possible for application for XRD and NPD methods, NPD methods tend to gain the most with increasing the electrode quantity, especially if the quantities of other components can be kept the same or even reduced. Methods that increase the content of electrode materials in a neutron beam include: coating both sides of current collectors and using more rolls per cylindrical battery or more layers per prism battery. In the case of *in situ* NPD experiments minimising hydrogen content also improves signal as it reduces contributions (often manifested by a large background) from the incoherent scattering from hydrogen. This is achieved by using deuterated carbonates instead of conventional carbonates found in the electrolyte and lower-hydrogen content separators such as polyvinyldiene fluoride or glass-fibre membranes rather than polyethylene-based membranes. Again both changes influence electrochemistry and applicable voltage and current ranges [7]. In the case of *in situ* XRD experiments with coin cells for example, the stainless steel casing has to be replaced or modified with an X-ray transparent window [15]. Therefore, careful optimisation of the electrochemistry applied and the materials inside the battery with respect to the data collection time and quality has to be undertaken.

Once these factors have been considered the angular range and type of analysis to be conducted has to be determined. This is often influenced by what question the researcher wants answered. A recent example is the work undertaken on the high current rate behaviour of $LiFePO_4$ cathodes where the authors show evidence for a metastable phase during higher current cycling that is difficult to observe with moderate-low current rates [36]. The authors choose to study only a small angular region (5° in 2θ), as determined by the detector coverage, but tune the synchrotron X-ray beam wavelength to provide the reflections of interest in the limited angular range. This effectively allows probing of a limited angular range but presumably at higher resolution than larger angular ranges, but pre-disposes the need to understand the material under study. For a new material, this is not advised but for something like $LiFePO_4$ where thousands of publications are found in the literature it is possible. Note, sufficient evidence has to be presented to show that the extra reflection is indeed due to the $LiFePO_4$ cathode and not to some other component in the *in situ* cell or even degradation of the cell (as undertaken in the report [36]). This is because the evidence is based on peak positions and intensities, and in this case it was only one extra reflection, so researchers have to be confident that the new reflection is in fact from the cathode. The work [36] does illustrate the level of detail that can be extracted from *in situ* diffraction data with sufficient time-resolution. In this work, a new phase was found at high current rates, which means the mechanisms that were previously assigned to $LiFePO_4$ at these current rates need to be re-assessed for validity.

SOME RECENT RESULTS

To illustrate some of the aspects discussed above, some recent results using time-resolved *in situ* neutron diffraction data are presented below. This work follows from the report in [27], where the studied battery was stored for 11 months and then further *in situ* neutron diffraction data were collected during electrochemical cycling. Data were collected on Wombat [33], the high-intensity powder diffractometer, at the Open Pool Australian Light-water (OPAL) reactor facility at the Australian Nuclear Science and Technology Organisation (ANSTO), similar to our previous work [7,9,10,12,13,16,17,27-29]. The battery details can be found in ref. [27] and this battery was placed in a neutron beam of wavelength (λ) = 2.4130(1) Å, and data were continuously collected in 5 minutes intervals in the angular range $25 \leq 2\theta \leq 135°$. Data correction, reduction, visualization and single peak fitting routines were performed in LAMP [37]. Rietveld refinements were undertaken using GSAS [38] with the EXPGUI [39] interface. Electrochemical cycling was performed using an Autolab potentiostat/galvanostat (PG302N).

A single 5 minute *in situ* neutron diffraction pattern in a limited angular range $35 \leq 2\theta \leq 45°$, is shown in Figure 1a. By collating 24 of these 5 minute patterns as a function of time during electrochemical cycling and representing intensity as a colour scale, we arrive at the 2D plot shown in Figure 1b. Clearly the reflection(s) change position (2θ value) indicating an increase or decrease of the lattice parameter and the changes in intensity (e.g. brightness of the red regions near $2\theta \sim 40°$) indicate atomic re-arrangements. Figure 1 represents 5 minute patterns collected on a relatively unique reactor-based neutron diffractometer. If the signal was inadequate in a 5 minute data collection and the best achievable compromise was a 2 hour data collection the resulting pattern would resemble Figure 2. There is a loss of time resolved information. It is worthwhile to note that a 2 hour pattern from neutron diffractometers at reactor sources is still considered relatively fast. However, this example shows that the loss of time resolution can influence the interpretation of the data – a 2 hour collection seems to show 4 reflections that might arise from 4 phases co-existing in the studied electrode simultaneously. The interpretation of the NPD pattern may suggest such a phase-mixture but it would have to be taken into account with the data collection time. The 5 minute collections suggest a different process for the 002 reflection of Li_xC_6, further details can be found in ref.'s [10,29]. This argument does not consider the fact that the battery has undergone nearly 2 charge/discharge cycles during 2 hours. Thus there is a need to account for the time resolution of the data collection strategy and its relationship to the electrochemistry. These results indicate that similar time resolved structural information on Li_xC_6 can be obtained using either *in situ* XRD[40,41] or NPD, but each technique is better suited for different sized batteries, with NPD requiring larger format batteries and XRD generally requiring smaller format batteries. The major advantage of *in situ* NPD is the potential ability to determine lithium atomic parameters.

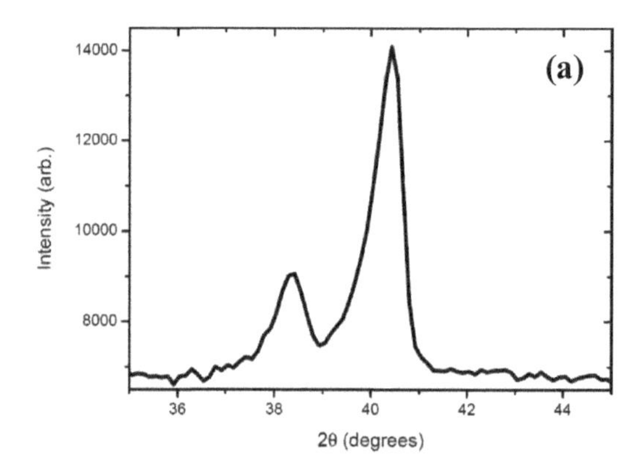

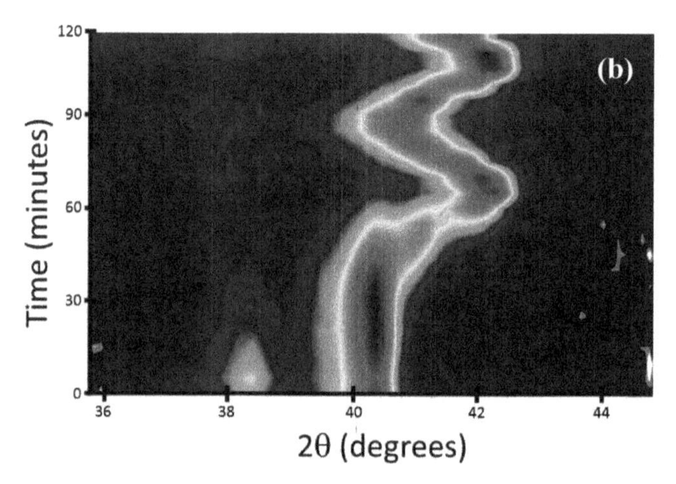

Figure 1. **(a)** A 5 minute NPD pattern of the battery at the initial stages of cycling (charged battery) in a limited angular range of $35 \leq 2\theta \leq 45°$, the reflection at $2\theta \sim 38°$ corresponds to the 001 reflection of LiC_6 and the reflection at $2\theta \sim 40°$ corresponds to the 002 reflection of LiC_{12}. **(b)** *In situ* NPD data shown in the region $35.5 \leq 2\theta \leq 45°$ as a function of time, the reflection highlighted is the 002 reflection of graphite and Li_xC_6 (or 001 reflection of LiC_6). The colour scale represents reflection intensity.

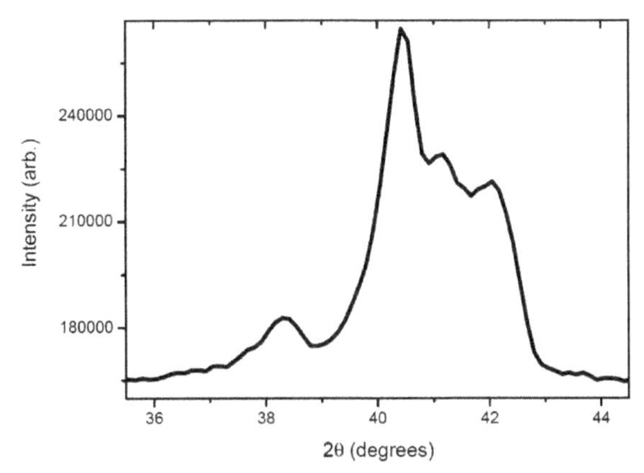

Figure 2. The same angular region ($35.5° \leq 2\theta \leq 45°$) shown in Figure 1 but summed in time. This represents a 2 hour NPD data collection relative to the 5 minute collections collated in Figure 1.

The advantages of acquiring time resolved data include the ability to track in real-time the evolution of the reflections and structural parameters of the electrodes. This in turn allows the determination of kinetic parameters associated with the evolution of the battery. For example, Figure 3 shows the evolution of the integrated intensities corresponding to the 001 reflection of LiC_6 at $2\theta \sim 38°$, 002 reflection of LiC_{12} at $2\theta \sim 40°$ and the 002 reflection of graphite at $2\theta \sim 42°$ that are present in Figure 1. The reflection assignments were determined based on our previous work and references therein [10,29,40,41]. LiC_6 is present in significant quantities illustrated by large integrated intensities around 10 minutes in the electrochemical cycling (corresponding to the charged state of the battery). Additionally, LiC_{12} (and LiC_6) show the opposite trend to graphite, as graphite increases LiC_{12} decreases agreeing with the previously proposed staged lithium insertion/extraction mechanisms. This information is derived from single peak fitting routines applied to the dataset shown in Figure 1b. Further analysis can be taken to determine the rates of phase disappearance and appearance by linear or more advanced fitting to the time-evolution information shown in Figure 3. For example, a linear fit to the disappearance of the LiC_{12} 002 reflection between 40 and 70 minutes yields a rate of disappearance of 240(30) arbitrary (intensity) units per minute (fitting R^2 value = 0.93). Whereas the rate of appearance from 70 to 95 minutes of the same reflection is 330(40) arbitrary (intensity) units per minute (fitting R^2 value = 0.92). There is a difference between these rates even though the charge and discharge processes represented by the appearance and disappearance rates were conducted with the same magnitude of applied current (1 A). This may suggest it is easier to form LiC_{12} than it is to remove it. However, this speculation needs further detailed analysis, e.g. investigation of the phase fraction of LiC_{12} using Rietveld methods. Comparatively the LiC_6 001 reflection disappearance from 10 to 65 minutes is 60(7) arbitrary (intensity) units per minute (fitting R^2 value = 0.86) which is less rapid than the LiC_{12} reflection and may require two rates to account for the behaviour due to the poorer fitting statistics. Note, ideally more data points are required within the charge and discharge process at the 1 A current rates to ensure better statistical outcomes for the linear or more advanced fitting routines. This can be achieved by reducing the current rate or shortening the NPD data collection times. Nonetheless, the ability to extract rates of reactions and processes occurring within electrodes is a powerful advantage of time-resolved *in situ* neutron and X-ray diffraction.

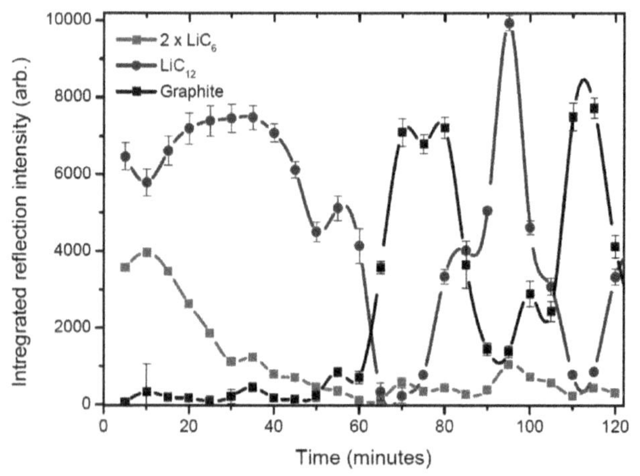

Figure 3. Variation in the integrated intensity of selected anode reflections present in Figure 1b. The reflections are the 002 reflection of graphite, 002 reflection of LiC_{12}, and 001 reflection of LiC_6. The lines between data points are a guide for the eye.

Single peak fitting routines can provide information regarding one or a number of reflections from a phase of interest, both in terms of peak position and intensity. This information can provide a guide to the structural evolution of the phase. However, whole pattern fitting methods such Le Bail or Rietveld methods generally provide more information about the structural evolution of the electrode of interest. In particular, Rietveld analysis of neutron diffraction data can illustrate the location and quantity of lithium in an electrode [27] due to the greater sensitivity afforded by neutron diffraction towards lithium atoms relative to X-ray or electron diffraction techniques. Time-resolved diffraction data can be used to extract detailed structural information of phases that are not stable under ambient conditions and are only produced in electrochemical processes, such as charge/discharge cycling. For example, at the charged state of the battery discussed in this work (corresponding to 10 minutes in Figures 1 and 3), the anode is a mixture of LiC_6 and LiC_{12} in a ratio of 0.34(2):0.66(8) assuming the anode is only composed of crystalline phases that are detectable by diffraction-based techniques. LiC_{12} was modelled adopting P_{63}/mmc symmetry with refined lattice parameters of a = 2.4805(7) and c = 7.0328(9) Å with one carbon (C1) at x=y = 0, z = 1/4 with refined atomic displacement parameter (ADP) of 0.063(9) $Å^3$ and the other carbon (C2) at x = 1/3, y = 2/3, z = 1/4 with refined ADP of 0.226(16) $Å^3$. LiC_6 was modelled adopting P_6/mmm symmetry with refined lattice parameters of a = 4.3113(28) and c = 3.7015(15) Å, carbon at x = 1/3, y = 0, z = 1/2 with ADP of 0.017(6) $Å^3$ and Li at x=y=z=0 with a fixed ADP of 0.02 $Å^3$. The cathode was composed of $Li_xMn_2O_4$ where x ~ 0 in *Fd*-3*m* symmetry 8.0576(9) Å and oxygen positional parameter (x=y=z) of 0.2699(10). The $Li_xMn_2O_4$ lattice parameter corresponds to the reported lattice of λ-MnO_2 of 8.0639(3) Å [42]. Copper and aluminium were also included in the model to account for the current collectors. The reflections of the Cu and Al do not change in position (2θ value) and intensity during electrochemical cycling and are modelled as such in the Rietveld analysis. Although some electrode reflections overlap with Cu and Al, they can be modelled using the whole pattern approach. The Rietveld refined fit to the 5 minute *in situ* NPD data using the above models is shown in Figure 4. The statistics of the fit were profile factor R_p = 1.75 %, weighted profile factor wR_p = 2.36 and goodness-of-fit term χ^2 = 1.91 using 25 parameters. These parameters illustrate that structural data from phases produced during electrochemical cycling can be extracted and compared to data collected *ex situ* to provide further details of the nature of reactions occurring at the electrodes.

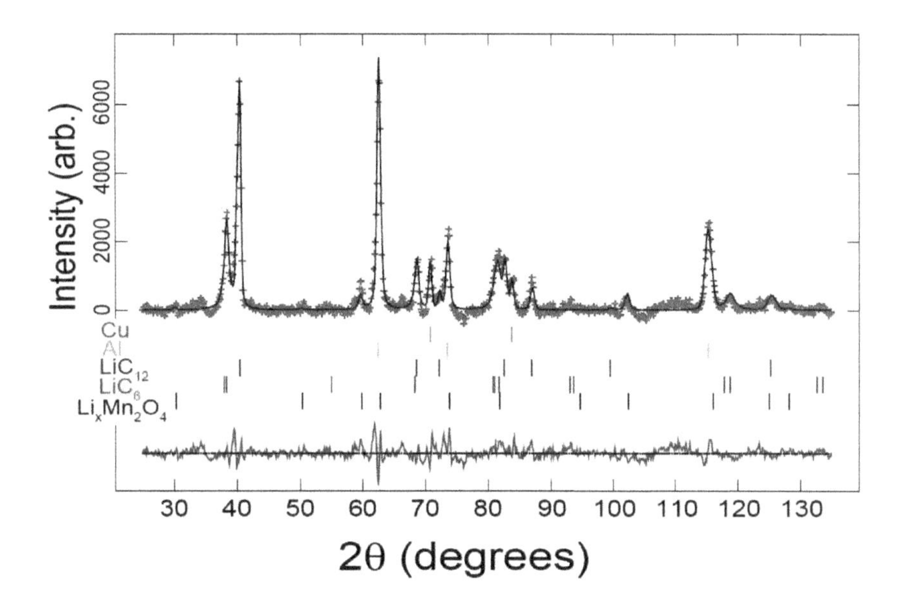

Figure 4. The Rietveld refined fit (black line) using the structural models of Cu, Al, LiC_{12}, LiC_6, and $Li_xMn_2O_4$ to the second (10 minute in Figures 1 and 3) *in situ* NPD data (red crosses). The difference between the fit and data is shown by the purple line at the bottom. Colour-coded vertical lines represent Bragg reflection markers of the structural models used.

Similar to the single peak fitting routines and determination of rates of appearance and disappearance of reflections, time-resolved *in situ* diffraction data can be used to show the evolution of lattice parameters and other structural parameters during charge/discharge. The information extracted from time-resolved *in situ* neutron diffraction data also enables the tracking of lithium during the charge/discharge process. Thus researchers can literally follow the lithium insertion/extraction processes and determine the rates of insertion and extraction into and out of compounds as recently demonstrated in ref. [27] using this battery.

Time-dependent information allows the development of materials or the manipulation of compounds with dopants, particle-size processing, and controlled morphologies to minimise the rate of lattice expansion and contraction during electrochemical processes. Recent work has shown a link (although further work is required for confirmation) between the rate of lattice expansion/contraction and the applied current [10,13,27]. Materials that have a low rate of lattice expansion/contraction at moderate current rates are likely to still have moderate rates of expansion/contraction at higher rates, assuming no other current-dependent processes occur. By minimising both the rate and magnitude of lattice expansion/contraction or increasing the ease of lithium insertion/extraction reduces the stresses applied to the electrodes. Thus it provides an opportunity to make better materials with minimal structural distortions which are seemingly a key aspect for higher rate capabilities.

Until now, no significant detail has been given in regards to the electrochemistry, the voltage or applied current. Thus structural information for the better part has been used to show the electrochemical processes in the battery, compared to using electrochemical information to suggest what happens to the structure. The classic example of the use of electrochemical information to infer structural information is the two-phase and single-phase reaction pathway of $LiFePO_4$, where authors often use a sloping voltage curve to indicate a single-phase transition, $Li_{1-x}FePO_4$ where x continuously decreases during charge, and a flat voltage profile or plateau in the voltage profile to indicate a two-phase reaction, $LiFePO_4 \rightarrow FePO_4$ during charge [1,13,43-46]. Our data allows the electrochemical processes to be directly correlated to *in situ* NPD data and the information presented above takes this into account. Many publications (e.g. ref. [27]) illustrate this link and the data shown here and the direct link between electrochemistry and structure will form part of a larger manuscript under preparation.

CONCLUSIONS

The work undertaken on *in situ* studies of battery materials while the battery is functioning sheds new light on the mechanisms occurring and allows researchers to further tune material properties to improve battery performance. There are important parameters that need to be considered prior to an *in situ* experiment, such as the structural detail required – is it essential to determine the lithium site occupancy factors in crystal structures and their evolution or is the lattice evolution sufficient? These structural details then influence whether synchrotron XRD or NPD experiments are undertaken, with the latter suited to determining the lithium site occupancy factors. In turn, the type of experiment influences whether modified coin-type cells, commonly used by researchers, are adequate or modified larger cells are required. At the instrument itself, consideration needs to be given to the type of detector and available time resolution. If kinetic information is required then the current rate used has to be matched with the available data collection speed. Further, the data collection times have to be optimised providing sufficient signal for meaningful analysis. The concept of the data collection time and structural evolution were highlighted using *in situ* NPD data on a prismatic lithium-ion battery with the Li_xC_6 002 reflection. A two hour collection for the current rate used averaged many structural processes occurring in the electrode, whereas five minute data collections allowed determination of the mechanisms of electrode evolution. Whether even faster data collections are required, still needs to be considered. The evolution of reflections from major phases found in the graphite electrode were followed and their transitions qualitatively shown. Kinetic information relating to the rates of transformation were also determined. Further, Rietveld analysis was used to characterise the structural parameters of the LiC_6 and LiC_{12} components in the anode and $Li_xMn_2O_4$ cathode. This example highlights some of the considerations that become important when conducting time-resolved *in situ* neutron and X-ray diffraction studies.

ACKNOWLEDGMENTS

NS would like to thank Y. Wu, Y. Shu (Fudan University), D. Yu, V. K. Peterson (ANSTO) for work in this fruitful collaboration and AINSE Ltd for providing financial assistance through the research fellowship scheme.

REFERENCES

[1]Goodenough, J. B.; Kim, Y., Challenges for rechargeable Li batteries, *Chem. Mater.*, **22**, 587-603 (2009).

[2]Tarascon, J.-M.; Armand, M., Issues and challenges facing rechargeable lithium batteries, *Nature*, **414**, 359-367 (2001).

[3]Winter, M.; Besenhard, J. O.; Spahr, M. E.; Novak, P., Insertion electrode materials for rechareable lithium ion batteries, *Adv. Mater.*, **10**, 725-763 (1998).

[4]León, B.; Corredor, J. I.; Tirado, J. L.; Perez-Vicente, C., On the Mechanism of the Electrochemical Reaction of Tin Phosphide with Lithium, *J. Electrochem. Soc.*, **153**, A1829-A1834 (2006).

[5]Liu, W.; Huang, X.; Wang, Z.; Li, H.; Chen, L., Studies of Stannic Oxide as an Anode Material for Lithium-Ion Batteries, *J. Electrochem. Soc.*, **145**, 59-62 (1998).

[6]Isnard, O., In situ and/or time resolved powder neutron scattering for materials science, *J. Optoelectron. Adv. Mater.* , **8**, 411-417 (2006).

[7]Sharma, N.; Reddy, M. V.; Du, G.; Adams, S.; Chowdari, B. V. R.; Guo, Z.; Peterson, V. K., Time-dependent *in-situ* neutron diffraction investigation of a $Li(Co_{0.16}Mn_{1.84})O_4$ cathode, *J. Phys. Chem. C* **115**, 21473–21480 (2011).

[8]Palacin, M. R.; Le Cras, F.; Seguin, L.; Anne, M.; Chabre, Y.; Tarascon, J.-M.; Amatucci, G.; Vaughan, G.; Strobel, P., In Situ Structural Study of 4V-Range Lithium Extraction/Insertion in Fluorine-Substituted LiMn2O4, *J. Solid State Chem.*, **144**, 361-371 (1999).

[9]Sharma, N.; Peterson, V. K., In situ neutron powder diffraction studies of lithium-ion batteries, *J. Solid State Electrochem.*, **16**, 1849-1856 (2012).

[10]Sharma, N.; Peterson, V. K., Current dependency of lattice fluctuations and phase evolution of electrodes in lithium-ion batteries investigated by in situ neutron diffraction, *Electrochim. Acta*, **In press**, 10.1016/j.electacta.2012.09.101 (2013).

[11]Baehtz, C.; Buhrmester, T.; Bramnik, N. N.; Nikolowski, K.; Ehrenberg, H., Design and performance of an electrochemical in-situ cell for high resolution full-pattern X-ray powder diffraction, *Solid State Ionics*, **176**, 1647-1652 (2005).

[12]Sharma, N.; Peterson, V. K.; Elcombe, M. M.; Avdeev, M.; Studer, A. J.; Blagojevic, N.; Yusoff, R.; Kamarulzaman, N., Structural changes in a commercial lithium ion battery during electrochemical cycling: An *in-situ* neutron diffraction study, *J. Power Sources*, **195**, 8258–8266 (2010).

[13]Sharma, N.; Guo, X.; Du, G.; Guo, Z.; Wang, J.; Wang, Z.; Peterson, V. K., Direct Evidence of Concurrent Solid-Solution and Two-Phase Reactions and the Nonequilibrium Structural Evolution of $LiFePO_4$, *J. Am. Chem. Soc.*, **134**, 7867–7873 (2012).

[14]Jones, J.; Hung, J.-T.; Meng, Y. S., Intermittent X-ray diffraction study of kinetics of deilthiation in nano-scale $LiFePO_4$, *J. Power Sources*, **189**, 702-705 (2009).

[15]Brant, W. R.; Schmid, S.; Du, G.; Gu, Q.; Sharma, N., A simple electrochemical cell for in-situ fundamental structural analysis using synchrotron X-ray powder diffraction, *J. Power Sources*, **In press**, 10.1016/j.jpowsour.2013.1003.1086 (2013).

[16]Sharma, N.; Du, G.; Studer, A. J.; Guo, Z.; Peterson, V. K., Structural change in the MoS_2 anode of a Li ion battery during discharge: *in-situ* neutron diffraction studies using an optimised cell design, *Solid State Ionics*, **199-200**, 37-43 (2011).

[17]Du, G.; Sharma, N.; Kimpton, J. A.; Jia, D.; Peterson, V. K.; Guo, Z., Br-doped $Li_4Ti_5O_{12}$ and composite TiO_2 anodes for Li-ion batteries: Synchrotron X-ray and *in situ* neutron diffraction studies, *Adv. Funct. Mater.*, **21**, 3990-3997 (2011).

[18]Bergstom, O.; Andersson, A. M.; Edstrom, K.; Gustafsson, T., A neutron diffraction cell for studying lithium-insertion processes in electrode materials, *J. Appl. Cryst.*, **31**, 823-825 (1998).

[19]Berg, H.; Rundlov, H.; Thomas, J. O., The $LiMn_2O_4$ to gamma-MnO_2 phase transition studied by neutron diffraction, *Solid State Ionics*, **144**, 65-69 (2001).

[20]Colin, J.-F.; Godbole, V.; Novák, P., In situ neutron diffraction study of Li insertion in Li4Ti5O12, *Electrochem. Comm.* , **12**, 804-807 (2010).

[21]Rodriguez, M. A.; Ingersoll, D.; Vogel, S. C.; Williams, D. J., Simultaneous in siu neutron diffraction studies of the anode and cathode in lithium-ion cell, *Electrochem. Solid-State Lett.*, **7**, A8-A10 (2004).

[22]Rosciano, F.; Holzapfel, M.; Scheifele, W.; Novak, P., A novel electrochemical cell for in situ neutron diffraction studies of electrode materials for lithium-ion batteries, *J. Appl. Cryst.* , **41**, 690-694 (2008).

[23]Senyshyn, A.; Muhlbauer, M. J.; Nikolowski, K.; Pirling, T.; Ehrenberg, H., "In-operando" neutron scattering studies on Li-ion batteries, *J. Power Sources*, **203**, 126-129 (2012).
[24]Godbole, V. A.; Hess, M.; Villevieille, C.; Kaiser, H.; Colin, J. F.; Novak, P., Circular in situ neutron powder diffraction cell for study of reaction mechanism in electrode materials for Li-ion batteries, *RSC Adv.*, **3**, 757-763 (2013).
[25]Roberts, M.; Biendicho, J. J.; Hull, S.; Beran, P.; Gustafsson, T.; Svensson, G.; Edstrom, K., Design of a new lithium ion battery test cell for in-situ neutron diffraction measurements, *J. Power Sources*, **226**, 249-255 (2013).
[26]X.-L., W.; An, K.; Cai, L.; Feng, Z.; Nagler, S. E.; Daniel, C.; Rhodes, K. J.; Stoica, A. D.; Skorpenske, H. D.; Liang, C.; Zhang, W.; Kim, J.; Qi, Y.; Harris, S. J., Visualizing the chemistry and structure dynamics in lithium-ion batteries by in-situ neutron diffraction, *Scientific Reports*, **2**, 1-7 (2012).
[27]Sharma, N.; Yu, D.; Zhu, Y.; Wu, Y.; Peterson, V. K., Non-equilibrium Structural Evolution of the Lithium-Rich $Li_{1+y}Mn_2O_4$ Cathode within a Battery, *Chem. Mater.*, **25**, 754–760 (2013).
[28]Hu, C.-W.; Sharma, N.; Chiang, C.-Y.; Su, H.-C.; Peterson, V. K.; Hsieh, H.-W.; Lin, Y.-F.; Chou, W.-C.; Shew, B.-Y.; Lee, C.-H., Real-time investigation of the structural evolution of electrodes in a commercial lithium-ion battery containing a V-added $LiFePO_4$ cathode using in-situ neutron powder diffraction, *J. Power Sources*, **In press**, 10.1016/j.jpowsour.2013.1002.1074 (2013).
[29]Sharma, N.; Peterson, V. K., Overcharging a lithium-ion battery: Effect on the LixC6 negative electrode determined by in situ neutron diffraction, *J. Power Sources*, **In press**, 10.1016/j.jpowsour.2012.1012.1019 (2013).
[30]Liss, K.-D.; Hunter, B. A.; Hagen, M. E.; Noakes, T. J.; Kennedy, S. J., Echidna, *Physica B*, **385-386**, 1010-1012 (2006).
[31]K. S. Wallwork; B. J. Kennedy; D. Wang, *AIP Conference Proceedings*, 879 (2007).
[32]Lee, P. L.; Shu, D.; Ramanathan, M.; Preissner, C.; Wang, J.; Beno, M. A.; Von Dreele, R. B.; Ribaud, L.; Kurtz, C.; Antao, S. M.; Jiao, X.; Toby, B. H., A twelve-analyzer detector system for high-resolution powder diffraction, *J. Synchrotron Rad.*, **15**, 427-432 (2008).
[33]Studer, A. J.; Hagen, M. E.; Noakes, T. J., Wombat: The high intensity powder diffractometer at the OPAL reactor, *Physica B*, **385-386**, 1013-1015 (2006).
[34]Hansen, T. C.; Henry, P. F.; Fischer, H. E.; Torregrossa, J.; Convert, P., The D20 instrument at the ILL: A versatile high-intensity 2-axis neutron diffractometer, *Meas. Sci. Technol.*, **19**, 034001 (2008).
[35]Hoelzel, M.; Senyshyn, A.; Juenke, N.; Boysen, H.; Schmahl, W.; Fuess, H.; A. Senyshyn, N. J., H. Boysen, W. Schmahl, H. Fuess, Nucl. Instr. A 667, 32-37 (2012). High-resolution neutron powder diffractometer SPODI at research reactor FRM II *Nucl. Inst. A*, **667**, 32-37 (2012).
[36]Orikasa, Y.; Maeda, T.; Koyama, Y.; Murayama, H.; Fukuda, K.; Tanida, H.; Arai, H.; Matsubara, E.; Uchimoto, Y.; Ogumi, Z., Direct Observation of a Metastable Crystal Phase of Li_xFePO_4 under Electrochemical Phase Transition, *J. Am. Chem. Soc.*, **135**, 5497-5500 (2013).
[37]Richard, D.; Ferrand, M.; Kearley, G. J., Large array manipulation program, *J. Neutron Research* **4**, 33-39 (1996).
[38]Larson, A. C.; Von Dreele, R. B.; Los Alamos National Laboratory Report LAUR 86-748: 1994.
[39]Toby, B. H., EXPGUI, a graphical user interface for GSAS, *J. Appl. Cryst.*, **34**, 210-213 (2001).
[40]Dahn, J. R., Phase diagram of Li_xC_6, *Phys. Rev. B*, **44**, 9170-9177 (1991).
[41]Dahn, J. R.; Fong, R.; Spoon, M. J., Suppression of staging in lithium-intercalated carbon by disorder in the host, *Phys. Rev. B*, **42**, 6424-6432 (1990).
[42]Fong, C.; Kennedy, B. J.; Elcombe, M. M., A powder neutron diffraction study of l and g manganese dioxide and of $LiMn_2O_4$, *Zeitschrift für Kristallographie*, **209**, 941-945 (1994).
[43]Bai, P.; Cogswell, D. A.; Bazant, M. Z., Suppression of phase separation in $LiFePO_4$ nanoparticles during battery discharge, *Nano Lett.*, **11**, 4890-4896 (2011).
[44]Laffont, L.; Delacourt, C.; Gibot, P.; Wu, M. Y.; Kooyman, P.; Masquelier, C.; Tarascon, J.-M.; , Study of the $LiFePO_4/FePO_4$ two-phase system by high-resolution electron energy loss spectroscopy, *Chem. Mater.*, **18**, 5520-5529 (2006).
[45]Srinivasan, V.; Newman, J., Existence of path dependence in the $LiFePO_4$ electrode, *Electrochem. Solid-State Lett.*, **9**, A110-A114 (2006).

[46]Wagemaker, M.; Singh, D. P.; Borghols, W. J. H.; Lafont, U.; Haverkate, L.; Peterson, V. K.; Mulder, F. M., Dynamic solubility limits in nanosized olivine $LiFePO_4$, *J. Am. Chem. Soc.*, **133**, 10222-10228 (2011).

ELECTROCHEMICAL PERFORMANCE OF $LiNi_{1/3}Co_{1/3}Mn_{1/3}O_2$ LITHIUM POLYMER BATTERY BASED ON PVDF-HFP/m-SBA15 COMPOSITE POLYMER MEMBRANES

Chun-Chen Yang[*,#], Zuo-Yu Lian
Department of Chemical Engineering, Mingchi University of Technology, New Taipei City 243, Taiwan, R.O.C.
[#]Battery Research Center of Green Energy, Ming Chi University of Technology, New Taipei City 243, Taiwan, R.O.C.
*Corresponding author, E-mail: ccyang@mail.mcut.edu.tw

ABSTRACT

This work reports the preparation of the composite polymer electrolyte for $LiNi_{1/3}Co_{1/3}Mn_{1/3}O_2$ lithium polymer battery. The poly(vinylidiene fluoride-hexafluoropropylene) (PVDF-HFP) is used as the polymer host and mesoporous SBA15 (silica) ceramic fillers used as the solid plasticizer are added into the polymer matrix. The SBA15 fillers with mesoporous structure and high specific surface can trap more liquid electrolytes to enhance the ionic conductivity. The ionic conductivity of the as-prepared PVDF-HFP/SBA15 composite polymer electrolyte is in the order of 10^{-3} S cm^{-1} at room temperature. The characteristic properties of the composite polymer membranes are examined by using FTIR spectroscopies, scanning electron microscopy (SEM), XRD and an AC impedance method. For comparison, the $LiNi_{1/3}Co_{1/3}Mn_{1/3}O_2$ composite batteries with the conventional polyethylene (PE) separator and pure PVDF-HFP polymer membrane are also prepared and studied. As a result, the $LiNi_{1/3}Co_{1/3}Mn_{1/3}O_2$ polymer battery comprised of the PVDF-HFP/3wt.%m-SBA15 composite polymer electrolyte achieves the discharge capacities of 156 and 88 mAh g^{-1} at 0.1C and 3C, respectively. It is shown good coulomb efficiency of ca. 95~99%. It is demonstrated that the PVDF-HFP/m-SBA15 composite membrane exhibits a good candidate for application to $LiNi_{1/3}Co_{1/3}Mn_{1/3}O_2$ polymer batteries.

1. INTRODUCTION

$LiCoO_2$ is widely used in commercial Li ion batteries due to its excellent electrochemical properties and easy preparation. However, the toxicity and high cost of Co restrict its application in future [1-5]. Compared with $LiCoO_2$ with a layered crystal structure, the ternary counterpart $LiNi_xCo_{1-2x}Mn_xO_2$ is a cheaper, safer and more environmentally friendly positive electrode material for Li ion batteries [7]. Among them, $LiNi_{1/3}Co_{1/3}Mn_{1/3}O_2$ (denoted as LNCM) has been widely studied [8]. They were reported to have some advantages over $LiCoO_2$ such as high discharge capacity, more structural and thermal stable, good rate capability, and excellent cycling performance. It was reported that $LiNi_{1/3}Co_{1/3}Mn_{1/3}O_2$ showed a specific discharge capacity of 160 mAh g^{-1} over 2.5-4.4 V [9]. It has a layered structure with a and c lattice parameters of 0.2862 nm and 1.427 nm, respectively, with a reversible capacity of more than 200 mAh g^{-1} with the voltage range of 2.5~4.6 V [10-20]. The Ni^{2+} and Co^{3+} cations in LNCM act as the electrochemically active species during the charge/discharge test. While the Mn^{4+} cations are electrochemically inactive, they are generally though to help the increasing structure stability of the material framework. Therefore, LNCM is considered to be one of the best candidates of positive electrode material to replace $LiCoO_2$. The $LiNi_{1/3}Co_{1/3}Mn_{1/3}O_2$ composite materials can be prepared via co-precipitation method [9,15], emulsion method [10], spray-dry method [17,18], sol-gel process [20], and a wet chemical process [16].

A solid polymer electrolyte (SPE) membrane is essential for the components of a Li-ion battery. The SPE required have higher ionic conductivity, excellent mechanical and interfacial properties, and stable electrochemical performance. The current polyolefin separators (PEs and PPs) are not fully suitable to use on the high power Li-ion battery applications, like electric vehicles (EVs). Therefore, it is imperative to develop a new SPE with a large pore size and volume, high porosity, and high ionic conductivity properties for application on the high power secondary Li-ion battery. PVDF-HFP has a high dielectric constant (ε =8.4) that facilitates a high concentration of charge carriers, and it has both an amorphous part (–HFP) and a crystalline phase (-VDF). The amorphous phase of the PVDF-HFP helps trapping large amounts of liquid electrolytes, whereas the crystalline phase acts as a mechanical support for the polymer electrolyte. PVDF-HFP composite polymer electrolyte membranes for Li ion batteries (LIBs) are usually prepared by a phase inversion process or a solution casting method [21-41]. Pu et al. [21] prepared microporous PVDF-HFP composite membranes for a Li-ion battery by a phase inversion method. Kim et al. [22] studied a composite porous PVDF-HFP/TiO_2 polymer electrolyte membrane for a Li-ion battery. Jiang et al. [26] reported the characteristics properties of the PVDF-HFP microporous polymer membranes with molecule sieves (MCM41 and SBA15).

The molecular sieve of SBA15 (silica) material has a mesoporous structure with a high specific surface area above 1000 m^2 g^{-1} [26]. Stephan et al. [23,24] studied PVDF-HFP composite polymer electrolyte membranes with two different sizes (14 μm and 7 nm) of aluminum oxyhydroxide ($AlO[OH]_n$) for Li-ion batteries. They found that the incorporation of nano-fillers greatly enhanced the ionic conductivity and compatibility of the composite polymer electrolyte. The addition of mesoporous SBA15 fillers may reduce the crystallinity of PVDF-HFP copolymer and help to trap more liquid electrolyte. Prior to adding SBA15 into the polymer matrix, the SBA15 powder was carried out by a surface modification with 3-Glycidoxypropyl-trimethoxysilane (GPTMS) compound [27,28]. The modified SBA15 material is denoted as m-SBA15. The characteristic properties of m-SBA15 powders and the composite polymer membrane were examined by nitrogen adsorption isotherm (BET), scanning electron microscopy (SEM) and FTIR spectroscopy.

In this work, the $LiNi_{1/3}Co_{1/3}Mn_{1/3}O_2$ composite battery [29], which is composed of PVDF-HFP/m-SBA15 composite polymer membrane, was assembled and examined. The electrochemical performance of the $LiNi_{1/3}Co_{1/3}Mn_{1/3}O_2$/Li battery was examined by a galvanostatic charge-discharge method at varied rates. The electrochemical performances of $LiNi_{1/3}Co_{1/3}Mn_{1/3}O_2$/Li batteries with a PE separator and a pure PVDF-HFP polymer membrane were also examined for comparison.

2. EXPERIMENTAL

2.1 Preparation of $LiNi_{1/3}Co_{1/3}Mn_{1/3}O_2$ cathode material

For the preparation of $LiNi_{1/3}Co_{1/3}Mn_{1/3}O_2$ material, an aqueous solution of 1 mol L^{-1} $NiSO_4$, $CoSO_4$, and $MnSO_4$ (use cation molar ratio of Ni:Co:Mn = 1:1:1) was pumped in a tank reactor with a continuous agitating under a nitrogen atmosphere. At the same time, 2 mol L^{-1} NaOH and 0.30 mol L^{-1} NH_4OH solutions were separately added in the CSTR reactor. After finished co-precipitation, the mixture solution was kept at 50°C with a continuous stirring for 12 h and the pH value of the mixture solution was maintained at 11.5. The above precipitation solid product was filtered to obtain the crude hydroxide solid product. The hydroxide solid product was then washed three times with deionized water and dried at 120°C for 12 h. The obtained $Ni_{1/3}Co_{1/3}Mn_{1/3}(OH)_2$ was directly mixed with a stoichiometric amount of $LiOH.H_2O$ and ball milled at 250 rpm for 5h. The dried mixture was then pre-calcined at 350°C for 4 h and then calcined again at 950°C for 10 h in air. The final oxide products of $LiNi_{1/3}Co_{1/3}Mn_{1/3}O_2$ were obtained and store in PE bag ready to the properties study.

2.2 Preparation of composite PVDF-HFP/m-SBA15 polymer membrane

The highly ordered large mesoporous SBA15 powders with a hexagonal structure were prepared by using a triblock copolymer ($EO_{20}PO_{70}EO_{20}$ or Pluronic 123, Aldrich)) as a template and Tetraethoxysilane (TEOS, Aldrich) as a silica source [15]. An amount of 10g of triblock copolymer was added to 300 mL of an aqueous solution of 2 M HCl and stirred at 50°C for 2 h. Subsequently, 20 g of TEOS was added slowly to the solution and the mixture solution was stirred at 50°C for 2 h. The as-prepared gels were maintained at 110°C for 24 h under a static condition. The final products were filtered, washed with distilled water, and dried at 25°C for 24 h, and then dried at 110°C for another 24 h. The as-prepared samples were sintered under N_2

atmosphere at 500°C for 6 h. The as-prepared SBA15 powders were further modified by using a 3-glycidoxypropyl-trimethoxysilane (GPTMS) compound. The suitable amounts of SBA15 powders were added to a 10wt.% GPTMS in anhydrous alcohol under stirring condition for 2 h, then washed by anhydrous alcohol. The modified SBA15 powder is denoted as m-SBA15 in this work. The m-SBA15 powders were examined by FE-SEM, BET, and FTIR spectroscopy.

An appropriate amount of PVDF-HFP was dissolved in NMP at 65°C for 1 h to form a viscous polymer solution. The suitable amounts of m-SBA15 powders (0 ~ 10 wt.%) were added to the viscous polymer solution and continuously stirred for 1 h. The viscous blend solution was cast on a clean glass plate and remained in air for 5 days at room temperature. The dried composite polymer membrane was immersed into 1 M $LiPF_6$ in EC/DEC (1:1 v/v) liquid electrolyte for 24 h to form a composite polymer electrolyte in a dry glove box. For comparison, conventional PE separator and pure PVDF-HFP polymer membrane containing lithium liquid electrolyte were also prepared and studied.

2.3 Characterization of $LiNi_{1/3}Co_{1/3}Mn_{1/3}O_2$ and SBA15 powders and the composite membranes

The surface morphologies of $LiNi_{1/3}Co_{1/3}Mn_{1/3}O_2$ and m-SBA15 material and PVDF-HFP/m-SBA15 composite polymer membranes were measured by a high-resolution transmission electron microscopy (HR-TEM, JEOL 2010F) and a scanning electron microscope (SEM, Hitachi), respectively. The crystal structure of $LiNi_{1/3}Co_{1/3}Mn_{1/3}O_2$ powders and PVDF-HFP/m-SBA15 composite membranes were examined by an X-ray diffraction (XRD) spectrometer (Philip, X'pert Pro System). The PVDF-HFP/m-SBA15 composite polymer membranes were analyzed by a Fourier transform infrared (FTIR) spectrometer (Perkin-Elmer Spectrum 100) equipped with an ATR accessory. Air was used as the reference background. The FTIR spectra were obtained in the wave numbers of 600 ~ 4000 cm^{-1} at an ambient temperature. The conductivity measurements were conducted for PVDF-HFP/m-SBA15 composite polymer electrolyte via an AC impedance method. The composite membranes electrolyte sample was clamped between stainless steel (SS304), as the ion-blocking electrodes, each with a surface area of 1.32 cm^2, in a spring-loaded glass holder. A thermocouple was positioned in close proximity to the composite polymer electrolyte for temperature measurement. Each sample was equilibrated at the experimental temperature for a minimum of at least 30 min before measurement. The AC impedance measurements were carried out by using an Autolab PGSTAT-30 equipment (Eco Chemie B.V., Netherlands). The AC spectra in the range of 1 MHz to 100 Hz at an excitation signal of 5 mV were recorded. AC impedance spectra of the composite polymer electrolyte membranes were recorded at a temperature range between 30 to 70°C. The experimental temperatures were maintained within ±0.5°C by a convection oven. These PVDF-HFP/m-SBA15 composite polymer electrolytes were examined a minimum of at least three times.

2.3 Electrochemical performance measurements

The electrochemical performance was studied on a two-electrode system that was assembled in an argon-filled glove box. The $LiNi_{1/3}Co_{1/3}Mn_{1/3}O_2$ composite cathodes (prepared by a co-precipitation and a solid state method) were prepared by mixing active materials, Super P, and poly(vinyl fluoride) (PVDF) in a weight ratio of 80:10:10, and pasted on an aluminum foil (Aldrich), and then dried in a vacuum oven at 110°C for 12 h. The lithium foil (Aldrich) was used as the counter and reference electrode. The PVDF-HFP/m-SBA15 composite polymer membrane was used as the separator. The electrolyte was 1 M $LiPF_6$ in a mixture of EC and DEC (1:1 in v/v). The $LiNi_{1/3}Co_{1/3}Mn_{1/3}O_2$/Li composite batteries were charged by a constant current and a constant voltage profile (CC-CV) and discharged by a constant current profile, over a potential range of 3.0 ~ 4.3 V (vs. Li/Li^+) at varied C rates with an Autolab PGSTAT302N potentiostat. The second CV charge step of 4.3 V was terminated when the charged current was below 0.1 C current.

3. RESULTS AND DISCUSSION

Fig. 1 shows the XRD pattern of the $LiNi_{1/3}Co_{1/3}Mn_{1/3}O_2$ materials prepared by a co-precipitation method. All diffraction peaks were indexed based on a hexagonal α-$NaFeO_2$ structure (Space group R-3*m*) and no impurity phase was detected in XRD peaks [9,11]. The peak splits of (006)/(102) and (018)/(110) are found, which indicated the compound with a characteristic of the layered structure. The integrated intensity ratio of the (003) to (104) peaks can be used to as an indicator for the degree of the cation mixing. In general, a value of $I_{003}/I_{104} > 1.2$ [17,29] have a well-crystallized phase. The intensity ratio of (003)/(104) peaks for our synthesized $LiNi_{1/3}Co_{1/3}Mn_{1/3}O_2$ material is 1.75, indicating the as-prepared samples have good crystallinity without cation mixing. The inset of Fig. 1 shows the surface morphology of $LiNi_{1/3}Co_{1/3}Mn_{1/3}O_2$ material. The samples have spherical morphology with an average secondary particle size of 10μm in diameter. The spherical particles consist of aggregated primary particles.

Figs. 2(a) and (b) show the FE-SEM and HR-TEM images of as-prepared SBA15 powders, respectively. It was found that SBA15 powders or aggregates with a hexagonal structure show a long cylindrical shape. The length of the SBA15 aggregate was approximately 300 ~ 500 μm. The SBA15 powder is a mesoporous material with a high specific surface area, large dimension open channels (5 ~ 50 nm), and excellent thermal and chemical stability [26]. The surface morphology of m-SBA15 powders was not altered; HR-TEM result shows the

SBA15 powder with a hexagonal structure. According to the results of FTIR analysis (result not shown here) of the as-prepared and modified SBA15 (m-SBA15) powders, there is an extra C-H stretch vibrational peak appeared in the spectra of m-SBA15 powders and it is due to the surface modification of GPTMS. In addition, the intensity of the O-H stretching peak at 3500 cm^{-1} for m-SBA15 sample decreased. The GPTMS modified SBA15 powders become more organic-philic [27,28]. The N_2 adsorption-desorption isotherm (or BET isotherm) at 77 K for the SBA15 sample was measured, data not shown here. The N_2 adsorption isotherm of SBA15 material is a typical reversible type IV adsorption isotherm, which has a characteristic of a mesoporous material. The BET results (not shown here) also indicated that the specific surface area of SBA15 was 571 $m^2\ g^{-1}$. The pore size distribution is approximately 2 ~ 200 nm.

The top surface and cross-sectional SEM images for PVDF-HFP/3wt.%m-SBA15 polymer membranes by a solution cast method are shown in Figs. 3 (a) and (b), respectively. A highly porous structure and significantly rough surface are observed on the composite polymer membrane samples. These m-SBA15 powders are not well dispersed, and thus a form of aggregates appears. The cross-sectional view shows several macro and micro-pores distributed irregularly on the membrane and also exhibits a high porosity structure. However, this may enhance the liquid electrolyte holding capacity [35,37,38,39,40]. Therefore, the higher liquid uptake was absorbed and the ionic conductivity can be increased.

Fig. 4 shows the FTIR spectra of PVDF-HFP/xwt.%m-SBA15 (where x: 0~5%) composite polymer membrane. The spectrum of PVDF-HFP composite film shows the vibrational peaks at wave numbers of 834, 879, 1065, 1194, 1276, and 1400 cm^{-1} which corresponds to CF_3 rocking, CF_2 rocking, CF_3 rocking, CF_2 symmetric stretching, CF_2 asymmetric stretching, and CF_2 symmetric stretching, respectively. The peaks at wave numbers of 2981 and 3021 cm^{-1} are attributed to CH_2 symmetric stretching and CH_2 asymmetric stretching, respectively [16]. In addition, FTIR spectrum of m-SBA15 (data not shown here) shows peaks at wave numbers of 800, 1087, and 3500 cm^{-1}, which corresponds to Si-O-Si symmetric stretching, Si-O-Si asymmetric stretching, and SiO-H stretching, respectively [35,40]. The peak position at 1276 cm^{-1} for pure PVDF-HFP membrane without SBA15 filler was shifted to 1173 cm^{-1}, i.e., a lower wavenumber direction, for the PVDF-HFP/10wt.%m-SBA15 composite polymer membrane. The shifting of peaks of the PVDF-HFP/m-SBA15 composite polymer membranes indicates a number of interactions among the constituents of the composite polymer membrane.

Fig. 5 shows the typical AC impedance spectra of PVDF-HFP/10 wt.%m-SBA15 composite polymer membrane at various temperatures. The AC spectra were typically non-vertical spikes for stainless steel (SS) blocking electrodes, that is, the SS | PVDF-HFP/3wt.%m-SBA15 SPE | SS cell. Analysis of the spectra yielded information about the properties of the PVDF-HFP/3wt.%m-SBA15 polymer electrolyte, such as the resistance, R_b. By considering the thickness of the composite electrolyte films, the R_b value was converted into the ionic conductivity value, σ, according to the equation: $\sigma = L/R_b \cdot A$, where L is the thickness (cm) of PVDF-HFP/3wt.%m-SBA15 composite polymer electrolyte, A is the area of the blocking electrode (cm^2), and R_b is the resistance (ohm) of the composite polymer electrolyte. Table 1 lists the ionic conductivity values of the PVDF-HFP/1~10wt.%m-SBA15 composite polymer membranes at different temperatures.

Typically, the R_b values of PVDF-HFP/3wt.%m-SBA15 composite polymer electrolytes are in the order of 3 ~ 5 ohm (inset of Fig. 5(a)) and are highly dependent on the contents of m-SBA15 ceramic fillers and the concentration of liquid electrolyte. Note that the composite polymer membrane was immersed in Li-salt electrolyte for 24 h before measurement. Table 2 displays the ionic conductivity values of the PE separator, pure PVDF-HFP polymer membrane, and PVDF-HFP/3wt.%m-SBA15 composite polymer membranes with electrolyte at various temperatures for comparison. As a result, the highest ionic conductivity value for PVDF-HFP/3wt.%m-SBA15 is 3.23×10^{-3} S cm^{-1} at 30°C. By contrast, the ionic conductivity value for PE separator and pure PVDF-HFP polymer membrane are 4.51×10^{-4} and 1.50×10^{-3} S cm^{-1} at 30°C, respectively. The result clearly indicates that the ionic conductivities of PVDF-HFP/m-SBA15 composite polymer membranes are much higher than those of PE separator and pure PVDF-HFP membrane. Table 3 lists the values of the liquid uptake and swelling ratio for the PE separator, pure PVDF-HFP film, and PVDF-HFP/3~10 wt.%m-SBA15 composite polymer membranes with 1 M $LiPF_6$ (in EC/DEC 1:1) electrolyte. It was found that the liquid uptake is varied in the range of 80~86%. By comparison, the liquid uptake values of PVDF-HFP/m-SBA15 composite polymer membranes are in the range of approximately 82.83~85.36%, which are higher than that those of PE separator (80.41%) and pure PVDF-HFP membrane (82.55%). It is due to the SBA15 mesoporous structure effect. The mesoporous structure can absorb more electrolyte acting as a micro reservoir inside composite membrane. Moreover, it was observed that the swelling ratio values of the PE separator and pure PVDF-HFP membrane are 410.53% and 427.13%, which are smaller than those of composite polymer membranes (482~578%).

However, the ionic conductivity of PVDF-HFP/10wt.%m-SBA15 composite polymer electrolytes decreases when the amount of added m-SBA15 fillers is above 10 wt.%, which is due to the occurrence of significant aggregation. The $\log_{10}(\sigma)$ vs. 1/T plots, i.e., Arrhenius plot, as shown in the inset of Fig. 5, obtains the activation energy (E_a) of the PVDF-HFP/3wt.%m-SBA15 polymer electrolyte, which is highly dependent on the contents of the liquid electrolyte uptake in the PVDF-HFP/3wt.%m-SBA15 matrix. In addition, the E_a value of the PVDF-HFP/m-SBA15 composite polymer membranes is in the order of 42~44 kJ mol^{-1}.

Typical initial charge-discharge curves of the $LiNi_{1/3}Co_{1/3}Mn_{1/3}O_2$ composite sample with PE separator, pure PVDF-HFP membrane, and PVDF-HFP/3wt.%m-SBA15 SPE at 0.1C rate are displayed in Fig. 6, respectively. As shown in Fig. 6, the $LiNi_{1/3}Co_{1/3}Mn_{1/3}O_2$/Li cell shows the typical sloping potential profile at 3.0 ~ 4.3 versus Li/Li^+. The specific discharge capacities of the $LiNi_{1/3}Co_{1/3}Mn_{1/3}O_2$/Li cells with PE separator, pure PVDF-HFP membrane, and PVDF-HFP/3wt.%m-SBA15 SPE are 145, 144, and 156 mAh g^{-1}, respectively. It was demonstrated that the electrochemical performance of $LiNi_{1/3}Co_{1/3}Mn_{1/3}O_2$/Li cell with PVDF-HFP/3wt.%m-SBA15 SPE is much better than those of $LiNi_{1/3}Co_{1/3}Mn_{1/3}O_2$/Li cells with PE separator and pure PVDF-HFP membrane,

Moreover, the $LiNi_{1/3}Co_{1/3}Mn_{1/3}O_2$/Li cell with PE separator delivers the specific capacities of 146, 142, 123, 30, 10, and 9 mAh g^{-1}, at C rates of 0.2, 0.5, 1, 3, 5, and 10, respectively. In contrast, the discharge capacity of the $LiNi_{1/3}Co_{1/3}Mn_{1/3}O_2$/Li cell with PVDF-HFP/3wt.%m-SBA15 membrane delivers 154, 144, 135, 88, 31 and 16 mAh g^{-1}, at C rates of 0.2, 0.5, 1, 3, 5 and 10, respectively. As a result, the capacities of the $LiNi_{1/3}Co_{1/3}Mn_{1/3}O_2$/Li cell with the PE separator and the PVDF-HFP/3wt.%m-SBA15 membrane decrease markedly with the increasing discharge rates from 0.1 to 10 C, as shown in Fig. 7.

Fig. 8 shows the cycling-life performance of the lithium ion cells with PE separator and PVDF-HFP/3wt.%m-SBA15 membrane at 0.2C/0.2C charge/discharge rate, in the voltage range of 3.0~ 4.3 V for 30 cycles at room temperature for comparison. It was found that the average discharge capacity of the $LiNi_{1/3}Co_{1/3}Mn_{1/3}O_2$/Li cells with PE separator delivers 143.47 mAh g^{-1}; in contrast, the $LiNi_{1/3}Co_{1/3}Mn_{1/3}O_2$/Li cell with PVDF-HFP/3wt.%m-SBA15 membrane delivers 151.58 mAh g^{-1}. As a result, the PVDF-HFP/3wt.%m-SBA15 membrane is superior to the conventional PE separator. It may be due to the mesoprous m-SBA15 fillers being able to retain large amount of electrolyte. It results in higher ionic conductivity. It also found that the charge transfer resistance (R_{ct} = 51.05 ohm) of the $LiNi_{1/3}Co_{1/3}Mn_{1/3}O_2$/Li cell with the PVDF-HFP/3wt.%m-SBA15 SPE is lower than that (R_{ct} = 61.65 ohm) of the $LiNi_{1/3}Co_{1/3}Mn_{1/3}O_2$/Li cells with PE separator by AC impedance analysis (not shown here). In short, this demonstrated that the PVDF-HFP/m-SBA15 SPEs exhibit better electrochemical property. The above result also indicates that the optimal amount of the m-SBA15 filler in the PVDF-HFP matrix is around 3wt.%. Therefore, the as-prepared composite membrane can be a good candidate for $LiNi_{1/3}Co_{1/3}Mn_{1/3}O_2$/Li polymer battery.

4. CONCLUSIONS

This study reports on the preparation of composite polymer electrolytes for lithium-ion batteries. The poly(vinylidiene fluoride-co-hexafluoropropylene) (PVDF-HFP) was used as the polymer host and mesoporous SBA15 (silica) filler as a solid plasticizer was added into the polymer matrix. The SBA15 fillers with the mesoporous structure and high specific surface are can trap more liquid electrolytes to enhance the ionic conductivity. The $LiNi_{1/3}Co_{1/3}Mn_{1/3}O_2$/Li composite batteries comprised of a lithium metal anode, a $LiNi_{1/3}Co_{1/3}Mn_{1/3}O_2$ composite cathode, and PVDF-HFP/m-SBA15 composite polymer membranes were assembled and studied. The performances of $LiNi_{1/3}Co_{1/3}Mn_{1/3}O_2$/Li polymer batteries with the PE separator and the pure PVDF-HFP composite membrane were examined and compared. Overall, it was found that the $LiNi_{1/3}Co_{1/3}Mn_{1/3}O_2$/Li polymer batteries with the PVDF-HFP/SBA15 membranes exhibit much better electrochemical performance, as compared with those with PE separator and pure PVDF-HFP membrane. In particular, the $LiNi_{1/3}Co_{1/3}Mn_{1/3}O_2$/Li polymer battery comprising of PVDF-HFP/3wt.%m-SBA15 membrane delivered the capacities of as much as 156 and 88 mAh g^{-1} at 0.1 and 3C rates, respectively. It was also shown the high coulomb efficiency of 95~99%. Experimental results demonstrate that the PVDF-HFP/m-SBA15 composite polymer membrane is a suitable candidate for application on the $LiNi_{1/3}Co_{1/3}Mn_{1/3}O_2$ polymer batteries.

5. ACKNOWLEDGEMENTS

Financial support from the National Science Council, Taiwan (Project No: NSC 101-2221-E-131 -037) is gratefully acknowledged.

6. REFERENCES

[1] J.M. Tarascon, M. Armand, "Issues and challenges facing rechargeable lithium batteries", Nature 414 (2001) 359-367.

[2] D. Fouchard, J.B. Taylor, "Spinel electrodes for lithium batteries-A review", J. Power Sources 21 (1987) 3-4.

[3] W.D. Johnston, R.R. Heikes, D. Sestrich, "The preparation crystallography and magnetic properties of the $Li_xCo_{(1-x)}O$ system", J. Phys. Chem. Solids 7 (1958) 1.

[4] T. Nagaura, K. Tozawa, "Lithium-ion rechargeable battery", Prog. Batts. Sol. Cells 9 (1990) 209-217.

[5] T. Ohzuku, A. Ueda, M. Nagayama, "Comparative study of $LiCoO_2$, $LiNi_{1/2}Co_{1/2}O_2$ and $LiNiO_2$ for 4 volt secondary lithium cells", Electrochim. Acta 38 (1993) 1159-1167.

[6] Z. Liu, A. Yu, J.Y. Lee, "Synthesis and characterization of $LiNi_{1-x-y}Co_xMn_yO_2$ as the

cathode materials of secondary lithium batteries", J. Power Sources 81 (1999) 416-419.

[7] S. Jouanneau, D.D. MacNeil, Z. Lu, S.D. Beattie, G Murphy, J.R. Dahn, "Morphology and safety of $Li[Ni_xCo_{1-2x}Mn_x]O_2$ ($0 \leqslant x \leqslant 1/2$)", J. Electrochem. Soc. 150 (2003) A1299-A1304.

[8] S.T. Myung, M.H. Lee, S. Komaba, N. Kumagai, Y.K. Sun, "Hydrothermal synthesis of layered $Li[Ni_{1/3}Co_{1/3}Mn_{1/3}]O_2$ as positive electrode material for lithium secondary battery", Electrochim. Acta 50 (2005) 4800-4806.

[9] X. Luo, X. Wang, L. Liao, X. Wang, S. Gamboa, P.J. Sebastian, "Effect of synthesis conditions on the structural and electrochemical properties of layered $Li[Ni_{1/3}Co_{1/3}Mn_{1/3}]O_2$ cathode material via the hydroxides co-precipitation method LIB SCITECH", J. Power Sources 161 (2006) 601-605.

[10] D.G. Tong, Q.Y. Lai, N.N. Wei, A.D. Tang, L.X. Tang, K.L. Huang, X.Y. Ji, "Synthesis of $Li[Ni_{1/3}Co_{1/3}Mn_{1/3}]O_2$ as a cathode material for lithium ion battery by water-ion-oil emulsion method, Mater. Chem. Phys. 94 (2005) 423-428.

[11] X. Luo, X. Wang, L. Liao, S. Gamboa, P.J. Sebastian, "Synthesis and characterization of high tap-density layered $Li[Ni_{1/3}Co_{1/3}Mn_{1/3}]O_2$ cathode material via hydroxide co-precipitation", J. Power Sources 158 (2006) 654-658.

[12] G.H. Kim, S.T. Myung, H.S. Kim, Y.K. Sun, "Synthesis of spherical $Li[Ni_{(1/3-z)}Co_{(1/3-z)}Mn_{(1/3-z)}Mg_z]O_2$ as positive electrode material for lithium-ion battery", Electrochim. Acta 51 (2006) 2447-2453.

[13] S. Yang, X. Wang, X. Yang, L. Liu, Z. Liu, Y. Bai, Y. Wang, "Influence of Li source on tap density and high rate cycling performance of spherical $Li[Ni_{1/3}Co_{1/3}Mn_{1/3}]O_2$ for advanced lithium-ion batteries", J. Solid Sate Electrochem. 16 (2012) 1229-1237.

[14] Z. Chang, Z. Chen, F. Wu, X.Z. Yuan, H. Wang, "The synthesis of $Li[Ni_{1/3}Co_{1/3}Mn_{1/3}]O_2$ using eutectic mixed lithium salts $LiNO_3$-LiOH", Electrochim. Acta 54 (2009) 6529-6535.

[15] S. Zhang, C. Deng, S.Y. Yang, H. Niu, "An improved carbonate-co-precipitation method for the preparation of spherical $Li[Ni_{1/3}Co_{1/3}Mn_{1/3}]O_2$ cathode material", J. Alloys and Compd. 484 (2009) 519-523.

[16] X. Li, Y.J. Wei, H. Ehrenberg, F. Du, C.Z. Wang, G. Chen, "Characterization on the structural and electrochemical properties of $Li[Ni_{1/3}Co_{1/3}Mn_{1/3}]O_2$ prepared by a wet-chemical process", Solid State Ionics 178 (2008) 1969-1974.

[17] Z. Chang, Z. Chen, F. Wu, H. Tang, Z. Zhu, X.Z. Yuan, H. Wang, "Synthesis and characterization of high-density non-spherical $Li[Ni_{1/3}Co_{1/3}Mn_{1/3}]O_2$ cathode material for

lithium ion batteries by two-step drying method", Electrochim. Acta 53 (2008) 5927-5933.

[18] B. Lin, Z. Wen, Z. Gu, S. Huang, "Morphology and electrochemical performance of $Li[Ni_{1/3}Co_{1/3}Mn_{1/3}]O_2$ cathode material by a slurry spray drying method", J. Power Sources 175 (2008) 564-569.

[19] B.-C. Park, H.-B. Kim, S.-T. Myung, K. Amine, I. Belharouak, S.-M. Lee, Y.-K. Sun, "Improvement of structured and electrocehmical properties of AlF_3-caoted $Li[Ni_{1/3}Co_{1/3}Mn_{1/3}]O_2$ cathode materials on high voltage region", J. Power Sources 178 (2008) 826-831.

[20] P. Samarasingha, D.H. Tran-Nguyen, M. Behm, A. Wijayasinghe, "$Li[Ni_{1/3}Co_{1/3}Mn_{1/3}]O_2$ synthesized by the Pechini method for the positive electrode in Li-ion batteries: Materials characterization and electrochemical behaviour", Electrochim. Acta 53 (2008) 7995-8000.

[21] W. Pu, X. He, L. Wang, C. Jiang, C. Wan, "Preparation of PVDF-HFP microporous membrane for Li-ion batteries by phase inversion", J. Membr. Sci. 272 (2006) 11-14.

[22] K.M. Kim, N.G. Park, K.S. Ryu, S.H. Chang, "Characteristics of PVDF-HFP/TiO_2 composite membrane electrolytes prepared by phase inversion and conventional casting methods", Electrochim. Acta 51 (2006) 5636-5644.

[23] A.M. Stephan, K.S. Nahm, T.P. Kumar, M.A. Kulandainathan, G. Ravi, J. Wilson, "Nanofiller incorporated poly(vinylene fluoride-hexafluoropropylene) (PVDF-HFP) composite electrolyte for lithium batteries", J. Power Sources 159 (2006) 1316-1321.

[24] A.M. Stephan, K.S. Nahm, M.A. Kulandainathan, G. Ravi, J. Wilson, "Poly(vinylene fluoride-hexafluoropropylene) (PVDF-HFP) based composite electrolytes for lithium batteries", Eur. Polym. J. 42 (2006) 1728-1734.

[25] C.G. Wu, M.I. Lu, C.C. Tsai, H.-J. Chuang, "PVDF-HFP/metal oxide nanocomposite: The matrices for high-conducting, low-leakage porous polymer electrolytes", J. Power Sources 159 (2006) 295-300.

[26] Y.X. Jiang, Z.-F. Chen, Q.-C. Zhuang, J.-M. Xu, Q.-F. Dong, L. Huang, S.-G. Sun, "A novel composite microporous polymer electrolyte prepared with molecule sieves for Li-ion batteries", J. Power Sources 160 (2006) 1320-1328.

[27] M. Stolarska, L. Niedzicki, R. Borkowska, A. Zalewska, W. Wieczorek, "Structure, transport properties and interfacial stability of PVDF/HFP electrolyte containing modified inorganic filler", Electrochim. Acta 53 (2007) 1512-1517.

[28] M. Walkowiak, A. Zalewska, T. Jesionowski, D. Waszak, B. Czajka, "Effect of chemically silica on the properties of hybrid gel electrolyte for Li-ion batteries", J. Power Sources 159

(2006) 449-453.

[29] H.S. Kim, C.-W. Lee, S.-I. Moon, "Electrochemical performances of lithium-ion polymer battery using $LiNi_{1/3}Co_{1/3}Mn_{1/3}O_2$ as cathode materials", J. Power Sources 159 (2006) 227-232.

[30] G.C. Li, P. Zhang, L.C. Yang, Y.P. Wu, "A porous polymer electrolyte based on P(VDF-HFP) prepared by a simple phase separation process", Electrochem. Commun. 10 (2008) 1883-1885.

[31] Z.H. Li, H.P. Zhang, P. Zhang, G.C. Li, Y.P. Wu, X.D. Zhou, "Effect of the porous structure on conductivity of nanocomposite polymer electrolyte for lithium ion batteries", J. Membr. Sci. 322 (2008) 416-422.

[32] X. Li, G. Cheruvally, J.K. Kim, "Polymer electrolytes based on an electrospun poly(vinylidene fluoride-co-hexafluoropropylene) membrane for lithium batteries", J. Power Sources 167 (2007) 491-498.

[33] Y.J. Hwang, S.K. Jeong, K.S. Nahm, "Electrochemical studies on poly(vinylidene-hexafluoropropylene) membranes prepared by phase inversion method", Europ. Polym. J. 43 (2007) 65-71.

[34] Y. Ding, P. Zhang, Z. Long, "The ionic conductivity and mechanical property of electrospun P(VdF-HFP)/PMMA membranes for lithium ion batteries" , J. Membr. Sci. 329 (2009) 56-59.

[35] M. Walkowiak, A. Zalewska, T. Jesionowski, "Stability of poly(vinylidene fluoride-co-hexafluoropropylene)-based composite gel electrolytes with functionalized silicas", J. Power Sources 173 (2007) 721-728.

[36] H. Xie, Z. Tang, Z. Li, "PVDF-HFP composite polymer electrolyte with excellent electrochemical properties for Li-ion batteries", J. Solid State Electrochem. 12 (2008) 1497-1502.

[37] P. Raghhavan, X. Zhao, J. Manuel, "Electrochemical performance of electrospun poly(vinylidene fluoride-co-hexafluoropropylene)-based nanocomposite polymer electrolytes incorporating ceramic fillers and room temperature ionic liquid", Electrochim. Acta 55 (2010) 1347-1354.

[38] H.S. Jeong, D.W. Kim, Y.U. Jeong, "Effect of microporous structure on thermal shrinkage and electrochemical performance of Al_2O_3/poly(vinylidene fluoride-co-hexafluoropropene) composite separators for lithium-ion batteries", J. Membr. Sci. 364 (2010) 177-182.

[39] H.S. Jeong, D.W. Kim, Y.U. Jeong, "Effect of phase inversion on microporous structure

development of Al_2O_3/poly(vinylidene fluoride-co-hexafluoropropene)-based ceramic composite separators for lithium-ion batteries", J. Power Sources 195 (2010) 6116-6121.

[40] H.S. Jeong, E.S. Choi, S.Y. Lee, "Composition ratio-dependent structural evolution of SiO_2/poly(vinylidene fluoride-hexafluoropropylene)-coated poly(ethylene terephthalate) nonwoven composite separators for lithium-ion batteries", Electrochim. Acta 86 (2012) 317-322.

[41] X. Li, J.K. Kim, J.W. Choi, "Polymer electrolytes based on an electrospun poly(vinylidene fluoride-co-hexafluoropropylene) membrane for lithium batteries", J. Power Sources 167 (2007) 491-498.

Table 1

The ionic conductivities (S cm^{-1}) of PVDF-HFP/x%m-SBA15 composite polymer electrolyte membranes containing varied amounts of SBA15 at various temperatures

Param. / T/ °C	PVDF-HFP/ 3%m-SBA15	PVDF-HFP/ 5%m-SBA15	PVDF-HFP/ 7%m-SBA15	PVDF-HFP/ 10%m-SBA15
20	3.23×10^{-3}	2.32×10^{-3}	3.48×10^{-3}	3.78×10^{-3}
30	3.29×10^{-3}	2.56×10^{-3}	3.56×10^{-3}	3.95×10^{-3}
40	3.54×10^{-3}	2.84×10^{-3}	3.74×10^{-3}	4.27×10^{-3}
50	3.87×10^{-3}	3.18×10^{-3}	3.81×10^{-3}	4.58×10^{-3}
60	4.12×10^{-3}	3.43×10^{-3}	3.24×10^{-3}	4.89×10^{-3}
Ea/ kJ mol^{-1}	42.55	42.42	44.75	41.36

Note: Merck LP40 composition: $LiPF_6$/EC+DEC+DMC（mole ratio＝35:50:15）

Table 2

Comparison of the ionic conductivities (S cm^{-1}) of PE separator, PVDF-HFP/3%m-SBA15 composite polymer electrolyte membranes at various temperatures

T/ °C \ Param.	PE separator	Pure PVDF-HFP	PVDF-HFP/ 3%m-SBA15
20	4.51×10^{-4}	1.59×10^{-3}	3.23×10^{-3}
30	4.85×10^{-4}	1.61×10^{-3}	3.29×10^{-3}
40	5.33×10^{-4}	1.69×10^{-3}	3.54×10^{-3}
50	5.86×10^{-4}	1.74×10^{-3}	3.87×10^{-3}
60	6.11×10^{-4}	1.76×10^{-3}	4.12×10^{-3}
Ea/ kJ mol^{-1}	38.5	51.83	42.55

Table 3

The A% and SW% of PVDF-HFP/x%m-SBA15 SPE

Samples	Thickness/ cm	A/ %	SW/ %
PE separator	0.0030	80.41	410.52
Pure PVDF-HFP	0.0110	82.55	473.27
PVDF-HFP/3%m-SBA15	0.0158	82.83	482.51
PVDF-HFP/5%m-SBA15	0.0108	83.09	491.50
PVDF-HFP/7%m-SBA15	0.0098	84.39	540.65
PVDF-HFP/10%m-SBA15	0.0122	85.36	578.57

Using Merck LP4 electrolyte.

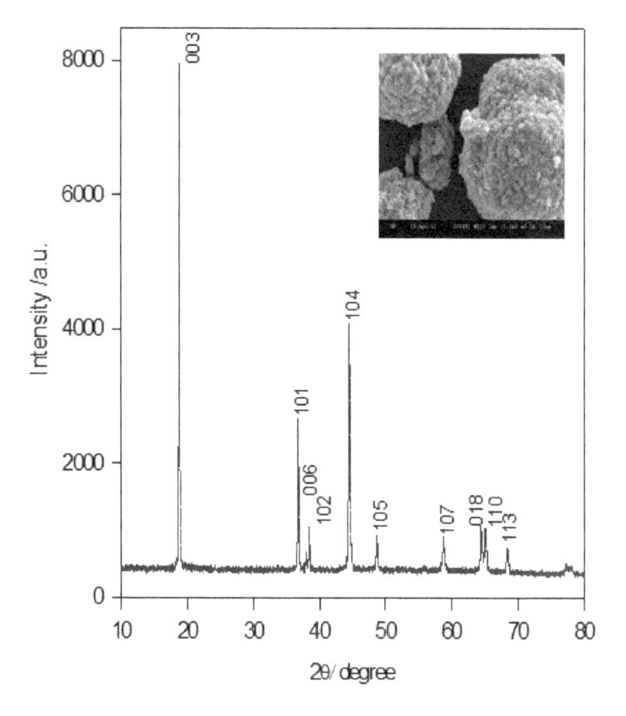

Fig. 1 XRD pattern of as-prepared $LiNi_{1/3}Co_{1/3}Mn_{1/3}O_2$ powders.

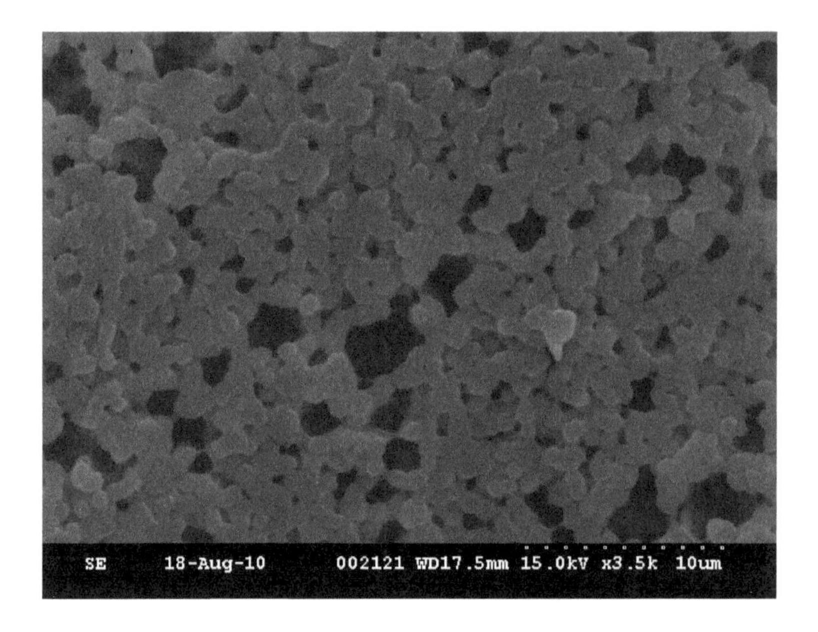

(a)

(b)

Fig. 2 FE-SEM (a) and HR-TEM (b) images of mesoporous m-SBA15 powders.

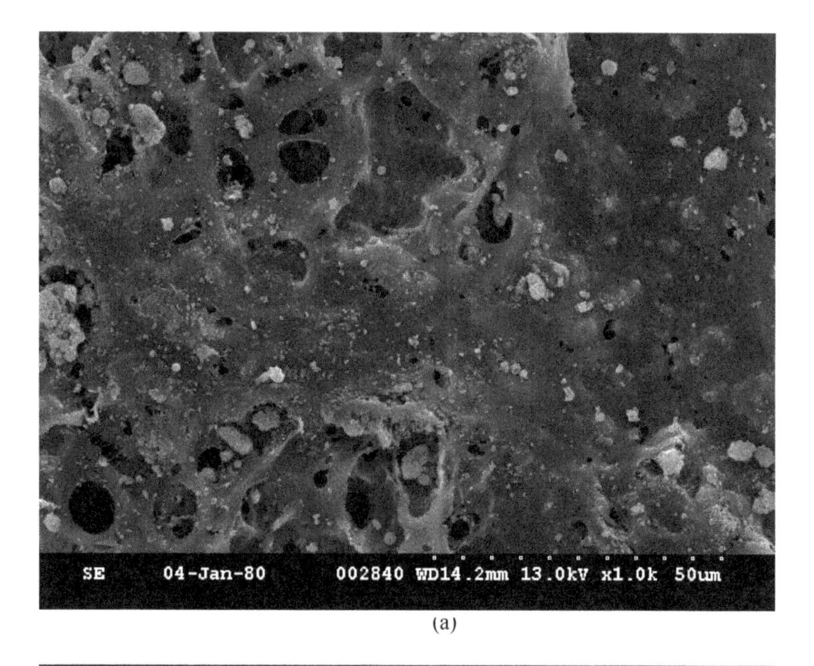

(a)

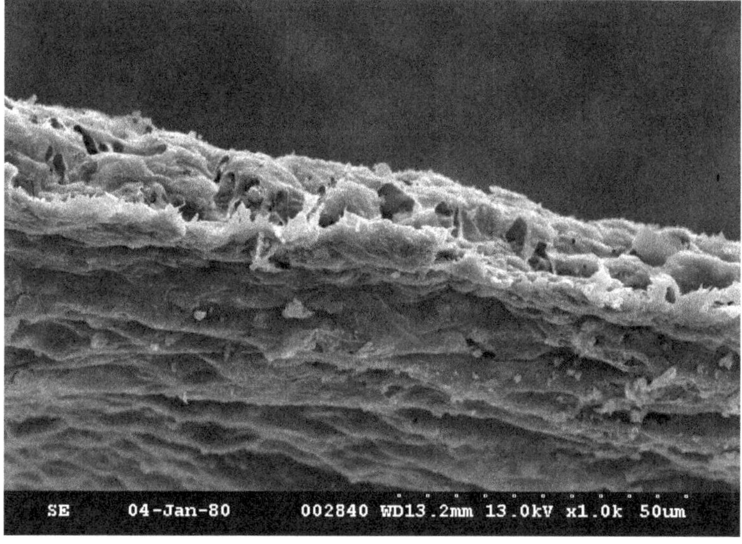

(b)

Fig. 3 SEM images of PVDF-HFP/3wt.%m-SBA15 composite polymer membrane; (a).Top view; (b). Cross-section view.

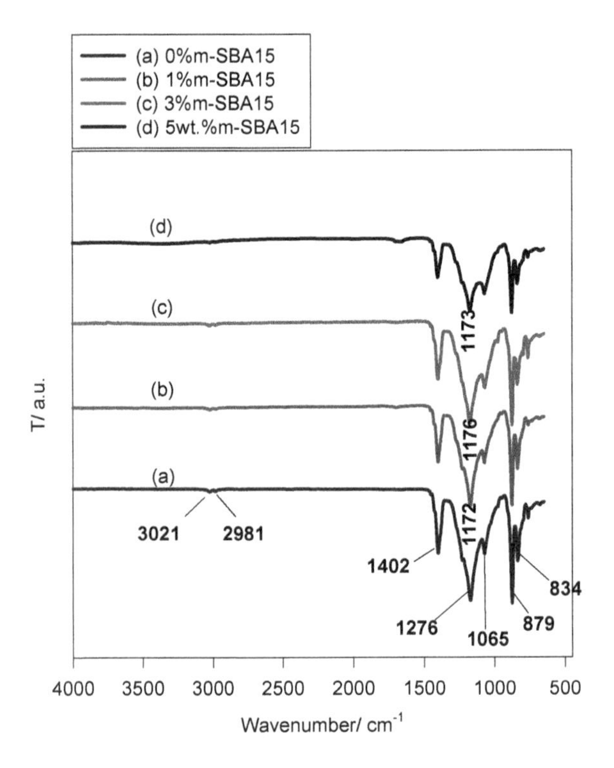

Fig. 4 FTIR images of PVDF-HFP/x%m-SBA15 composite polymer electrolyte membranes.

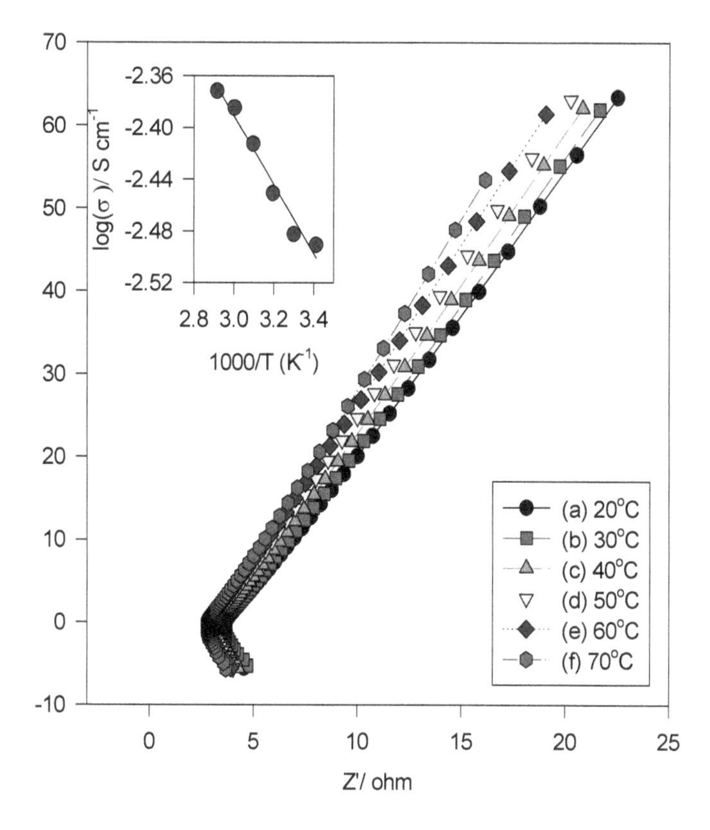

Fig. 5 Nyquist plot for PVDF-HFP/3%m-SBA15 composite polymer electrolyte at various temperatures; the inset for Arrhenius plot.

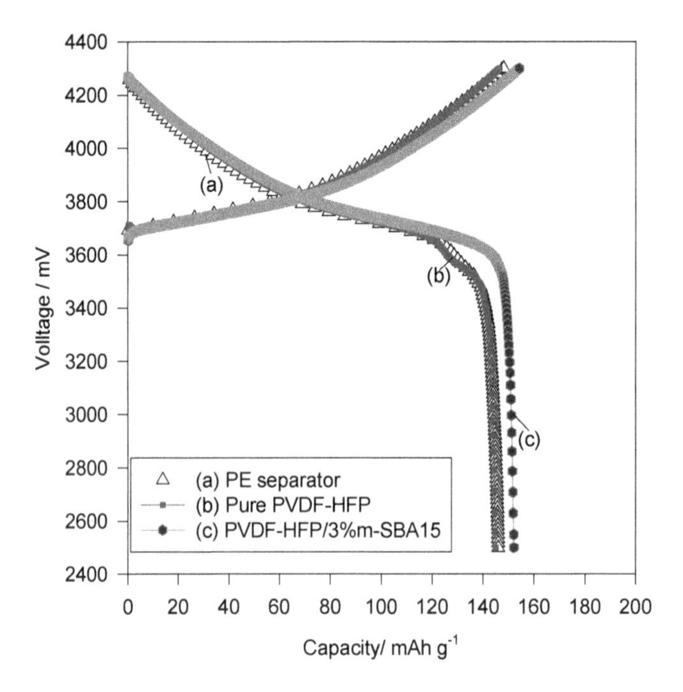

Fig. 6 The initial charge-discharge curves of $Li/LiNI_{1/3}Co_{1/3}Mn_{1/3}O_2$ lithium polymer batteries with PE separator, pure PVDF-HFP film, and PVDF-HFP/3wt.%SBA15 composite SPEs at 0.1C rate.

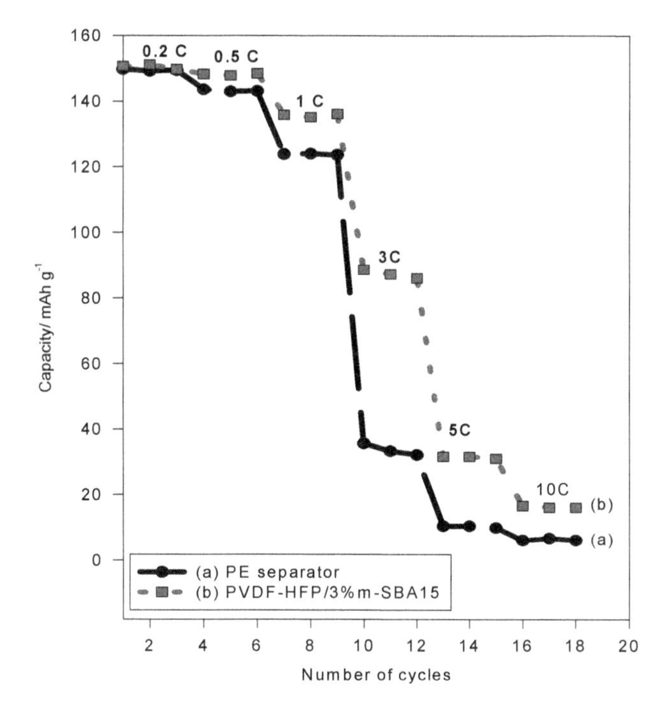

Fig. 7 The rate capability comparison for Li/ $LiNi_{1/3}Co_{1/3}Mn_{1/3}O_2$ lithium polymer batteries with (a) PE separator; (b) PVDF-HFP/3%m-SBA15 SPE at 0.2C-10C rates.

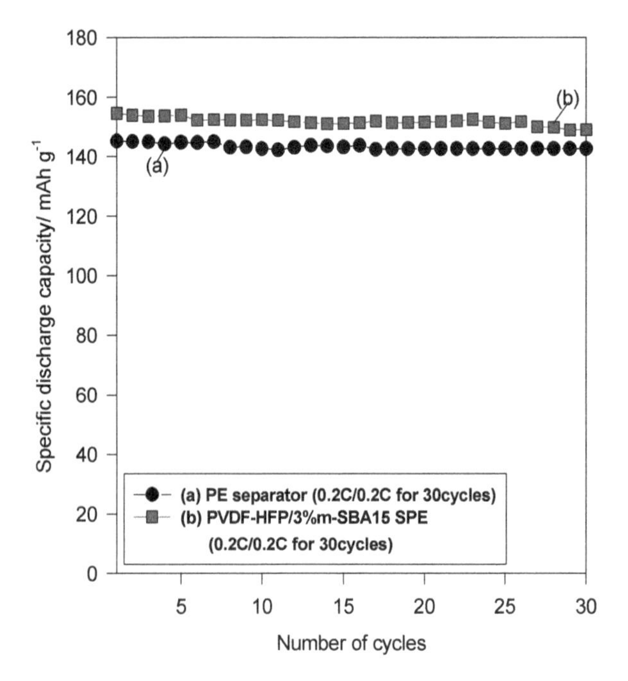

Fig. 8 The cycle-life performance of Li/ $LiNi_{1/3}Co_{1/3}Mn_{1/3}O_2$ lithium polymer batteries composed of PE separator and PVDF-HFP/3%m-SBA15 SPE at 0.2C/0.2C rate.

Glasses and Ceramics for Nuclear and Hazardous Waste Treatment

BOROSILICATE GLASS FOAMS FROM GLASS PACKAGING RESIDUES

R.K.Chinnam[1], Silvia Molinaro[2], Enrico Bernardo[3], Aldo R. Boccaccini[1 (*)]

1. University of Erlangen-Nuremberg, Department of Materials Science and Engineering, Cauerstr- 6, 91058 Erlangen, Germany.
2. Stevanato Group s.p.a., via Molinella, 17, 35017 Piombino Dese (PD), Italy.
3. University of Padova, Department of Industrial Engineering, Via Marzolo 9, 35131 Padova, Italy.

(*) Corresponding author: A. R. Boccaccini; e-mail: aldo.boccaccini@ww.uni-erlangen.de

ABSTRACT

In the present study, pharmaceutical waste derived borosilicate glass (BSG) was investigated to fabricate foams without the addition of external foaming agents. Foaming was shown to occur when glass powder compacts were sintered at 950°C and above. It was found that foaming ability depended on the initial powder particle size and it was higher when the glass particle size was <38μm, while samples with a particle size of <100μm and <250μm exhibited bulk nature (no foaming). The exact reasons for the self-foaming behavior of BSG waste were not determined; however a hypothesis was proposed to explain the foaming mechanism. The novel borosilicate glass foams developed here are of interest for acoustic and thermal insulation structures exhibiting high thermal shock resistance.

INTRODUCTION

Borosilicate glass (BSG) usage in pharmaceutical packaging is a considerably growing market. The manufacturers of such vials, flasks and cartridges need to maintain accurate dimensions and tolerances and any piece exhibiting mismatch is discarded as waste. The amount of rejected borosilicate glass in a pharmaceutical packaging manufacturing industry can reach more than 1000 tonnes/year. Transporting such waste into landfills represents an economic burden for the industry. Hence, ways to safely transform such BSG waste into application oriented materials will be advantageous to industry and also will contribute to a more efficient natural resources usage leading to environmental benefits.

Foam glass is an attractive material with desirable attributes for structural applications such as high strength to weight ratio [1, 2]. Such porous glasses are promising because of their unique combination of properties, such as compressive strength, thermal insulation, freeze tolerance, non-flammability, chemical inertness and non-toxicity, rodent and insect resistance, bacteria resistance and water and steam resistance [3, 4].

Most glass foams are fabricated by sintering glass powders mixed with foaming agents. Foaming agents are substances releasing gasses in the pyroplastic mass formed by the softening glass matrix [5, 6]. The gas release is associated with oxidation or decomposition reactions. Oxidation reactions lead to the release of CO_x gas (carbon monoxide or carbon dioxide) from C-containing compounds, e.g. carbon black, graphite, SiC, organic substances, reacting with oxygen from the atmosphere. Decomposition reactions are those provided by carbonates (mainly Na- and Ca-carbonates) or sulphates, leading to the release of CO_2 or SO_x [7]. Minay et al [8] used microwave heating to foam steel fiber reinforced BSG. It was reported that ~50% porosity was obtained in the glass due to localized overheating of glass. Pore formation in borosilicate glass by sintering under microwave heating has been also reported [9] indicating the influence of fast heating rate on pore formation in BSG. Borosilicate foam glass preparation using argon gas dissolution was reported by Wang et al. [10]. In their study, argon gas was initially dissolved into molten BSG and later by controlled heat treatment the dissolved argon was released in the glass-matrix, which resulted in BSG foaming.

Given the availability of BSG waste from the pharmaceutical packaging industry, in the present study the possibility of developing BSG foams from such residues was considered. Using powder sintering technique, borosilicate glass foams were produced without the addition of external foaming agents. The experimental method is presented and a hypothesis to explain the possible mechanisms involved in self-foaming of BSG is discussed.

MATERIALS AND METHODS

The pharmaceutical waste borosilicate glass (BSG) (provided by NUOVA OMPI, Italy) (see Table 1) was crushed in a Retsch BB51 crushing machine repeating the process for 4 times with a decreasing gap between the crushers. The obtained fine powders were graded according to their particle size distribution as follows: <38μm, >38μm to <100μm, and >100μm to <250μm. The different particle sizes were achieved by sieving.

Table 1 - Composition of borosilicate glass waste

Components	Content in BSG (wt%)
Al_2O_3	5
B_2O_3	10.5
CaO	1.5
Na_2O	7
SiO_2	75

Power compacts were prepared by cold compaction in a cylindrical die. Typically, powder was mixed with ~5wt% distilled water and the resulting wet powder was uniaxially pressed using a pressure of 30MPa into pellets of nominal dimensions 20mm in diameter x ~3mm in height. Powder compacts obtained in this way were dried in an oven at ~70°C overnight. The dry compacts were carefully transferred to an electric furnace at 950°C. Followed by direct heating, the samples were sintered for 30min and cooled in air to room temperature. Samples were characterized by scanning electron microscopy (SEM). Back scattered SEM images were taken using a ESEM, Quanta 200 (FEI, NL). Samples were also inspected visually to assess morphology changes which were documented by digital camera imaging.

RESULTS AND DISCUSSION

In this study, the heating profile used to sinter BSG powder compacts comprises direct heating (950°C), annealing for 30min and cooling in air to room temperature. Different particle sized compacts when sintered under the given conditions showed different behavior. Specifically, the sample made with <38μm particles has foamed, while compacts made with powders of size <100 and <250μm resulted in bulk (dense) BSG (see figure 1).

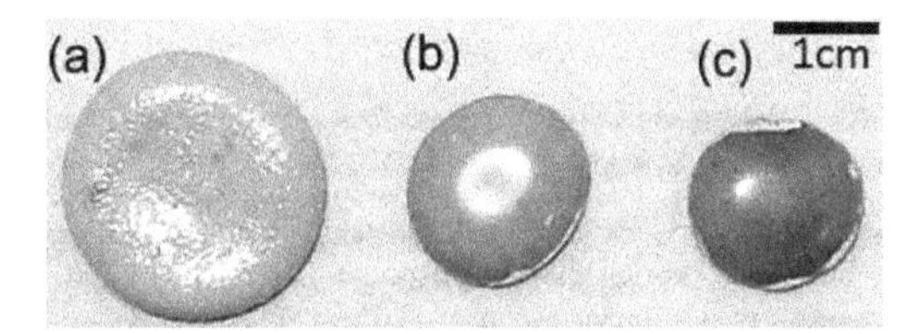

Figure 1. Change in morphology of (a) <38μm, (b) <100μm and (c) <250μm particle sized BSG powder compacts sintered at 950°C for 30min.

The different behavior of BSG powder compacts made from different size powders upon sintering at 950°C can be explained as follows. The compact of fine (<38μm) BSG powder has higher surface area compared to that of coarse particles (<100μm and <250μm). Direct heating of BSG compact to 950°C, which is well above the softening temperature of BSG (~820°C) [11], can activate the formation of a low viscosity BSG layer on the surface at the initial stages of sintering. This low viscosity BSG surface layer is vulnerable to nucleation / crystallization, and any such event will increase the viscosity of a low viscosity phase [12]. In the later stages of sintering, because of a temperature homogenization effect in the sample, the interior region of the compact will start to soften and form a low viscosity phase. Hence, a closed system with a high viscosity BSG layer at the outer surface and a low viscous BSG core (bulk) will be formed (see figure 2).

In the green state, coarse particle compacts contain relatively large pores (in comparison to fine particle compacts) because of particle packing limitations. At the initial stages of sintering, when the low viscosity BSG is forming on the outer surface of the pellets, the pore size of the green compact plays an influential role on the densification behavior because of different capillary forces, e.g. pores in fine particle green compacts exhibit higher capillarity. Due to direct heating, the viscous phase formed on the outer surface of the compact may seal more effectively the small pores forming a viscous and non-porous surface layer which inhibits the escape of gases from the bulk of the sample during sintering (Figure 2). Hence, the occluded gas expands and forms closed pores leading to close-cell foam like structures as observed in the SEM image in Figure 3. Pores formed by the occluded gases in BSG were analyzed to have an average size of ~300μm however pores as large as 500 μm in diameter were also found. These values are comparable to other studies in literature on foams fabricated from waste glasses, for example Bernardo et al [1] analyzed the pore size of foam glass prepared from soda-lime glass residues with SiC additive and reported pore sizes <500μm. On the other hand, the spherical pores of foamed borosilicate glass made by microwave heating [9] were an order of magnitude smaller (up to ~ 50 μm).

In the case of coarse particle compacts, because of the reduced surface area, the amount of bulk viscous BSG formed on the surface will be reduced. In the green state, because of the larger pore size, capillary forces will be low. Hence, in coarse particle compacts, at the initial stages of sintering, the viscous BSG layer formed on the outer surface cannot completely seal the pores. The presence of open porosity even at the later stages of sintering helps the escape of gases from the bulk of the sample. Therefore, coarse particle compacts (<100μm and <250μm) do not show foaming ability and rather sinter to a dense bulk sample (see figure 2). Even if the self-foaming behavior of BSG was not further investigated, several mechanisms can be suggested to be involved in activating the release of gases from the bulk of the sample to form close porosity in the soft viscous matrix (as discussed above) during sintering. In the literature, boron escape from BSG melts while processing at high temperatures (1000-1200 °C) has been reported [13, 14]. This may be also the case in the fine particle compacts (discussed here), e.g. evaporation of boron due to the increased surface area is possible. Moreover, while processing at higher temperatures, the glass tends to dissolve gases. Wang et al [10] used this effect of BSG to produce foams. A similar mechanism could have been active in the present recycled BSG powder. Along with the above possibilities, the entrapped air / moisture in the pores formed during powder compaction could also lead to pore formation, for example these inclusions might have occluded in the viscous glass and expanded while sintering. Finally, it should be noted that the BSG waste from the pharmaceutical packaging plant was rejected for further processing prior to the cleaning of cartridges. Though the surface of glass looked clear (by simple visual inspection), there is a possibility of the presence of residual oils other microscopic organic contaminants on the glass walls which evaporate while sintering acting also as pore formers.

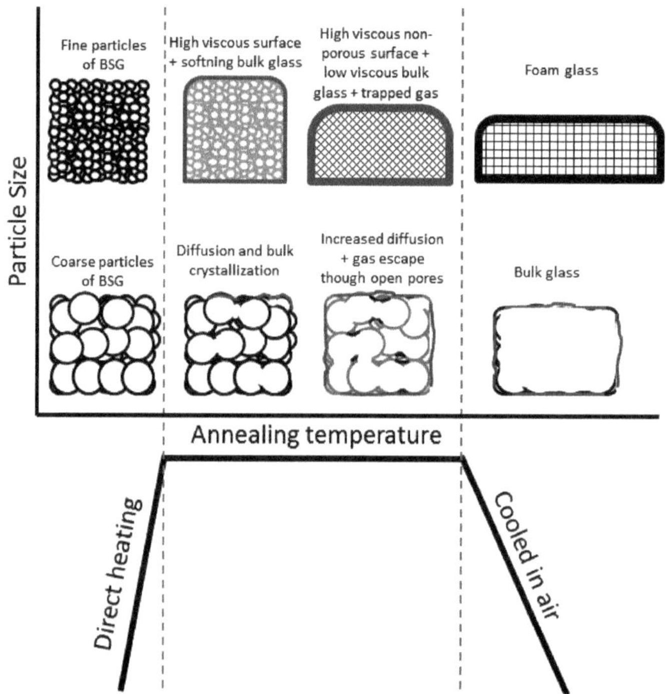

Figure 2. Schematic diagram illustrating the suggested influence of BSG particle size on the evolution of densification and foaming of glass powder compacts during heat treatment.

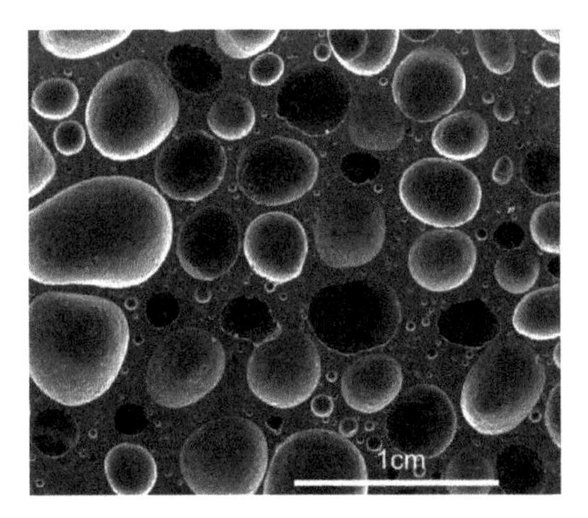

Figure 3. Back scattered SEM image of <38μm particle compact sintered at 950°C for 30min showing closed porosity in a densified glass matrix.

CONCLUSIONS

Fine particle BSG powder (<38μm) compacts were found to foam by heating at 950°C for 30min, while an increase in the particle size to 100μm produced dense bulk glass. The highly viscous non-porous BSG layer formed on the surface of the powder compact could have inhibited the release of occluded gases from the bulk of the compact. The release of gas due to evaporation of contaminants from the surface of waste glass, evaporation of boron, release of dissolved gasses or the occlusion of air / moisture trapped during powder compaction are possible reasons for foaming of BSG pellets. In the future, a detailed probe into each suggested factor responsible for self-foaming BSG must be carried out. Foaming by a single heat-treatment procedure without using porogens or special additives to produce BSG foams with large pores (~ 300 μm) represents an attractive approach for reusing discarded BSG from pharmaceutical packaging industry. The high temperature resistant foams are attractive materials for efficient sound and thermal insulation applications.

ACKNOWLEDGEMENTS

The support of the European Community's Seventh Framework Programme through a Marie-Curie Research Training Network ("GlaCERCo" PITN-GA-2010-264526) is gratefully acknowledged.

REFERENCES

[1] E. Bernardo, G. Scarinci, P. Bertuzzi, P. Ercole, L. Ramon, Recycling of waste glasses into partially crystallized glass foams, *J. Porous Mater.*, 17, 359–365 (2009).

[2] Y. Attila, M. Güden, A. Taşdemirci, Foam glass processing using a polishing glass powder residue, *Ceram. Int.*, 39, 5869-5877 (2013).

[3] S. Hasheminia, A. Nemati, B. Eftekhari Yekta, P. Alizadeh, Preparation and characterization of diopside-based glass–ceramic foams, *Ceram. Int.*, 38, 2005–2010 (2012).

[4] G. Scarinci, G. Brusatin, E. Bernardo, Glass Foams, in M. Scheffler, P. Colombo, Cellular Ceramics: Structure, Manufacturing, Properties and Applications, Wiley–VCH, Weinheim, 158-176 (2005).

[5] Bo Chen, K. Wang, X. Chen, Lu A., Study of Foam Glass with High Content of Fly Ash Using Calcium Carbonate as Foaming Agent, *Mater. Lett.*, 79, 263–265 (2012).

[6] Yi-Chong Liao, Chi-Yen Huang, Glass foam from the mixture of reservoir sediment and Na_2CO_3, *Ceram. Int.*, 38, 4415–4420 (2012).

[7] R.K. Chinnam, A.A. Francis, J. Will, E. Bernardo, A.R. Boccaccini, Review. Functional glasses and glass-ceramics derived from iron rich waste and combination of industrial residues, *J. Non-Cryst. Solids*, 365, 63–74 (2013).

[8] E. J. Minay, P. Veronesi, V. Cannillo, C. Leonelli, A. R. Boccaccini, Control of pore size by metallic fibres in glass matrix composite foams produced by microwave heating, *J. Eur. Ceram. Soc.*,24, 3203–3208 (2004).

[9] A.R. Boccaccini, P. Veronesi, C. Leonelli, Microwave processing of glass matrix composites containing controlled isolated porosity, *J. Eur. Ceram. Soc.*, 21, 1073-1080 (2001).

[10] Bo Wang, K. Matsumaru, J. Yan, Z. Fu, K. Ishizaki, Mechanical behavior of cellular borosilicate glass with pressurized Ar-filled closed pores, *Acta Mater.*, 60, 4185–4193 (2012).

[11] A.A. El-Kheshen, M.F. Zawrah, M. Awaad, Densification, phase composition, and properties of borosilicate glass composites containing nano-alumina and titania, *J. Mater. Sci: Mater. Electron.* 20, 637–643 (2009).

[12] E.E. Hamzawy, A.A. El-Kheshen and M.F. Zawrah, Densification and properties of glass/cordierite composites, *Ceram. Int.*, 31, 383–389 (2005).

[13] R.G.C. Beerkens, Modelling the Kinetics of Volatilization from Glass Melts, *J.Am.Ceram.Soc.*, 84, 1952-60 (2001).

[14] M. Asano, T. Harada, Y. Mizutani, Vaporization of glasses in the R_2O-Cs_2O-B_2O_3-SiO_2 (R = Li, Na, K and Rb) system, *J. Mater. Sci.*, 26, 399–406 (1991).

THE DURABILITY OF SIMULATED UK HIGH LEVEL WASTE GLASS COMPOSITIONS BASED ON RECENT VITRIFICATION CAMPAIGNS

Mike T. Harrison
National Nuclear Laboratory
Sellafield, Seascale, CA20 1PG, United Kingdom

Carl J. Steele
Sellafield Ltd
Sellafield, Seascale, CA20 1PG, United Kingdom

ABSTRACT

The Waste Vitrification Plant (WVP), operated by Sellafield Ltd, converts Highly Active Liquor (HAL) from the reprocessing of spent nuclear fuel in the UK into a vitrified product suitable for safe interim storage with a view to its ultimate disposal in a geological disposal facility. One of the important aspects to understand in relation to the disposal of HLW glass is its behavior over long timescales in aqueous conditions and, to this end, a program of non-active laboratory-based leach tests is currently being undertaken. The aim of these tests is to determine the envelope of behavior of the expected range of WVP glass compositions using a standard test protocol, i.e. the ASTM Product Consistency Test (PCT).

This paper describes the results from several series of tests that investigate the PCT response over 182 days of a number of glass compositions typical of those currently produced on the WVP. The waste incorporations of these glasses range from ~26 to 31 wt% with lithia contents of ~3.8 to 4.5 wt%. In addition to these laboratory glasses, a number of samples taken from the glass residue remaining at the base of a drained melter from the full-scale inactive Vitrification Test Rig (VTR) were tested. The aim of this was to investigate the effect on durability of settled solids and the potential enrichment of certain species in the heel of glass at the base of the melter.

INTRODUCTION

In the UK highly radioactive liquors from the reprocessing of spent nuclear fuel are converted into a borosilicate glass at the Sellafield Waste Vitrification Plant (WVP). The glass matrix provides a stable and durable waste form suitable for safe long-term storage and subsequent disposal. Two types of highly active liquors (HAL) are vitrified, which result from the reprocessing of Magnox and Oxide (Advanced Gas-cooled Reactor, AGR, and Pressurised Water Reactor, PWR) spent fuel. The Magnox HAL is relatively low in fission products but contains significant amounts of Al and Mg from the fuel cladding. Oxide HAL is of higher burn-up and has higher fission product content, but lacks any Al or Mg. In order to produce a product glass of acceptable quality the Oxide feed has to be blended with Magnox. Historically, the standard Blend ratio is 75o:25m by weight, where 'o' and 'm' refer to Oxide and Magnox respectively. Lower Oxide to Magnox ratios, e.g. 50o:50m, are likely to be implemented on WVP in the near future depending on HAL availability. However, if the Oxide content is >80% then the lower Al concentration in the glass will reduce the aqueous durability.

Since the WVP product glass cannot be sampled and analysed it is manufactured to a Quality Assured Process Specification, with historic targeted waste oxide incorporation in 'MW' base glass* of 25 wt% for both Magnox and 75o:25m Blend products. Recently, operational envelopes for higher incorporations have been determined using the full-scale inactive Vitrification Test Rig (VTR). This will result in fewer containers being produced, which will provide considerable savings not only operationally, but also for storage, transport and final disposal. VTR-fabricated Magnox and Blend product glasses up to 38 wt% waste oxides have been shown to have similar product quality and, as a result of this qualification work, higher incorporations have been implemented on WVP. The targeted waste oxide loadings for Magnox and 75o:25m Blend glasses in recent campaigns have been up to 32 and 28 wt% respectively.

The underpinning of product quality by long-term durability testing is also required for the final disposal of vitrified HLW glass in the UK. In addition, in order to allow WVP to move to higher waste incorporations in the short term it is necessary to demonstrate that new compositions have no worse durability under standard test conditions than the envelope of existing vitrified product, Hence, the range of behaviour of existing vitrified product under the selected test conditions needs to be defined for comparison.

The majority of the simulated HLW glasses fabricated in the lab and at full-scale on the VTR uses inactive HAL with compositions based on representative Magnox or Oxide fuel reprocessing flowsheets. The current compositions are defined as 'WRW17' for Magnox (4.8 GWd/teU), and 'WRW16' for a 75o:25m Blend with 26.56 GWd/teU Oxide (AGR, BWR, PWR) and 4.8 GWd/teU Magnox. Previous studies using these simulants[1, 2] have shown that the long-term durability of simulated UK vitrified HLW in de-ionised water (DIW), as measured by the ASTM Product Consistency Test (PCT)[3], depends on a number of compositional factors, i.e.

- waste incorporation,
- waste type (Magnox or Blend) and blend ratio, and
- lithia content.

The lithia content in particular has a significant influence on the durability, and since its concentration can be quite variable in the actual HLW vitrified product it is important to ensure this is as representative as possible in any inactive simulant.

In addition to the major compositional variations listed above, the HAL fed to WVP for vitrification will also deviate from the representative WRW-based simulants as a result of variations in the spent fuel type, burn-up, cooling time before reprocessing, etc. Hence, in order to investigate the effect on durability of potential deviations from the standard simulants, a series of five non-active glasses based on the calculated compositions of recent WVP campaigns underwent standard 182 day duration PCTs.

There may also be some significant deviations from the WRW simulant compositions as a result of settling of insoluble material in the melter. The presence of such material in WVP vitrified product could affect its long term durability. Hence, some samples of highly-enriched residue material taken from used VTR melters also underwent standard 182 day duration PCTs.

BACKGROUND

Recent WVP Campaign Glass Compositions

The five glass compositions from recent WVP campaigns were calculated using data provided by Sellafield Ltd for elements present in the HA liquor feed along with the average calculated incorporation rate. In order to simplify the composition from ~100 elements down to ~30 for the non-active analogues, anything present in insignificant quantities ($\leq$0.018% oxide fraction) was neglected with Nd replacing any actinides still present, and Re substituting for Tc on a molar basis. The resulting compositions are shown in the table below. The main intention of these studies was to ensure that the glasses tested had Li, Al and Mg, which are known to have a significant impact on durability, tailored towards their concentration in actual current waste feeds rather than in the WRW16 and 17 simulants. 28 wt% Magnox and 75o:25m Blend glass based on the WRW17 and WRW16 simulants respectively are also shown in Table I for comparison.

Table I. Summary of the compositions used for the non-active analogues of vitrified HLW produced during recent WVP Campaigns (1 to 5). WRW17- and WRW16-based 28 wt% Magnox and 75o:25m Blend compositions are also shown for comparison.

WVP Campaign Waste Oxide (wt%)	1 Magnox	2 Blend	3 Magnox	4 Blend	5 Magnox	WRW17 Magnox	WRW16 Blend
Al_2O_3	6.67	1.59	6.08	1.43	5.02	5.18	2.00
B_2O_3	15.37	16.48	16.11	16.73	16.42	16.26	16.26
BaO	0.73	0.92	0.68	1.03	0.67	0.56	1.06
CeO_2	1.21	1.24	0.94	1.41	0.98	1.33	1.55
Cr_2O_3	0.53	0.58	0.49	0.50	0.54	0.70	0.52
Cs_2O	1.02	1.12	0.98	1.24	0.97	1.25	1.74
Eu_2O_3	0.00	0.06	0.00	0.07	0.00	-	-
Fe_2O_3	3.33	3.10	3.22	2.57	3.22	3.44	2.24
Gd_2O_3	0.58	2.77	0.49	2.30	0.78	-	4.50
La_2O_3	0.52	0.57	0.49	0.55	0.45	0.68	0.76
Li_2O	4.45	4.23	3.93	4.07	3.76	3.90	3.90
MgO	6.02	1.41	5.48	1.33	4.52	5.49	1.76
MnO_2	0.09	0.00	0.08	0.00	0.06	-	-
MoO_3	1.59	2.14	1.41	2.14	1.68	1.69	2.48
Na_2O	7.47	8.01	7.83	8.14	7.98	8.02	7.95
Nd_2O_3	2.03	2.56	1.78	2.36	1.93	2.19	2.54
SO_3	0.28	0.11	0.17	0.12	0.00	0.12	0.03
P_2O_5	0.28	0.41	0.29	0.49	0.37	0.19	0.48
NiO	0.38	0.48	0.37	0.41	0.39	0.43	0.33
PdO	0.38	0.66	0.24	0.80	0.40	-	-
Pr_6O_{11}	0.51	0.59	0.48	0.67	0.44	0.65	0.75
Rb_2O	0.14	0.17	0.13	0.18	0.12	-	-
ReO_2	0.67	1.02	0.76	0.96	1.09	-	-
Rh_2O_3	0.33	0.58	0.37	0.49	0.22	-	-
RuO_2	0.93	1.21	0.95	1.22	0.93	0.91	0.82
SiO_2	41.85	44.60	43.80	45.29	44.56	44.17	43.93
Sm_2O_3	0.38	0.43	0.36	0.48	0.33	0.45	0.53
SrO	0.30	0.35	0.29	0.40	0.30	0.33	0.48
TeO_2	0.21	0.26	0.20	0.29	0.19	0.20	0.28
Y_2O_3	0.23	0.26	0.21	0.28	0.19	0.21	0.33
ZrO_2	1.53	2.10	1.36	2.05	1.50	1.65	2.79
Total	100.0	100.0	100.0	100.0	100.0	100.0	100.0
Waste Oxide Incorporation (wt%)	30.9	26.7	28.3	25.8	27.3	28.0	28.0

Heel Residue Glass

The WVP melter crucible is designed to operate with a heel of about 70 kg of glass that is not poured under normal operations. The purpose of the heel is to help with heat transfer to the next feed cycle, and is achieved by locating the entrance of the pour nozzle ~14 cm above the base of the melter crucible.

Over a number of feed cycles and depending on melter operating conditions, e.g. temperature and sparge rate, some of the insoluble material (platinoids, spinels, etc.)** in the melt can settle into the heel increasing its viscosity and potentially causing problems with glass pouring. In addition, the phase-separated molybdate-based material known as 'yellow phase',*** which tends to float on the surface of the melt due to its lower density, can also build up in the heel over time. Hence, periodic 'drain pours' through the melter drain nozzle located at the base of the melter crucible were historically carried out on WVP. Note that following the introduction of multi-sparge melters on the WVP, the issue of settled solids in the heel has largely been removed, and periodic drain pours are no longer carried out.

However, from inspection of used VTR melters, there is evidence of solids settling into the heel. For example, the single-sparged VTR melter used for early campaigns contained a quantity of material that had not drained with the final pour, see Figure 1. The analysis of this heel residue (HR1), see Table II, showed very high waste incorporation (~60 wt%) and increased concentrations of Cr, Mo, Ni and Ru, which indicated the settling of undissolved spinels, yellow phase and RuO_2. It appears that these species accumulated in 'static' regions of the melter where the single sparge agitation was less effective, increasing the viscosity up to the point where the material could not be drained.

Figure 1. Glass heel residue (HR1) remaining around the pour nozzle cloche in the sectioned VTR Melter 1.

A similar inspection was carried out on a later VTR melter, which also revealed the presence of heel residue glass on its base following the final drain pour despite being triple-sparged, see Figure 2.

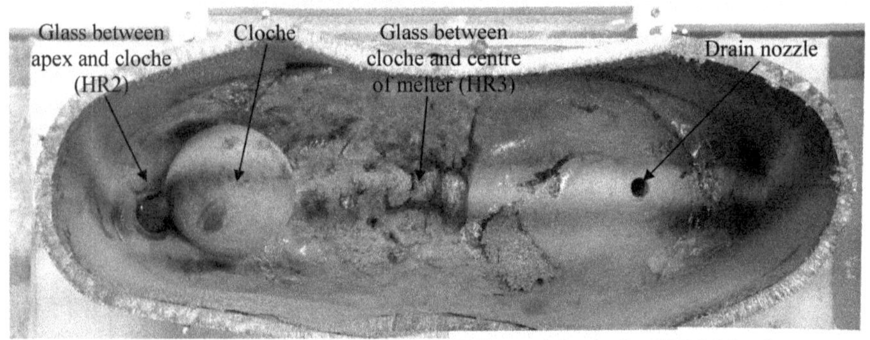

Figure 2: Glass heel residues (HR2 and HR3) remaining in the VTR Melter 2.

The majority (~9.5 kg) of this residue was between the pour nozzle cloche and the centre of the melter in an area that was relatively unsparged. Analysis of the glass between the cloche and centre of the melter (HR3) showed a significant enrichment in Ru and high waste incorporation, see Table 2, plus small increases in Fe, Cr and Ni compared to the preceding drain pour.

In addition, there was a small amount (~2.4 kg) of heel residue glass between the melter apex and cloche (HR2), see Figure 2. Analysis of this material showed a very similar composition and waste incorporation to the drain pour from the melter with only a slight enrichment in Ru. Hence, it was considered that this glass did not drain due to the enriched heel residue material (HR3) blocking its path rather than increased viscosity.

One of the concerns about these heel residues from a product quality point of view is that some of this material may be present in drain pours and waste resulting from melter disposal, which could affect the aqueous durability of the glass. In particular, the long-term durability under PCT conditions could be degraded by the extremely high waste incorporation and the presence of increased concentrations of Cr, Mo, Ni and Ru. Hence, standard 182 day PCT experiments were carried out on the three samples of heel residues (HR1 to HR3) described above, and summarised in Table II.

Table II. Summary of the heel residue samples (HR1 to HR3) with compositional analysis. Enriched species are indicated *.

Heel Residue Sample Waste Oxide (wt%)	HR1 *Melter 1 below surface*	HR2 *Melter 2 between apex and cloche*	HR3 *Melter 2 between cloche and centre*
Al_2O_3	1.5	4.7	8
B_2O_3	11	15	10.7
BaO	1.2	0.48	0.81
CeO_2	1.6	1.3	2.5
Cr_2O_3	8.9*	1.9	1.8
Cs_2O	1.4	1.1	1.8
Fe_2O_3	4.2	4.2	6.5
Gd_2O_3	2.9	0.18	0
La_2O_3	0.78	0.6	0.99
Li_2O	2.3	4.1	3.3
MgO	1.6	5.3	9.1
MoO_3	3.4*	1.5	2.5
Na_2O	4.2	8	6
Nd_2O_3	2.2	1.8	2.9
NiO	15.5*	1.5	1.2
Pr_2O_3	0.69	0.53	0.88
RuO_2	7.2*	1.7	5.2*
SiO_2	26.4	43.3	31.6
Sm_2O_3	0.5	0.4	0.66
SrO	0.53	0.29	0.46
TeO_2	0.3	0.18	0.32
Y_2O_3	0.3	0.18	0.3
ZrO_2	2.6	1.7	3
Total	101.2	99.94	100.52
Waste Incorporation (wt%)	57.3	29.5	48.9

EXPERIMENTAL

The five recent WVP campaign glasses were fabricated according to the target compositions in Table I. The appropriate quantities of oxides, carbonates, nitrates, phosphates, sulphates and 'MW-½Li' base glass frit for 400 g melts were weighed out and thoroughly mixed using a rotary powder mixer. Table III shows the oxide components in the glasses and the corresponding reagent used for the fabrication.

Note that 'MW-½Li', which has a composition of 62.7 SiO_2, 23.2 B_2O_3, 2.75 Li_2O, and 11.3 Na_2O wt%, is the standard base glass feed used on the WVP. It contains only half the lithia content required in the final product glass, with the rest added as $LiNO_3$ solution with the HAL feed. Historic development work showed that this strategy minimised the formation of unreactive refractory oxides and reduced the amount of dust formed. For 25 wt% incorporation product, a 50:50 split of the Li between the HAL and the MW-½Li base glass gives a 1:1 Li:Na mole ratio and a final Li_2O content of 4 wt%. For higher incorporation product, the amount of $LiNO_3$ added is ideally that required to maintain equi-molar Li and Na in the product, resulting in less being added to the HAL and a final Li_2O content of <4 wt%. However, due to WVP operational requirements, variable $LiNO_3$ dosing produces a range of lithia contents and Li:Na ratios in the product glass. Historically, the dosing strategy for higher incorporations (>25 wt%) has generally yielded Li_2O contents >4 wt% and Li:Na molar ratios >1, e.g. 1 and 2 in Table I. However, the $LiNO_3$ dosing strategy in the most recent WVP campaigns has targeted the ideal 1:1 Li:Na ratio in the product, resulting in the lower Li_2O contents for 3 to 5 shown in Table I.

Table III: List of oxide components with the corresponding reagents used for the five recent WVP campaign glasses.

Oxide	Reagent	Oxide	Reagent	Oxide	Reagent
Al_2O_3	Al_2O_3	MgO	$MgCO_3$	ReO_2	Re_2O_7
BaO	$BaCO_3$	MnO_2	MnO_2	Rh_2O_3	$Rh_2O_3.5H_2O$
CeO_2	CeO_2	MoO_3	MoO_3	RuO_2	RuO_2
Cr_2O_3	$Cr(NO_3)_3.9H_2O$	Nd_2O_3	Nd_2O_3	SiO_2	SiO_2
Cs_2O	Cs_2CO_3	SO_3	$(NH_4)_2SO_4$	Sm_2O_3	Sm_2O_3
Eu_2O_3	Eu_2O_3	P_2O_5	$(NH_4)H_2PO_4$	SrO	$SrCO_3$
Fe_2O_3	Fe_2O_3	NiO	NiO	TeO_2	TeO_2
Gd_2O_3	Gd_2O_3	PdO	PdO	Y_2O_3	Y_2O_3
La_2O_3	La_2O_3	Pr_6O_{11}	Pr_6O_{11}	ZrO_2	ZrO_2
Li_2O	Li_2CO_3	Rb_2O	Rb_2CO_3		

The mixed powders were melted in silica crucibles at 1050 °C for 4 hours with stirring after 3 hours. The melts were then poured into brass moulds to form ingots, and then annealed at 500 °C for 3 hours before cooling to room temperature at 0.5 °C/min.. The bulk densities of the glasses were measured according to the Archimedes displacement method in water, with samples of glasses 1 and 2 sent for external analysis in order to verify the target compositions shown in Table I. The heel residue samples (HR1 to HR3) were used as received following their mechanical removal from the used VTR melters, with the analysis shown in Table II having been carried out previously.

Leaching experiments were carried out on all of the glasses described above using a protocol based on the ASTM Standard Product Consistency Test (PCT).[3] Powders were prepared by first crushing the glass ingots or pieces using a stainless steel percussion mortar, and then milling using a Cole Palmer Analytical Mill with a silicon carbide blade. The resulting powders were sieved to obtain the 75 – 150 μm size fraction and then washed to remove fines adhering to the surface. Note that the glass samples sent for compositional analysis were taken from the unwashed <75 μm sieved size fraction.

The sieving was performed on a mechanical sieve shaker (Endecotts Octagon Digital) with 200 mm (8") diameter sieves using a 30 min programme at amplitude 9. This was repeated a number of times (at least 3) until the weight of material passing through the lower 75 μm sieve was <0.2 g.

The washing process, which was slightly modified from the ASTM procedure in order to avoid any pre-leaching of the glasses, consisted of forcibly squirting de-ionised water (DIW) on to the powder in a beaker to a volume ~2.5 times that of the powder and stirring with a nickel spatula.[****] This was then placed in an ultrasound bath for 3 mins before decanting off the DIW. Absolute ethanol was then forcibly squirted into the beaker, stirred with a nickel spatula, ultra-sounded for 3 minutes, and decanted, with two further repeats. Finally, the cleaned powder was dried in an oven at 90 °C overnight.

All of the leach tests in this study were performed in PFA TFE-fluorocarbon screw-lidded 60 ml jars (Savillex Corporation), cleaned according to the ASTM procedure.[3] For previously-used jars, glass residues from previous wasteform testing were removed by rinsing with DIW using at least 3 vessel volumes for each container. The vessels and lids were soaked in 0.16 M (1 wt%) HNO_3 at 90 ± 10 °C for 1 h, and then rinsed with at least 3 vessel volumes of fresh DIW at ambient temperature. The vessels and lids were then soaked in fresh DIW at 90 ± 10 °C for 1 h before being filled 80-90% full of fresh DIW at ambient temperature, the lids screwed tightly closed, and left in an oven at 90 ± 10 °C for a minimum of 16 h. The jars were then removed from the oven and the pH of an aliquot of the DIW measured. If the pH was between 5 and 7, then the vessels and lids were dried at 90 ± 10 °C for a minimum of 16 h and stored ready for us. If the pH was outside this range, then the cleaning process was repeated.

The tests were set up by weighing 4.0 g of the glass powder into the PFA test vessel, adding 40.0 ml DIW (18 MΩ·cm) via pipette, tightly screwing on the lid, and then gently swirling to ensure full wetting of the glass sample. The test vessels were then weighed prior to placing in an oven at 90 °C. The screw lids were periodically tightened over the first 24 hours in order to minimise the potential for evaporative losses. The tests were carried out in triplicate along with a 'blank', i.e. a jar with just ~40 ml DIW and no glass powder, for standard durations of 7, 14, 21, 28, 42, 56, 84, 112 and 182 days. At each duration, the leach test vessels were removed from the oven, allowed to cool, weighed to check the evaporative losses, and then two ~10 ml aliquots removed for analysis. The first was used for measuring the pH of the leachate solution. The second was filtered using a 0.45 μm syringe filter and acidified with concentrated nitric acid to reduce the pH to <7 for elemental analysis. For most of the samples, 1 volume % (~0.1 ml) was sufficient, but for some of the highly concentrated Magnox glass leachates, double this was required.

All of the leachate solutions were analysed by ICP-OES for Al, B, Cr, Li, Mg, Mo, Na and Si, and normalised elemental mass losses NL(i) calculated according to equation (1):

$$NL(i) = \frac{c_i}{f_i(S/V)} \tag{1}$$

where c_i is the concentration of element i in the leachate in ppm, f_i is the mass fraction of element i in the unleached wasteform, and S/V is the surface area to volume ratio in m^{-1} calculated according to the ASTM procedure[3] using the actual glass densities (as measured using the Archimedes method), weights, and leachate volumes. The normalised boron mass loss, NL(B), was used as an indicator for general glass dissolution.

RESULTS AND DISCUSSION

Recent WVP Campaign Glasses

Table IV compares the target and analysed compositions for the WVP recent campaign simulated glasses 1 (Magnox) and 2 (Blend) fabricated in the laboratory. Note that B, Li and Ru were analysed via ICP-OES, all other elements by XRF.

Table IV: Comparison of the target and analysed compositions for WVP recent campaign simulated glasses 1 (Magnox) and 2 (Blend) fabricated in the laboratory.

WVP Campaign		1			2	
Waste Oxide (wt%)	Target	Analysis	%	Target	Analysis	%
Al_2O_3	6.67	7.02	105	1.59	1.77	111
B_2O_3	15.37	15.60	101	16.48	16.70	101
BaO	0.73	0.75	103	0.92	0.94	103
CeO_2	1.21	1.27	105	1.24	1.29	104
Cr_2O_3	0.53	0.65	122	0.58	0.71	122
Cs_2O	1.02	0.75	73	1.12	0.72	65
Eu_2O_3	0.00	0.01	-	0.06	0.06	103
Fe_2O_3	3.33	3.55	107	3.10	3.39	109
Gd_2O_3	0.58	0.57	99	2.77	2.28	82
La_2O_3	0.52	0.49	95	0.57	0.55	97
Li_2O	4.45	4.10	92	4.23	3.86	91
MgO	6.02	5.31	88	1.41	1.25	88
MnO_2	0.09	0.10	107	0.00	0.02	-
MoO_3	1.59	1.53	96	2.14	2.03	95
Na_2O	7.47	7.83	105	8.01	8.37	104
Nd_2O_3	2.03	1.53	75	2.56	2.50	98
SO_3	0.28	0.05	18	1.21	0.00	0
P_2O_5	0.28	0.26	94	0.41	0.38	93
NiO	0.38	0.39	102	0.48	0.55	115
PdO	0.38	0.08	21	0.66	0.26	39
Pr_6O_{11}	0.51	0.43	85	0.59	0.32	54
Rb_2O	0.14	0.15	106	0.17	0.16	94
ReO_2	0.67	0.21	31	1.02	0.32	31
Rh_2O_3	0.33	0.15	46	0.58	0.43	74
RuO_2	0.93	0.73	79	1.21	1.09	90
SiO_2	41.85	43.44	104	44.60	46.49	104
Sm_2O_3	0.38	0.32	84	0.43	0.42	97
SrO	0.30	0.32	106	0.35	0.39	110
TeO_2	0.21	0.19	91	0.26	0.21	82
Y_2O_3	0.23	0.23	101	0.26	0.27	105
ZrO_2	1.53	1.61	105	2.10	2.20	105
Total	100.0	99.7	100	100.0	100.0	100
Waste Loading (wt%)	30.9	28.7	93	26.7	24.5	92

There is generally good agreement between the target and analysed compositions (between 90 and 110%), although there are a number of elements that show systematic discrepancies. For example, the Cs, Pd, Pr, Re, Rh, and S analyses are low for both glasses 1 and 2, which has resulted in the analysed waste incorporation being only 92-93% of that targeted. For the platinoids (Pd and Rh), which are present as metallic inclusions due to their insolubility in the glass, the discrepancy could be due either to some settling occurring during melting, or incomplete dissolution of the sample prior to the analysis carried out.

The analysed values of Cs_2O and ReO_2 are consistently lower than the calculated value in all the glasses. Both Re, which is used as a surrogate for Tc, and Cs are known to volatilise from alkali borosilicate glass melts forming $CsReO_4$ / $CsTcO_4$ as one of their volatile products.[6] Hence, for the Cs and Re the discrepancy is likely to be due to volatilisation during melting at 1050 °C.

The analysed Li_2O and MgO contents are also only ~90% of the target value, which may be due to systematic errors in the analysis of these elements (Li in particular can be difficult to analyse correctly). However, if correct, then this will have a significant effect on the PCT response as Li and Mg are key elements for determining the long term dissolution rates of UK Magnox glasses, and the compositions of glasses 1 to 5 may not be completely representative.

Figure 3 compares the normalised boron mass loss, NL(B), up to 182 days for the recent WVP campaigns glasses 1 to 5 (waste incorporation and Li_2O contents are the targeted values).

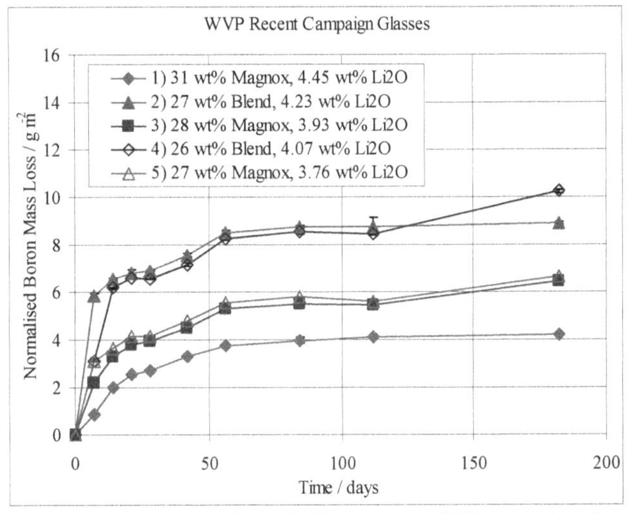

Figure 3: Normalised boron mass loss, NL(B), for recent WVP campaign glasses 1 to 5.

The NL(i) values for the three Magnox compositions (1, 3 and 5) are, in general, lower than those for the two 75o:25m Blends (2 and 4). In addition, of the three Magnox glasses, composition 1, which has the highest incorporation, has lower NL(B) than 3 and 5, which have similar PCT responses and similar waste loadings. For comparison, the PCT response under identical conditions (DIW at 90 °C, S/V ~2,000 m^{-1}) for a 25 wt% Magnox glass made at full-scale on the VTR using WRW17 simulant is shown in Figure 4. The WVP representative glasses 1, 3 and 5 all have lower NL(B) values than this 'standard' UK composition, indicative of superior aqueous durability under these test conditions up to 182 days. Note that a general observation from previous studies is that the durability of UK Magnox glasses tends to increase with incorporation, decrease with Li_2O concentration, and is quite sensitive to small compositional variations.[1, 2]

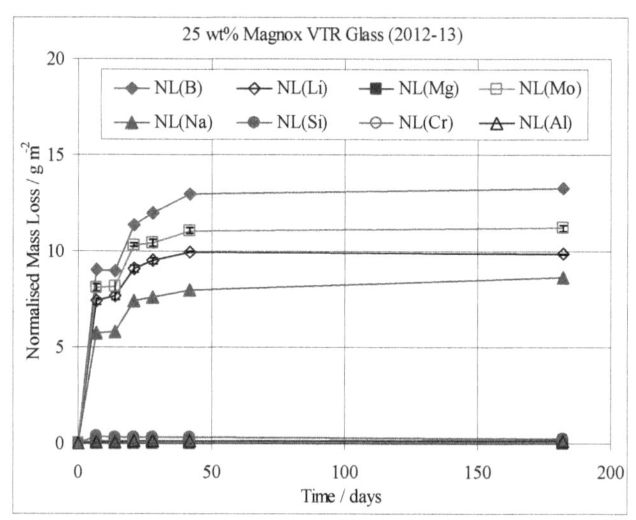

Figure 4: PCT response of a 25 wt% Magnox glass fabricated at full scale on the VTR using WRW17 simulant.

In addition, the Magnox glasses 1, 3 and 5 have lower NL(B) values to similar equivalent representative glasses made in previous studies using the WRW17 simulant composition. For example, Figure 5 compares glass 3 to one with similar waste incorporation and Li_2O content made in the laboratory with WRW17 calcine from the VTR. It can be seen from Figure 5 that not only does composition 3 generally have lower NL(i) values up to 182 days than the equivalent WRW17 glass, but the relative releases of Mo and Cr are also significantly higher in the latter. There is a similar pattern for compositions 1 and 5, which also have relatively low Mo and Cr releases when compared to WRW17-based glasses.

The origin of this difference is currently unclear. It may be the result of small differences in composition between the WRW17 simulant and the composition of HA liquor fed to WVP in recent campaigns. For example, glass 3 has slightly higher Al_2O_3 and MgO than an equivalent WRW17-based glass. Similarly, if the analysis of glass 1 shown in Table IV is correct, then the lower MgO and Li_2O than the target values could, in theory, yield a more durable glass than expected. Indeed, if the glass 1 target composition is correct, then the presence of ~4.5 wt% Li_2O would be expected to increase the NL(i) values significantly, although this would be off-set by the higher waste loading.[1, 2] For composition 3, if the actual Li_2O content is only ~90% of that targeted, i.e. ~3.5 wt%, then this could account for the lower NL(i) values shown in Figure 5.

However, these compositional variations are not systematic and it is difficult extracting from the data which species could be causing the observed lower NL(i) values. It could also be a result of differences in glass properties as a result of using separate chemical components vs a calcine, e.g. the amount of agitation from the decomposition of carbonates, the reactivity of waste components with the melt, the amount of crystalline species forming (spinels), etc.

For the two Blend glasses (2 and 4), the PCT responses are very similar to each other, and also to those reported in previous studies[1, 2], i.e. a 'levelling-off' of NL(B) after ~50 days at ~8 - 10 g m^{-2}. Figure 6 compares composition 4 with a similar (in terms of waste loadings and Li_2O content) glass made with WRW16-based calcine.

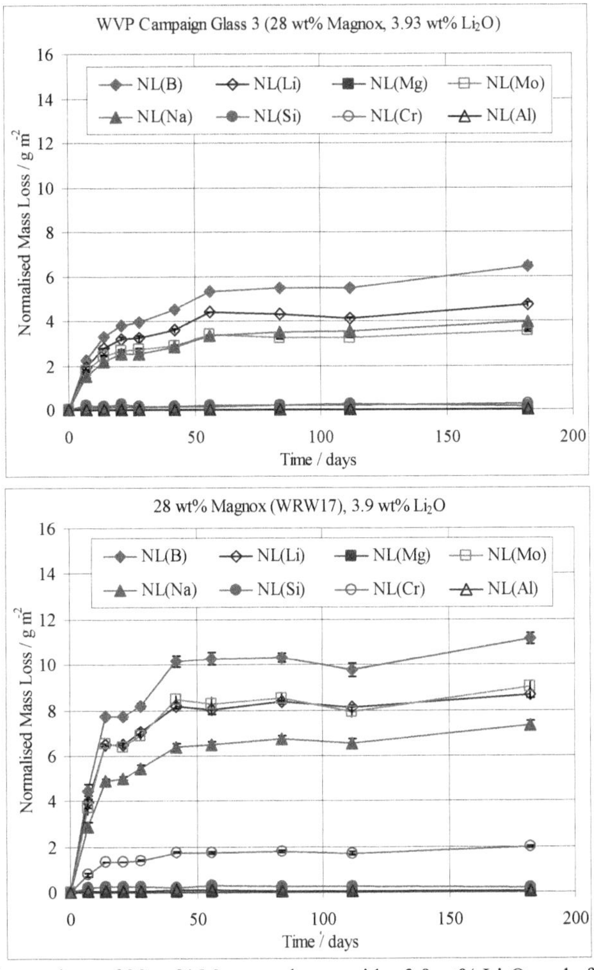

Figure 5: Comparison of 28 wt% Magnox glasses with ~3.9 wt% Li_2O made from separate components (composition 3) and WRW17 calcine.

The two glasses show very similar NL responses for most of the elements tested (with the exception of Cr) despite the differences in composition shown in Table I, e.g. the WRW16 glass contains significantly higher Gd, higher Al and Mg, but lower Fe. Note that the higher NL(Cr) values for the WRW16 glass may be a result of different Cr speciation in the calcine compared to using separate components. However, the general observation, which is that 75o:25m Blend glasses are less sensitive to variations in composition, fabrication and test conditions than their Magnox counterparts, is consistent with previous studies.[1, 2]

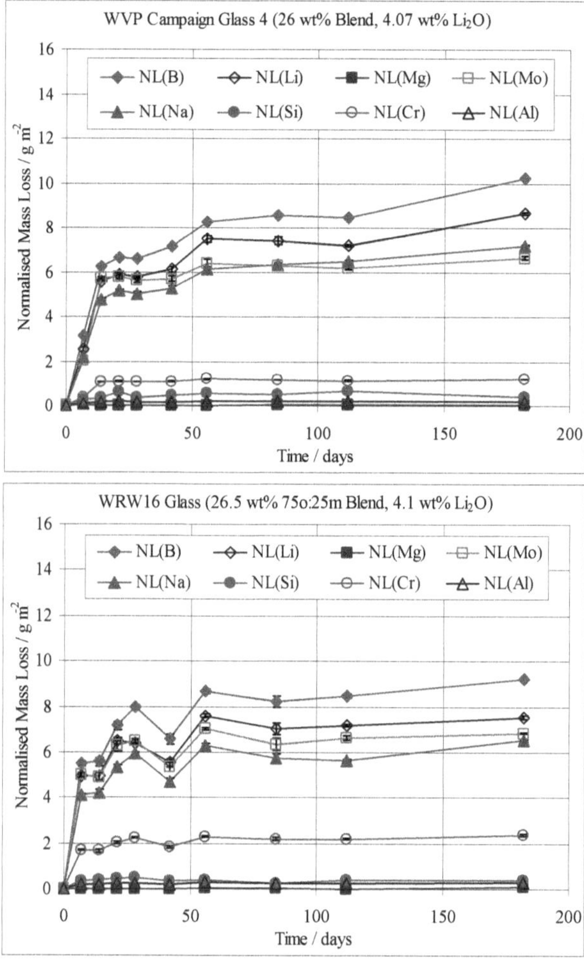

Figure 6: Comparison of similar 75o:25m Blend glasses made from separate components (glass 4) and WRW16 calcine.

Heel Residue Glasses

Figure 6a shows the PCT response (NL(i) vs duration) up to 182 days for the three heel residue glasses (HR1 to HR3) taken from VTR melters 1 and 2, see Table II. From Figure 6a, it can be seen that for the standard range of elements measured in the current leach test studies, none of the heel residues exhibit unusually high NL(i) values up to 182 days in DIW at 90 °C. There are, however, a number of other noteworthy features:

- HR1, which is the most highly enriched heel residue sample with the highest waste loading, has almost congruent release of B, Li, Na and Mo, but zero release of Cr despite it being present in extremely high concentrations. This is likely to be due to Cr only being present in relatively high durability crystalline phases rather than the glass.

- HR2, which is considered to be the least enriched of the three heel residue samples, has a PCT response very similar to a 'standard' 25 wt% Magnox glass, albeit with slightly higher NL(i), see Figure 4 and previous studies.[1, 2, 4, 5] This is consistent with the suggestion that this residue was actually relatively normal glass that didn't drain due its pathway being blocked by the high viscosity HR3 material.
- HR3, which is only significantly enriched in RuO_2, has a relatively normal PCT response, with the exception of NL(Mo), which is, unusually, similar to NL(Si). This could be due to the presence of relatively high durability Mo-containing phases, e.g. powellites.

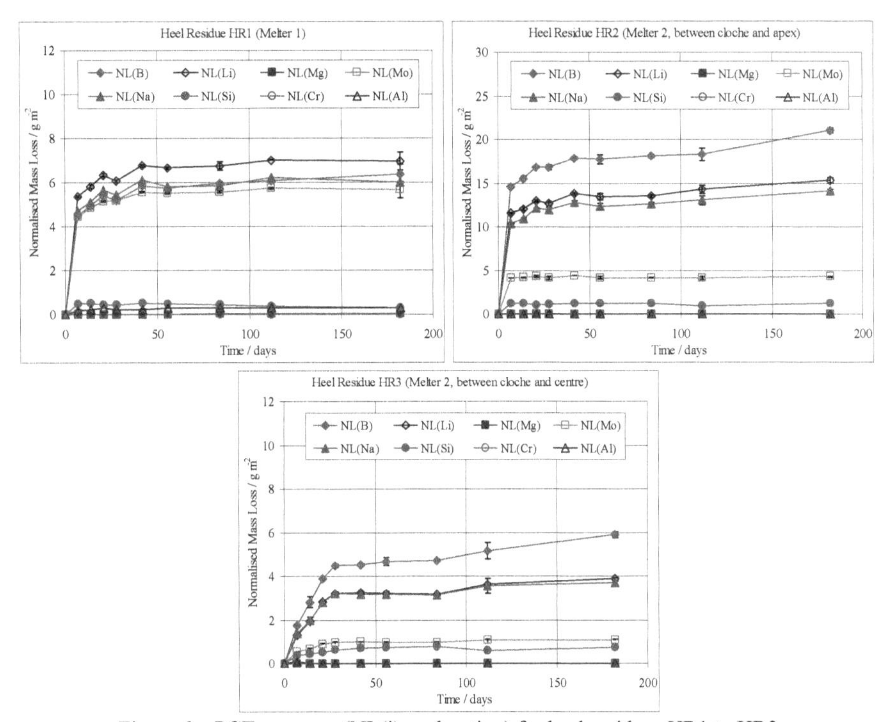

Figure 6a: PCT response (NL(i) vs duration) for heel residues HR1 to HR3.

These results indicate that, on the basis of a 182 day PCT at 90 °C in DIW, any heel residue material present in WVP vitrified product is unlikely to significantly increase elemental releases from the glass.

CONCLUSIONS

A number of non-active UK HLW glass compositions that are more representative of WVP product than ones based on standard simulants have been studied using static powder-based leach tests at 90 °C in de-ionised water for up to 182 days.

The test responses for the Magnox and 75o:25m Blend glasses with compositions based on recent WVP campaigns are all similar or better (as indicated by their lower NL(i) values up to 182 days) than equivalent glasses made with WRW16 and 17 VTR calcine. The Magnox glasses in particular show the greatest discrepancy. This is likely to be due either to the small differences in composition between the representative WVP glasses and their WRW17-based counterparts, deviations of the actual glass compositions from those targeted, or differences in the properties of the melt arising from using separate components rather than a pre-prepared calcine.

The PCT responses over 182 days of heel residue material from VTR melters do not show any significantly increased releases of the elements tested compared to 'standard' 25 wt% Magnox product glass based on WRW17 simulant. A number of lower elemental releases can be explained by the presence of relatively high durability crystalline phases. Future work will include powder X-ray diffraction in order to attempt to identify these phases.

ACKNOWLEDGMENTS

The NNL gratefully acknowledges the funding for this project provided by Sellafield Ltd and the NDA.

FOOTNOTES

* The 'MW' base glass used for the vitrification of HLW in the UK has a typical composition (in oxide weight %) of 21.9 B_2O_3, 5.3 Li_2O, 11.0 Na_2O and 61.8 SiO_2.

** Spinels of the type $M^{II}M^{III}{}_2O_4$, where M^{II} = Mg or Ni and M^{III} = Al, Fe or Cr, form in the melt due to the low solubility of Cr and are present in small quantities in WVP product glass. The noble platinoids (Ru, Rh, Pd) have very low solubilities in the melt and therefore tend to form metallic or oxide inclusions.

*** Yellow phase is a complex mixture of water soluble alkali, plus insoluble alkaline earth and rare earth molybdates, that can act as hosts for Tc-99 and Cs-137.

**** The initial PCT trials found that forcible squirting alone did not wet the powder at the bottom of the beaker adequately, resulting in residual fines and a 'crusty' finish to the powder.

REFERENCES

[1]C. Brookes, M. T. Harrison, A. Riley, and C. Steele "The effect of increased waste loading on the durability of high level waste glass", *Mat. Res. Soc. Symp. Proc.* **1265**, 109-114 (2010).

[2]Mike T. Harrison, Carl J. Steele & Andrew D. Riley "The effect on long term aqueous durability of variations in the composition of UK vitrified HLW product", *Glass Technol.: Eur. J. Glass Sci. Technol. A*, **53** (5), 211–216 (2012).

[3]ASTM C1285-02: Standard Test Methods for Determining Chemical Durability of Nuclear, Hazardous, and Mixed Waste Glasses and Multiphase Glass Ceramics: The Product Consistency Test (PCT).

[4] H. U. Zwicky, B. Grambow, C. Magrabi, E. T. Erne, R. Bradley, B. Barnes, Th. Graber, M. Mohos and L. O. Werme "Corrosion behaviour of British Magnox waste glass in pure water", *Mat. Res. Soc. Symp. Proc.* **127**, 129-136 (1989).

[5] E. Curti, J. L. Crovisier, G. Morvan, and A. M. Karpoff "Long-term corrosion of two nuclear waste reference glasses (MW and SON68): A kinetic and mineral alteration study", *Applied Geochemistry* **21**, 1152-1168 (2006).

[6] J. G. Darab and P. A. Smith, "Chemistry of technetium and rhenium species during low-level radioactive waste vitrification", *Chem. Mater* **8**, 1004-1021 (1996).

SCALED MELTER TESTING OF NOBLE METALS BEHAVIOR WITH JAPANESE HLW STREAMS

Keith S. Matlack[1]; Hao Gan[1]; Ian L. Pegg[1]; Innocent Joseph[2], Bradley W. Bowan[2], Yoshiyuki Miura[3], Norio Kanehira[3], Eiji Ochi[3], Tamotsu Ebisawa[3], Atsushi Yamazaki[3], Toshiro Oniki[4], Yoshihiro Endo[4]
1. Vitreous State Laboratory, The Catholic University of America, Washington, DC, United States.
2. Energy*Solutions*, Columbia, MD, United States.
3. JNFL, Rokkasho-mura, Aomori, Japan.
4. IHI Corporation, Yokohama-shi, Kanagawa, Japan.

ABSTRACT

The Rokkasho Reprocessing Plant, located in northern Japan, includes facilities for reprocessing spent nuclear fuel and conversion of the high-level waste (HLW) into a stable glass waste form by vitrification. The vitrification process centers on two 2.6 m^2 joule heated ceramic melters. A major consideration in the design of these melters is the large concentrations of noble metal fission products (Ru, Rh, Pd) in the projected HLW streams. These streams also contain significant amounts of molybdenum, which can lead to the formation of a molybdate salt phase ("yellow phase") during vitrification. In addition to the operational issues that can be caused by settling and accumulation of noble metals, noble metals can also influence yellow phase formation. In this study, an extensive series of tests was conducted at two melter scales (~1/100 (DM10) and ~1/20 (DM100)) to investigate various factors affecting noble metals settling and their influence on yellow phase formation. In particular, tests in the DM100 melter were able to replicate the "cold-bottom" mode of operation that is planned for the Rokkasho melters to reduce noble metals settling. Samples collected throughout the tests were used to monitor the noble metals distribution over time. Melt pool bubbling was found to be reasonably effective in suspending noble metals and the presence of noble metals tended to suppress yellow phase formation.

INTRODUCTION

The Rokkasho Reprocessing Plant, located in northern Japan, includes facilities for reprocessing spent nuclear fuel and conversion of the high-level waste (HLW) into a stable glass waste form by vitrification. The vitrification process centers on two 2.6 m^2 joule heated ceramic melters and is supported by a full scale pilot melter (KMOC), which is illustrated in Figure 1. The joule heated melters feature a conical bottom designed to minimize the accumulation of secondary phases, particularly those associated with noble metals. Glass is discharged through a bottom drain, which facilitates the removal of such phases. The melters employ a dual feeding system that introduces the liquid waste and additives in the form of glass beads separately. Heaters located in the vapor space maintain a target plenum temperature of 600°C. Separate power electrode in the upper and lower portions of the melter are used to establish high and low temperature zones in the glass pool; the temperature of the glass in the upper portion of the melter is about 1150°C while that in the lower portion is about 950°C to reduce the settling of noble metals. The HLW stream contains high concentrations of molybdenum, which can form secondary phases ("yellow phase") during vitrification[1-4], and noble metals, which tend to settle and deposit towards the bottom of the melter[5-6].

The formation of a molybdate yellow phase has been observed in certain tests performed with high level waste (HLW) simulants at IHI Corporation's mock-up vitrification test facility (KMOC) as well as in active commissioning tests at Rokkasho. Since the formation of such a phase is undesirable, IHI commissioned Energy*Solutions* and the Vitreous State Laboratory (VSL) of The Catholic University of America (CUA) to develop strategies to suppress the formation of yellow phase during waste vitrification[7-9]. Crucible melts and horizontal and vertical gradient furnace test were performed and followed by melter testing on the VSL DuraMelter 10 (DM10) to explore methods to suppress yellow phase formation. A mitigation approach was developed that involves the redistribution of some of the additives from the frit to the liquid waste (while maintaining a low suspended solids content in the blended waste stream) and modifying the frit composition in order to maintain the same glass product composition. This approach was demonstrated to be successful in both crucible-scale tests and DM10 melter tests. Subsequently, this approach was scaled up to the VSL DM100 melter system, which not only has a larger surface area than the DM10 (0.108 vs. 0.021 m^2) but is also capable of producing a similar vertical temperature profile in the melt pool to that of the Rokkasho melter. The strategy was also further developed to achieve significantly higher waste loading in the glass when the redistribution of frit components to the liquid waste was combined with redesign of the glass composition.

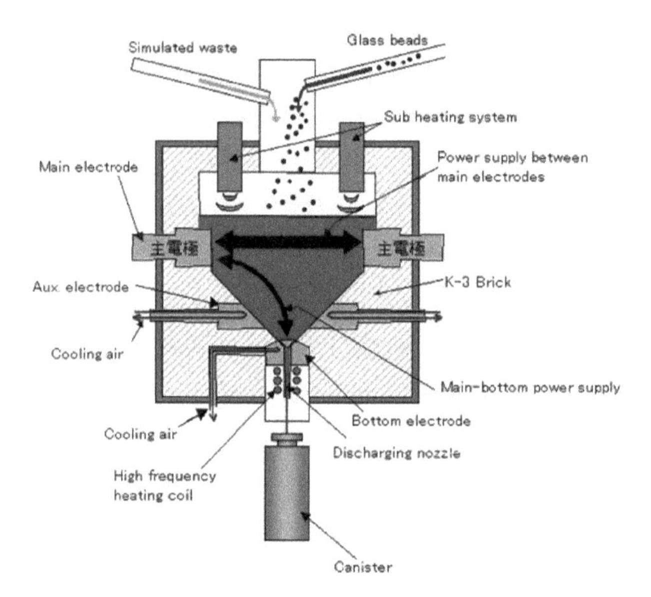

Figure 1. Schematic of IHI Corporation's full-scale mock-up (KMOC) melter system.

Testing on continuously fed melters at VSL has been used to address other issues associated with processing high level waste (HLW) simulants at the KMOC mock-up vitrification test facility as well as in active commissioning tests at Rokkasho. The use of melt pool bubbling was shown to increase glass production rates by as much as five fold while also having the potential to mitigate yellow phase formation on the melt pool surface. A series of tests was conducted on the DM10 melter to determine factors contributing to foaming and its relationship to the presence of di-*n*-butyl-phosphate (DBP) in the melter feed. During these tests, foaming observed during melter operations at the KMOC full-scale pilot facility was reproduced. The test also investigated potential foaming mechanisms and methods of foaming mitigation.

The ability to generate a vertical temperature profile in the DM100 melter that is similar to that in the full-scale KMOC and Rokkasho melter is important for studying yellow phase formation and noble metals settling. Two separate series of DM100 melter tests were conducted to determine the tendency towards secondary phase formation, noble metals behavior in the melt pool, and potential remediation strategies.

DM100 TEST METHODS

The DM100 melter used for these tests is a ceramic refractory-lined melter fitted with a pair of opposing Inconel 690 plate electrodes as well as a bottom electrode. The mode of electrode powering was determined based on pre-tests performed prior to the actual tests. Based on those tests, the side-to-side configuration, in which the bottom electrode is not powered, was selected in order to keep the temperature at the bottom of the melter closer to that found in the KMOC melter. The bubbler used for stirring the melt pool enters from the top, is removable and was only installed for turnover portions of the tests. The glass product is removed from the melter by means of an air-lift discharge system. The DM100 has a melt surface area of 0.108 m^2. The nominal depth of the melt pool is about 19 inches, which gives a typical glass inventory of about 115 kg. In the present tests, a dual feed system was employed to deliver the melter feed into the DM100 melter, analogous to KMOC melter. The liquid waste simulant and the glass frit are introduced into the melter through the same water-cooled vertical feed tube, the former via peristaltic pump and the latter through a mechanical screw feeder. The plenum is 27.5 inches in height and is fitted with heaters to maintain the target temperature. Temperatures are monitored by means of a series of thermocouples located in the melt pool, the electrodes, the plenum space, and the discharge chamber.

A variety of samples were taken throughout the tests to detect secondary phases and noble metals accumulations. Glass was discharged throughout the tests, sampled, analyzed, and inspected for secondary phases and noble metals. The surface of the melt pool was sampled by dipping a threaded rod into the melt pool. The floor and distinct horizons within the melt pool were sampled using suction tubes of various diameters depending on the desired amount of glass to be removed.

DM100 YELLOW PHASE MITIGATION TESTS

Four melter tests of 50 to 100 hours in duration were conducted on the DM100 melter using the modified frit composition and waste simulant compositions combined with a portion of the additives redistributed from the glass frit to the waste solution, as summarized in Table 1. Tests were conducted with and without noble metals to determine the effects on waste processing and the tendency to form yellow phase. The tests represent scale up from previously conducted tests on the DM10 melter system that demonstrated mitigation of yellow phase formation using a redistribution strategy, which involves replacing some of the components originally present in the glass beads by dissolved constituents in the waste. The simulants used in these tests are based on the actual waste composition ("AT waste") used in the commissioning tests of the full-scale vitrification facility at Rokkasho and employ rhenium as a surrogate for technetium. Unlike the smaller DM10, the DM100 was configured to simulate the low melt pool bottom temperatures typical of the full-scale vitrification facility at Rokkasho.

Over 960 liters of waste simulant with redistributed additives and 307 kg frit were processed to evaluate yellow phase formation with and without noble metals included in the waste simulant. Feed rates were scaled down from the 2.65 m^2 melt surface area of the full-scale vitrification facility at Rokkasho to the 0.108 m^2 surface area of the DM100 melter. Prior to each test, water was fed into the melter to achieve a plenum temperature of 600°C before starting the waste simulant and frit feed. In all but the first test, water was fed simultaneously with frit and waste then tapered off to maintain a 600°C plenum temperature as the cold cap was being established. Plenum heaters were used to prevent the plenum from dropping below 600°C during each test. The glass pool temperature was maintained at 1150°C throughout the tests, as indicated by thermocouples placed either eight or twelve inches from the melt pool floor. No bubbler was installed, therefore there was no glass pool bubbling during the four tests. After each test, the melt pool was extensively sampled for yellow phase, both on the melt pool surface and on the melt pool floor. In between each of the tests a variety of activities were performed including melt pool turnovers, extensive sampling to rid the melter of yellow phase, and review and analysis of test data; in addition, data were collected during idling periods of between about ten to sixty days.

Table 1. AT Waste Simulant Constituents, Contributions from Additives and Glass Frit, and Concentrations in Glass Product for Key Components.

Waste Simulant	Additives to Liquid Waste	Oxides from Glass Frit	Concentrations of Key Components in Product Glass (wt%)
Nitrates of Ag, Ba, Ce, Co, Cr, Cs, Eu, Fe, Gd, La, Na, Nd, Ni, Pd, Pr, Rb, Rh, Ru Sm, Sr, Y, Zn, Zr, perhenic acid, sulfuric acid, sodium molybdate, di-*n*-butyl phosphate	$Al(NO_3)_3 \cdot 9H_2O$ and silicic acid	Al_2O_3 B_2O_3 CaO Li_2O Na_2O SiO_2 ZnO	Cs_2O-0.98% MoO_3-1.50% Na_2O-9.89% PdO-0.49% Re_2O_7-0.42% Rh_2O_3-0.13% RuO_2-0.64% SO_3-0.26%

The tests demonstrated varying degrees of yellow phase formation and other observed behavior in response to changes in the feed compositions and operational conditions tested. Foaming was observed both in the feed tank upon mixing and on the cold cap while processing wastes that did not include noble metals. Waste with noble metals formed a more complete cold cap with fewer openings when comparing feeds processed at the nominal rates, indicating that the feed with noble metals is slower melting than that without. Only minor or trace amounts of yellow phase were identified on some dip samples from each test, indicating a general lack of any significant yellow phase on the melt pool surface. The minor amounts of yellow phase on the melt pool surface at the end of the tests suggest that yellow phase forms transiently in the cold cap but does not accumulate on the melt pool surface when assimilated into the glass at a rate comparable to the feed rate. No yellow phase was observed on any of the discharged glass. However, yellow phase was observed in suction samples taken at the conclusion of each of the four tests: tens of grams of yellow phase was collected after the first test without noble metals but trace amounts were observed in samples from the other three tests. Suction samples taken after subsequent idling periods showed more yellow phase, suggesting that either not all the yellow phase present after the tests was collected or that yellow phase settled during the idling period. The presence of noble metals in the feed and higher bottom glass pool temperatures decreased the amount of yellow phase detected on the bottom of the melter. In all previous tests on the DM10 with the same waste, redistribution strategy, frit composition, and glass composition, no yellow phase was observed on the bottom of the melter while the entire glass pool temperature was close to the 1150 and 1100°C target temperatures employed in those tests. Taken together, these results strongly suggest that the low temperature (< 950°C) at the bottom of the melter tends to promote the accumulation of yellow phase. However, these results demonstrate scale up of the redistribution approach, which prevented the formation and accumulation of yellow phase even when the bottom temperature is low when noble metals are present in the waste feed.

Extensive analysis of discharge glasses and samples from the surface and bottom of the melt pool was performed. The majority of the glass sample analysis indicates that both the discharge and melt pool samples approximate the intended composition. Yellow phase collected from the bottom of the melter was determined to be about half molybdenum oxide, fifteen percent soda, eight percent SO_3, two percent each of barium, calcium, chromium and cesium oxides, and two to twenty three percent rhenium oxide, depending on the concentration of noble metals in the feed, by mass on an oxide basis. Noble metals concentrations increased towards the target values in the melt pool during testing and were collectively as high as eight percent total oxides in samples removed from the bottom of the melter. Measured noble metals concentrations in the various samples indicate that noble metals are primarily in suspension in the glass while processing the waste and can be kept in suspension with bubbling; however, noble metals quickly deposit on the melter floor during idling without bubbling, as illustrated in Figure 2. After the test conducted with noble metals between 10/17-21, noble metals remained suspended in the melt pool until bubbling was discontinued. A resumption of bubbling resulted in the noble metals being re-suspended. Samples with high concentrations of noble metals were analyzed by SEM-EDS and XRD to document mineralogy and morphology of the noble metals; typical noble metal morphologies are shown in Figure 3.

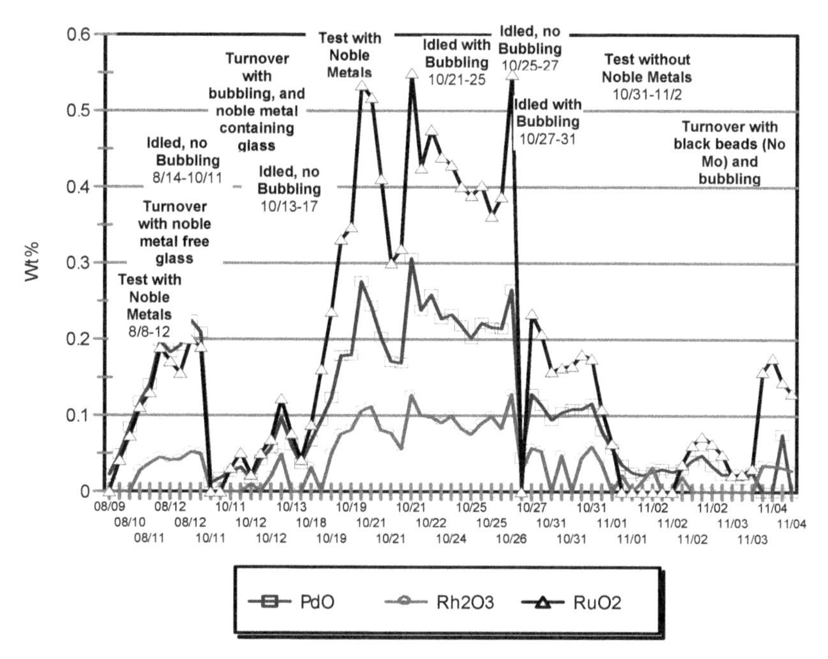

Figure 2. Noble metal concentrations measured by XRF in the melt pool glasses (discharge and dip) during and after tests with and without noble metals.

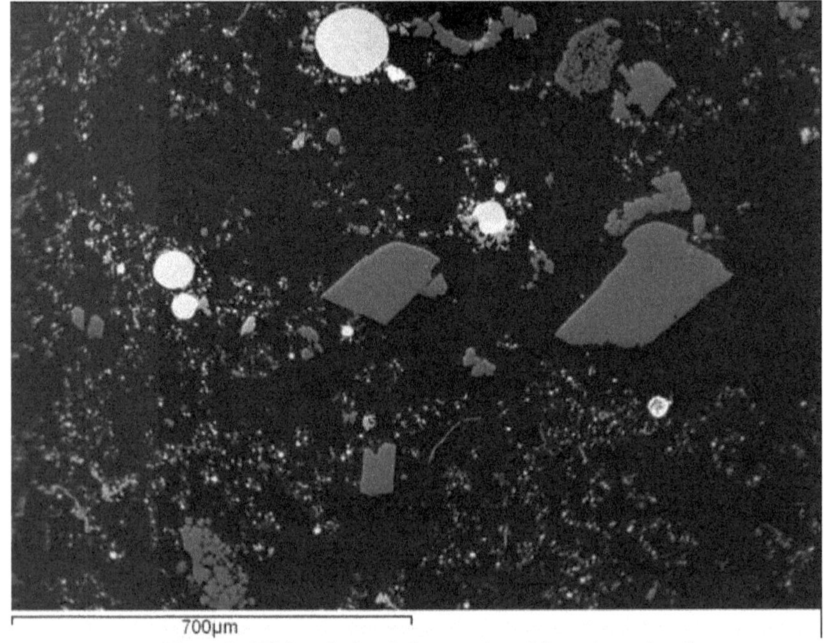

Figure 3. SEM analysis of glass suctioned from the melter floor.
Spherical Pd particles containing variable amounts of Rh and Te, mostly 10-100 micron size.
Ru is commonly elongated or needle shaped, 1-3 by 10-15 micron size, most likely RuO_2.
Some Ru oxides are subhedral equant clustered in patches. Ru oxide also is concentrated around Pd particles.

DM100 NOBLE METALS TESTS

Three melter tests each consisting of a 100-hour feeding period followed by idling periods and turnover with glass from previous KMCO tests ("black beads") were conducted on the DM100 melter using four percent soda glass beads and KMOC8 HB waste simulant containing noble metals combined with varying amounts and types of DBP and MBP (300 ppm DBP, no DBP, 300 ppm MBP/DBP). After each of the feeding and idling periods during the three tests, the melt pool was extensively sampled for secondary phases and to provide samples for noble metals determinations. The three tests processed 779.3 liters of KMOC8 HB waste simulant and 278 kg of four percent soda glass beads while sampling the melt pool for noble metals before and after each feeding and idling period. Over 960 kg of black beads (glass without noble metals and molybdenum) were processed during the turnover periods. Prior to each test, water was fed into the melter to achieve a plenum temperature of 600°C before starting the waste simulant and frit feed. Water was fed simultaneously with frit and waste then tapered off to maintain a 600°C plenum temperature as the cold cap was being established. Plenum heaters were used to prevent the plenum from dropping below 600°C during each test. The waste simulant and glass frit feed rates were scaled from the Rokkasho melter to the DM100 using the surface areas of the two melters. The glass pool temperature was maintained at 1150°C throughout the tests, as indicated by the thermocouple placed twelve inches from the melt pool floor. No bubbler was installed during the waste and frit feeding or during most of the idling periods, therefore there was rigorously no glass pool bubbling during these periods. The bubbler was installed for turnover with black beads and the melt pool was sparged with a separate lance prior to melt pool turnovers. In each test, there were at least two distinct turnover periods with black beads: the first 200 kg and the second 100 kg, each separated by idling periods.

Table 2. KMOC8 HB Waste Simulant Constituents, Contributions from Glass Frit and Target Oxide Compositions.

Waste Simulant	Oxides from Glass Frit	Concentrations of Key Components in Product Glass (wt%)
Nitrates of Ag, Ba, Ce, Co, Cr, Cs, Fe, Gd, K, La, Mn, Na, Nd, Ni, Pd, Pr, Rh, Ru, Sm, Sr, Y, Zn, Zr, perhenic acid, sodium sulfate, sodium phosphate, sodium molybdate, di-*n*-butyl phosphate	Al_2O_3 B_2O_3 CaO Li_2O Na_2O SiO_2	Cs_2O-1.1% MoO_3-1.23% Na_2O-10.03% PdO-0.42% Rh_2O_3-0.14% RuO_2-0.64%

Collectively, the data from these tests indicate the progressive accumulation of noble metals on the melt floor and associated increased joule heating near the bottom over the course of testing. Higher bottom glass and electrode temperatures coupled with decreases in glass pool resistance are clear indications of increased joule heating due to noble metals deposition. Typical behavior is illustrated in Figure 4: glass and electrode temperatures decrease over the 100-hr feeding period, increase by 200 to 300°C while idling (100 hrs to about 180 hrs) as the glass resistance drops, and are stabilized during the turnover with black beads while bubbling the melt pool (starting at about 180 hrs). However, the increases in power over the course of testing are more modest. Subsequent air sparging of the melt pool, turnover with black beads while bubbling, and scraping of the sides of the melter did result in the mobilization of much of the noble metals and also produced some recovery in glass temperatures and resistance during idling and turnover periods although not to the conditions at the onset of testing.

Extensive analysis of discharge glasses and samples from the surface, center, and bottom of the melt pool was performed. Dip and suction samples were analyzed with SEM-EDS to identify crystal morphologies associated with noble metals but no clear patterns were discerned from the SEM data with respect to crystal morphologies and test conditions. The analyzed compositions of the dip and suction samples largely reflected the discharge glass, although most suction samples had higher concentrations of noble metals and chromium. Samples taken from the scraping tool also contained higher concentrations of noble metals and chromium than other dip samples indicating that noble metals were deposited on the sides of melter; the chromium likely originated from the melter refractories or the Inconel electrodes. The higher noble metal and chromium concentrations are consistent with observations of noble metals and high-chromium spinels observed in the SEM analysis conducted on many of the samples.

Noble metal concentrations in the glass samples mostly increased with closer proximity to the melt pool floor. Noble metals did not appear to preferentially deposit in a particular location on the melt floor, as indicated by a lack of a clear trend in the five samples taken at the end of each feeding period. It should be noted that, despite the turnovers that were performed in between tests, at the start of each test there were still some noble metals in the melter from preceding tests, particularly on the melter floor. In Tests 1 and 2, more deposition of noble metals occurred during the first idling period after waste simulant feeding while in Test 3 the concentrations of noble metals in the samples from the melter floor were higher prior to the idling period. This difference may be related to large amounts of noble metals that were re-suspended into the melt pool and subsequently discharged in glass during the turnover in Test 3. The concentration of noble metals in samples from the melter floor generally decreased in response to the 200 kg turnover with bubbling. Although noble metal concentrations in samples from the bottom of the melter decreased throughout the turnovers during Test 1, the largest macroscopic noble metals deposit was recovered after the turnover prior to Test 2.

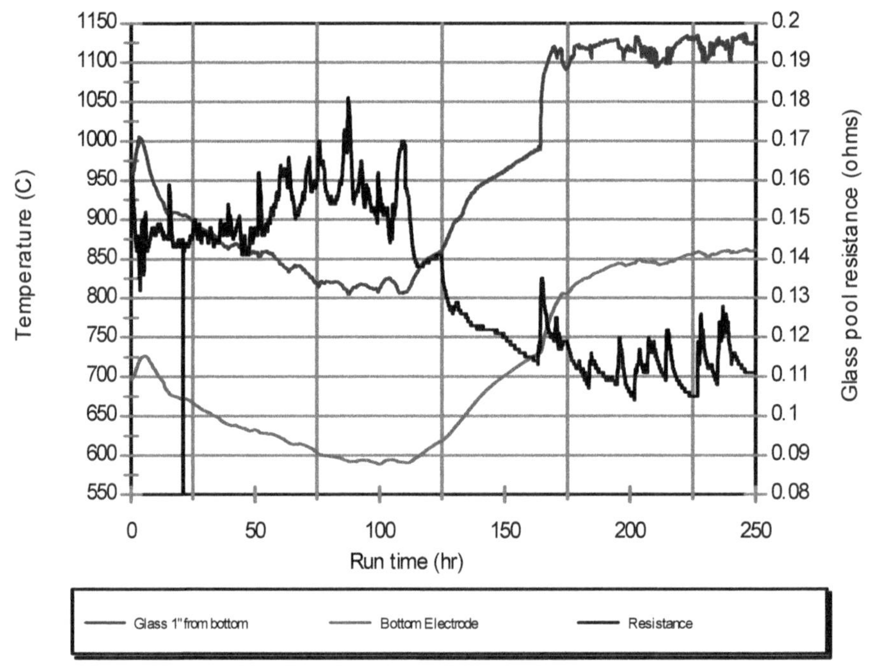

Figure 4. Bottom electrode temperature, glass temperature, and glass pool resistance while feeding noble metals, idling, and turnover with black beads while bubbling.

Noble metals concentrations were determined for all glasses for the purposes of determining the extent of settling and determining a mass balance. The results are summarized in Table 3 and illustrated for Test 1 in Figure 5. All three noble metals increase in concentration while feeding the waste simulant; however, the amounts observed in the discharged glass vary between the tests even though the concentrations in the waste simulant and the feed rate were the same for all three tests. The concentrations of noble metals in the discharged glass after waste simulant feeding in Test 1 are about twice those for the two subsequent tests. The total amounts of noble metals measured in discharge glasses during waste simulant feeding decreases with successive tests from 36 to 56% in Test 1 to 17 to 24% in Test 3. In response to sparging and bubbling during turnover with 200 kg black beads, noble metals concentrations increase by factors of two to three in Test 3 while only declines in concentrations were observed in the other two tests. As a result, the total amount of noble metals discharged in glass increased by a factor of about two over the 200 kg turnover with black beads in the first two tests while increasing by factors of four to five in Test 3. An increase in noble metals concentration was observed in Test 2 with sparging and bubbling during the second turnover with 100 kg black beads. After 300 kg turnover with black beads, the vast majority of the noble metals fed to the melter were recovered in the discharge glass. Over all three tests, 94% of the ruthenium, 104% of the rhodium and 73% of the palladium were recovered in the discharge glass. The near-complete recovery of ruthenium and rhodium coupled with the observed changes in bottom melter temperature and glass resistance suggests that either the missing palladium or very small amounts of the other noble metals were responsible for the observed changes melter conditions.

Table 3. Noble Metals Distributions in Discharged Glass.

		PdO	Rh_2O_3	RuO_2
Test 1	% Feed noble metals in discharge glass during 100 hour test	36	56	51
	% Feed noble metals in discharge glass during 100 hour test + 200 kg glass turnover	75	108	97
	% Feed noble metals in discharged glass during 100 hour test + 300 kg glass turnover	81	116	100
Test 2	% Feed noble metals in discharge glass during 100 hour test	29	36	35
	% Feed noble metals in discharge glass during 100 hour test + 200 kg glass turnover	54	70	60
	% Feed noble metals in discharge glass during 100 hour test + 300 kg glass turnover	64	85	75
Test 3	% Feed noble metals in discharge glass during 100 hour test	17	24	19
	% Feed noble metals in discharge glass during 100 hour test + 200 kg glass turnover	67	99	97
	% Feed noble metals in discharge glass during 100 hour test + 300 kg glass turnover	74	109	105

CONCLUSIONS

A series of melter tests were conducted with simulated high-level waste representative of that from the Rokkasho Reprocessing Plant. The DM100 test melter was operated in a manner to reproduce the "cold-bottom" mode of operation used in the Japanese melters to evaluate the formation of secondary phases such as molybdate salt (yellow phase) and the behavior of noble metals. Samples collected throughout the tests, monitored temperatures, and measured glass resistance collectively show the settling out of noble metals to the floor of the melter and joule heating associated with noble metal deposition. Melt pool bubbling was found to be effective in suspending noble metals and relatively effective at mobilizing noble metals and yellow phase from the floor of the melter. Noble metals mass balances across each test and its associated two turnover periods showed reasonably good closure for ruthenium and rhodium, suggesting that it was possible to mobilize and discharge the majority of these components. In contrast, the mass balance for palladium showed a consistent shortfall for all three tests, indicating that a significant fraction of the palladium was retained in the melter despite the turnover periods. Yellow phase formation was significantly reduced in the presence of noble metals as compared to feeds without noble metals. Consistent with the results from smaller scale tests, only traces of yellow phase were observed with the redistributed feed containing noble metals; however, the present test results are significant in that they included the more prototypical melt pool temperature gradient in addition to the nearly six-fold scale-up.

Parallel studies have been conducted to confirm the viability of implementing the redistribution approach in the Rokkasho facility. Since this approach will slightly increase the solids content in the liquid feed, suspension, settling, mixing, and transport tests have been performed. In addition, since the composition of the glass beads is modified somewhat, the effects on beads fabrication and bead properties have also been determined. No issues have been identified in these tests. In addition, the results of the present tests do not suggest any deleterious impact on feed rates or glass production rates. Therefore, based on the favorable results obtained to date, plans are in place to further confirm the viability of the redistribution approach for the Rokkasho facility through full-scale non-radioactive testing.

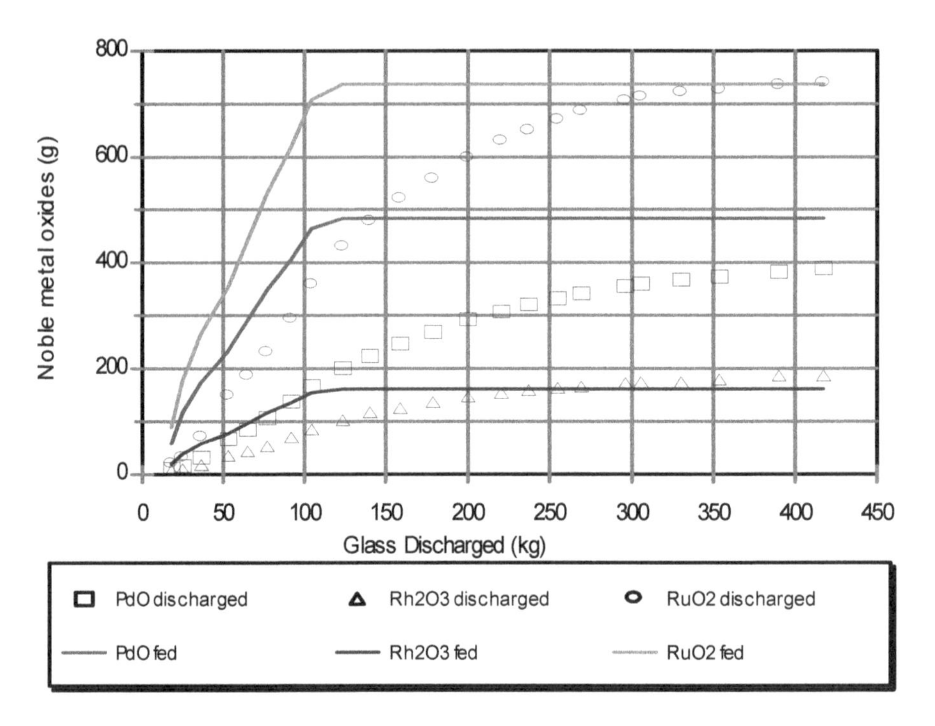

Figure 5. Noble metals measured by XRF in glasses discharged during Test 1.

ACKNOWLEDGMENTS

This work was carried out as a part of the research supported by Japan Nuclear Fuel Limited with Grant-in-Aid by the Ministry of Economy, Trade and Industry.

REFERENCES

[1]E. Schiewer, H. Rabe and S. Weisenburger, in *Scientific Basis for Nuclear Waste Management* V, Vol. 11, p 289, Materials Research Society (1982).

[2]W. Lutze, in *Radioactive Waste Forms for the Future*, Edited by W. Lutze and R.C. Ewing, Elsevier Science Publishers B.V. (1988).

[3]O. Pinet, J.L. Dussossoy, C. David, and C. Fillet, "Glass Matrices for Immobilizing Nuclear Waste Containing Molybdenum and Phosphorus," *J. Nucl. Mat.*, 377, 307 (2008).
[4]R.J. Short, R.J. Hand, and N.C. Hyatt, "Molybdenum in Nuclear Waste Glasses – Incorporation and Redox State," in *Scientific Basis for Nuclear Waste Management XXVI*, vol. 757, p. 141 (2002).
[5]G. Roth, S. Weisenburger, "Role of Noble Metals in Electrically Heated Ceramic Melters," Proceedings of SPECTRUM '90, p. 26-30 1990, American Nuclear Society, La Grange Park, IL.
[6]C. Chapman, "Reducing Noble Metals Precipitation and Extending Melter Life," Internal Report, CH2MHill Hanford Group, Inc., Richland Washington, February 2001.
[7]I.L. Pegg, H. Gan, K.S. Matlack, Y. Endo, T. Fukui, A. Ohashi, I. Joseph and B.W. Bowan, Mitigation of Yellow Phase Formation at the Rokkasho HLW Vitification Facility, Paper 10107, *Proc. Waste Management Symposium*, Phoenix Arizona, USA. (2010).
[8]H. Gan, K.S. Matlack, I.L. Pegg, I. Joseph, B.W. Bowan, Y. Miura, N. Kanehira, E. Ochi, T. Oniki, Y. Endo, "Suppression of Yellow Phase Formation during Japanese HLW Vitrification," *Ceramic Transactions*, this volume.
[9]Y. Miura, N. Kanehira, E. Ochi, T. Oniki, A. Yamasaki, Y. Endo, I.L. Pegg, H. Gan, K.S. Matlack, I. Joseph, B.W. Bowan, "Development of New Method for Yellow Phase Suppression by Redistribution of Frit Components," Proceedings of GLOBAL 2011, Makuhari, Japan, Dec. 11-16, 2011, Paper No. 441526.

SUPPRESSION OF YELLOW PHASE FORMATION DURING JAPANESE HLW VITRIFICATION

Hao Gan[1], Keith S. Matlack[1], Ian L. Pegg[1], Innocent Joseph[2], Bradley W. Bowan[2], Yoshiyuki Miura[3], Norio Kanehira[3], Eiji Ochi[3], Toshiro Oniki[4], Yoshihiro Endo[4],
1: Vitreous State Laboratory, The Catholic University of America, Washington, DC, USA
2: Energy*Solutions* Federal EPC, Inc., Columbia, MD, USA
3: Japan Nuclear Fuel Limited, Aomori, Japan
4: IHI Corporation, Kanagawa, Japan

ABSTRACT

The presence of significant amounts of molybdenum in high level waste (HLW) streams can lead to the formation of a molybdate salt phase ("yellow phase") during vitrification. These salts are undesirable from both melter operations and product quality perspectives, such that yellow phase formation can become the primary factor that limits waste loading in the HLW glass product. The Rokkasho Reprocessing Plant, located in northern Japan, includes facilities for reprocessing commercial spent nuclear fuel and conversion of the HLW into a stable glass waste form by vitrification. The primary objectives of the present study were to characterize yellow phase formation with projected Japanese HLW streams and to develop suppression strategies that could permit both more robust operations and the potential for increased waste loadings. The testing was divided between approaches that would not change the present baseline product glass composition and those that involve the modification of that composition. The former approach involves the least change and therefore could be implemented more quickly. Testing was performed using a combination of crucible scale glass melts, vertical gradient furnace tests, and scaled joule heated ceramic melter tests to develop viable strategies for yellow phase control that are being subjected to full-scale testing prior to deployment in the Rokkasho facility.

INTRODUCTION

The Rokkasho Reprocessing Plant is located in Rokkasho Village in northern Japan. The site is operated by Japan Nuclear Fuel, Ltd. (JNFL). Facilities at the Rokkasho site include the reprocessing plant, spent fuel receiving and storage facilities, the uranium enrichment plant, the low level waste disposal center, the vitrified waste storage center, and the MOX fabrication plant. The nominal capacity of the reprocessing plant is 800 metric tons of uranium per year, which is equivalent to the spent fuel from about forty 1000-MW power plants. High level waste (HLW) from the reprocessing plant will be converted to stable glass in the integrated waste vitrification facility.

The Rokkasho vitrification process centers on two, 2.6 m^2 joule heated ceramic melters. These melters were designed to treat the specific waste chemistry and high noble metal concentration of the Rokkasho flow-sheet. The waste stream contains significant amounts of molybdenum and it is known that when the concentration of molybdenum is sufficiently high, a molybdate salt phase can form during nuclear waste vitrification[1-4]. Such molybdate salts are typically referred to as "yellow phase," since in addition to a variety of other elements they also incorporate chromium, which gives them their yellow coloration. These salts are typically low melting, very fluid at typical melter operating temperatures, of variable density depending on their composition, and undesirable from both melter operations and product quality perspectives. Full-scale mock-up testing with waste simulants suggested that operating protocols that had been developed would be adequate to manage yellow phase formation. However, during the early part of the active test campaigns, yellow phase formation was found to be much more prevalent than was expected on the basis of the mock-up test results. This is believed to be due to significant differences in waste compositions used in the active and mock-up tests, including the presence of sulfur in the actual waste. During the latter part of the active tests, yellow phase formation was eliminated by diluting the HLW with another waste stream that had low concentrations of molybdenum and sulfur in order to decrease the levels of components contributing to yellow phase formation. However, this increased the amount of glass product made from the waste. In order to reduce the amount of waste glass product, it is preferable to develop high waste loading glass formulations that do not show the tendency for yellow phase formation.

Previous work investigated potential strategies to suppress the formation of the molybdate yellow phase during the vitrification of Japanese high-level waste without changing the composition of the final glass product[5]. It is likely that a solution that does not change the product glass composition could be implemented more easily and more quickly. The yellow phase mitigation strategy involved redistributing some of the components in the frit so that they are instead added to the liquid waste stream. The basis for this approach is the expectation that the incorporation of certain components in the liquid waste will make them available early in the melting process so that they are able to combine with and "tie up" components that would otherwise form the yellow phase. The reformulation was done in such a way that the general melting properties of the frit, including the overall glass production rate, are not deleteriously impacted. This approach proved successful in reducing the tendency for yellow phase formation[5].

In the current work, the same approach was used, with the exception that the final glass composition and glass frit composition were allowed to change. This permits considerably more flexibility in the glass formulation strategy, such that it is possible to significantly increase the waste loading in the glass product while still preventing yellow phase formation. A variety of screening tests were used to assess yellow phase formation, product glass processability and durability, and the properties of the glass frit. These results were used iteratively to refine the formulations to optimize the waste loading and glass properties. The selected glass formulation and feed blend were then subjected to confirmatory testing on a small-scale continuously fed joule heated ceramic melter system at different waste loadings. During the mock-up tests, it was observed that yellow phase was more likely to develop from melter processing of HLW simulants using noble metal (NM) surrogates (without noble metals) in comparison to HLW that contained Ru, Rh and Pd. While the origin of this effect is of interest, this observation is of considerable importance for full-scale mock-up tests in view of the high cost of noble metals. Thus, one part of this work was directed toward exploration of the association between yellow phase formation and the nature and concentrations of these three platinoids.

YELLOW PHASE FORMATION

The HLW stream investigated in the present work is rich in alkali and alkaline earth oxides (Na, Cs, and Ba), molybdenum oxide, rare earth oxides, and several other transition metal oxides (Fe, Mn, Ni, Zr). On an oxide basis one of the waste simulants used in the present work was composed of approximately 23 wt% Na_2O, 11 wt% each of CeO_2 and ZrO_2, 9 wt% each of MoO_3, Gd_2O_3, and Nd_2O_3, 5 wt% each of Fe_2O_3 and Cs_2O, and lesser amounts of a variety of other constituents, including about 3 wt% NiO, 2 wt% MnO, and 0.3 wt% Cr_2O_3. The currently adopted processing method introduces glass formers in the form of glass beads.

A sample of the yellow phase collected from a melter discharge during a mock-up campaign in the JAEA test facility at Tokai (KMOC) was analyzed by x-ray fluorescence spectroscopy (XRF) and dissolution followed by direct current plasma atomic emission spectroscopy (DCP-

AES). On an oxide basis, the sample was composed of about 70 wt% MoO_3, 13 wt% Na_2O, 4 wt% Li_2O, and about 2 wt% each of oxides of Ba, Ca, and Cs, and lesser amounts of a variety of other constituents. The major crystalline phases in the yellow phase identified by x-ray diffraction (XRD) and scanning electron microscopy with energy dispersive x-ray spectroscopy (SEM/EDS) were sodium and calcium molybdates, with minor amounts of cesium and barium molybdates. In addition, a yellow phase composed predominantly of calcium molybdate was observed on the bottom of the KMOC melter as a residue after the glass had been discharged. It is noteworthy that these yellow phases formed in a glass melt in which the MoO_3 content was considerably below its saturation limit, indicating the importance of kinetic effects.

Sodium molybdate melts at 686°C, cesium and calcium molybdates melt or decompose at 936°C and 985°C, respectively, and barium molybdate melts at a much higher temperature of 1480°C. Since sodium molybdate would be the most abundant salt from the HLW waste, extensive eutectic melting will occur for barium and calcium salts at around 700°C[6]. As a result, a molten molybdate salt phase incorporating various metals can develop in the melter, even though the overall glass system is not saturated with MoO_3. Once separated, the slow kinetics inhibits the incorporation of the molten molybdate phase into the silicate melt. We have observed and extensively investigated very similar phenomena during the vitrification of high-sulfur wastes, such as the low-activity fraction of the Hanford tank wastes[7-11]. Indeed, molybdate and sulfate show similar behavior in this respect and both are found in significant amounts in the yellow phase that forms when sulfur is present in the waste.

Since kinetic effects are important in the formation of a low-melting molybdate phase, its formation can be significantly affected by the way in which the sodium-molybdenum-rich waste simulant and highly viscous glass frit are introduced. As discussed previously, a constraint on the previous work[5] was that the composition of the final glass product not be changed. Consequently, the strategy that was investigated employed modification of the waste-only simulant part by redistributing some of the components in the glass frit so that they are instead added to the liquid waste stream. The basis for this approach is the expectation that the incorporation of certain components in the liquid waste will make them available early in the melting process so that they are able to combine with and tie up components that would otherwise form the yellow phase. However, the reformulation of the glass frit and simulant must be done in such a way that the general melting properties of the frit, including the overall glass production rate, are not deleteriously impacted. While this approach has proved very successful, the results from the present work demonstrate that significant further improvements are possible if the glass product and glass frit compositions are allowed to change.

SCREENING TESTS

Screening tests were performed using a mixture of dried simulants and their matching glass frits (i.e., as modified by the particular redistribution/reformulation scheme) in a vertical gradient furnace. The vertical gradient furnace creates a thermal environment similar to that in the cold cap that forms in a joule-heated ceramic melter (JHCM), in which the feed materials at different temperatures change, evolve with time, and interact with neighboring materials that have been evolving at different temperatures. Thus, the reacted feed materials after gradient furnace heat treatment simulate the cross section that might be observed in cold cap samples. In addition, samples generated from the gradient furnace experiments cover wide ranges of temperature in one test. Although horizontal gradient furnace tests and isothermal tests were also used in the early stages of this work, the vertical gradient furnace test was selected as the preferred standardized screening test in this work.

In all tests, the waste simulant was mixed with appropriate amounts of the redistributed glass former additives, dried overnight at 110°C, ground to pass 80 mesh sieve, and then mixed with the appropriate amount of the matching glass frit. This provided a simplified representation of the JHCM melter feed as it dries in the cold cap.

The temperature gradient inside the vertical gradient furnace (VGF) is maintained by two separate sets of heating elements, both of which are arranged in cylindrical form and aligned along their axis. The inner heater is set at 1150°C, to represent that of the glass pool, and the ambient heater is set at 600°C, representing the melter plenum temperature. A ceramic crucible of about 6 cm diameter is used to contain the reacting feed blend. Feed reactions under the controlled temperature gradient are allowed to continue for the designated test duration (typically 30, 45, 60, and 90 minutes) and then stopped by rapid cooling in room temperature air. The top surface of the reacted feed blends are then inspected and photographed. After collection of samples of salt and partially reacted feed blend for further characterization, the crucibles with their feed contents are cross-sectioned (dry) to reveal the conversion progress of feed blends. Samples are analyzed by optical microscopy, SEM/EDS, and x-ray diffraction (XRD).

SCREENING TESTS RESULTS

Original Feed Blend

Figure 1 shows the formation of yellow phase in VGF tests using the original feed blend (i.e., original glass frit composition without any glass former redistribution or glass formulation modification). The waste oxide loading in the product glass is 20.8 wt%. The yellow phase is a molten salt mixture that is typically less dense than the feed batch and therefore tends to accumulate on the surface. The yellow coloration is due to the accumulation of chromium in the salt mixture. Significant amounts of yellow phase were also seen in horizontal gradient furnace tests and in isothermal tests[1]. Figure 2 shows a typical SEM micrograph of the yellow phase material. XRD analysis shows the presence of Na, Ca, Cs, and Ba molybdates as well as rare earth molybdenum oxides.

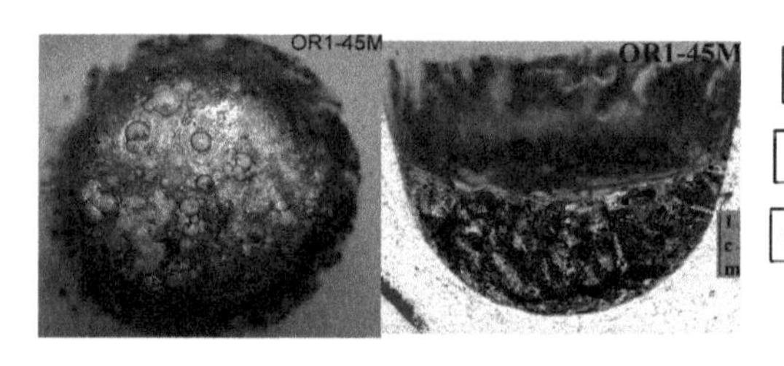

Figure 1. Top and cross-section views of original feed blend after VGF test (45 minutes).

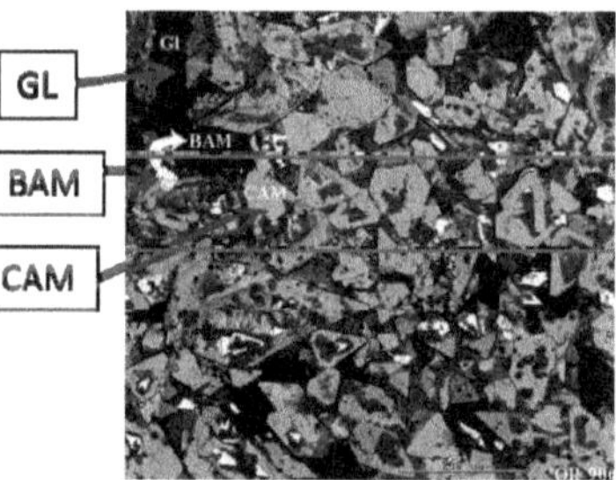

Figure 2. SEM image of the top surface of the reacted original feed blend after one hour heat treatment at 900°C. (CAM = $CaMoO_4$; BAM = $BaMoO_4$; GL = vitreous phase).

High Waste Loading Formulation

In the present work, new glass formulations were developed to increase waste loading and tolerance to higher concentrations of molybdenum and sulfur, while meeting all of the product quality and processability constraints. As was the case for the original formulation, the major constituents of these glasses are SiO_2, B_2O_3, Na_2O, Al_2O_3, and Li_2O, together with the numerous constituents originating in the waste, including molybdenum and sulfur. The concentrations of all of the waste constituents, including molybdenum and sulfur, were higher in the new glasses as a result of their higher waste loadings. For each new product glass composition, the reformulation process requires the development of a new glass frit that does not phase separate and a liquid waste that still has suitable rheological properties after part of the frit components are redistributed to the liquid waste. Glass property databases accumulated from many years of waste glass development work, glass property-composition models, and past experience in glass formulation development were used to define specific compositions that were projected to have acceptable processability and product quality. These compositions were then tested first for yellow phase formation, which was expected to be the most constraining property. Formulations that effectively suppressed yellow phase formation, as indicated by VGF tests, were then characterized for properties relevant to processability and product quality. The results from one set of tests were used as feedback for the next set of tests to iteratively refine the glass formulation. The changes that were made at each stage were guided by glass-property composition models and relevant exiting waste glass property data. The outcome of this process was a set of compositions with optimized waste loadings and that suppress yellow phase formation while meeting all product quality and processability constraints. The principal properties measured to assess processability were glass melt viscosity, glass melt electrical conductivity, and the crystal content of the glass melt under melter idling conditions. Product quality was assessed using the Product Consistency Test (PCT) glass leaching procedure (ASTM C 1285). Compositions that showed suitable tolerance to yellow phase formation were melted and tested for processing properties. Glasses that had measured processing properties within acceptable ranges were then tested for chemical durability using the PCT procedure. Glass compositions that met all processing and product quality requirements while maintaining high waste loadings without formation of yellow phase were candidates for further testing on a small-scale continuously fed joule-heated ceramic melter.

Figure 3 shows the results from VGF tests with one of the new formulations with a waste oxide loading of 34 wt%; there is only a very minor sign of yellow phase. Comparison with the VGF results for the original formulation at 20.8 wt% waste loading (Figure 1) demonstrates the significant improvement in yellow phase suppression even though the waste loading and the concentrations of the major yellow phase constituents, notably molybdenum and sulfur, are significantly higher. The measured properties of the high waste loading formulation are within acceptable ranges and generally comparable to those of the original glass composition; Figure 4 compares the melt viscosities and electrical conductivities of the two glasses.

SMALL-SCALE MELTER TESTS

DM10 Melter System

Based on the results from the screening tests, small-scale continuously fed joule-heated ceramic melter tests were performed with both the original feed blend (as a control) and one of the new high waste loading formulations. The glass frits used for each of these tests were sized to match the size range employed at the Rokkasho facility. The tests were performed on a DM10 melter system.

Figure 3. Top and cross-section views of new high waste loading formulation (34 wt% waste loading) after 30 minutes VGF test.

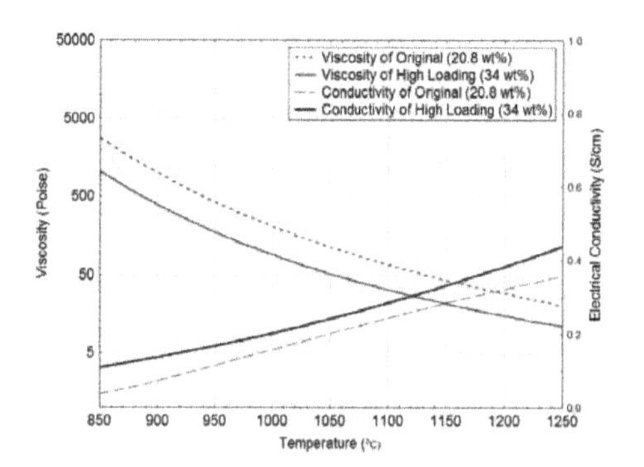

Figure 4. Melt viscosity and electrical conductivity of new high waste loading (34 wt%) glass formulation (solid curves) and original glass formulation (20.8 wt% waste loading, dashed curves).

The ceramic refractory-lined melter has an effective melt surface area of 0.021 m^2, and two Inconel 690 plate electrodes that are used for joule-heating of the glass pool. The glass product is removed from the melter by means of an air-lift discharge system. A dual feed system was used to deliver the melter feed into the DM10 melter, where separate containers were used for the HLW simulant and the glass frit. The HLW simulant container was mounted on load cells for weight monitoring, and feed was delivered to the feed tube using a peristaltic pump. Pre-weighed amounts of glass frit were periodically added to a calibrated hopper that was connected to a mechanical screw feeder. The waste simulant and the glass frit were both introduced into the melter through the same water-cooled, vertical feed tube. For operational simplicity, the DM10 is equipped with a dry off-gas treatment system involving gas filtration operations only. Exhaust gases leave the melter plenum through a film cooler device that minimizes the formation of solid deposits. The film-cooler air has constant flow rate and its temperature is thermostatically controlled. A transition line connects the melter with the first filtration device. Immediately downstream of the transition line are cyclonic filters followed by conventional pre-filters and HEPA filters. The temperature of the cyclonic filters is maintained above 150°C while the HEPAs are held above 100°C to prevent moisture condensation. The entire train of gas filtration operations is duplicated and each train is used alternately. An induced draft fan completes the system.

Since a key objective of these tests was to assess the extent of yellow phase formation and mitigation, a series of samples were taken before, during, and after the melter tests to determine the extent of secondary phase formation. Separated secondary phases on the melt pool surface were detected by inserting a threaded metal rod into the molten glass pool. The rod is then pulled out and the rod and attached material are allowed to cool. Any secondary phases floating on the surface of the glass melt adhere to the metal rod and can be readily visually distinguished from the attached bulk glass due to their different color and morphology. Several of these "dip" samples were taken at the end of each test at different locations across the melt pool. Separated secondary phases on the melter floor were detected by inserting a ceramic tube to the floor of the melter. The tube is connected through a three-way valve to an air supply and a vacuum pump. Slight positive pressure is supplied while the tube is inserted into the melt to prevent glass from entering the tube. The applied gas pressure is then switched to suction once the tube is in the sampling position, thereby pulling glass and any secondary phases from the bottom of the melter into the tube. Once the tube is removed from the melt pool and cooled, glass removed from the inside of the tube is representative of glass suctioned from the melt pool floor, whereas glass from the outside of the ceramic tube is representative of the surface of the melt pool. Multiple samples of each type were taken.

Original Feed Blend

Following a period to turn over the melt pool inventory, a 65-hour test was performed with the HLW simulant and the original glass frit composition (with no redistributed glass formers) at the nominal waste oxide loading of 20.8 wt%. Melt pool temperatures throughout the test were close to the target of 1150°C while plenum temperatures ranged from 400 to 550°C once the cold cap was established.

Samples taken prior to the test and midway through were essentially free of secondary phases. Visual observations of these glasses indicated no macroscopic secondary phases were present on the melt pool surface or floor. In contrast, many of the samples taken at the end of the test indicated that secondary molybdate phases were present on the melt pool surface and floor. Figure 5 shows a suction tube sample with the bottom portion of the ceramic tube removed to reveal the glass drawn inside from the bottom of the melter. Yellow streaks are evident along the exterior of the sample. This secondary phase originated from the surface of the melt pool and adhered to the outside of the tube as the tube was removed from the melter. In contrast, the orange-brown material at the far-right tip originated from the melter floor. SEM/EDS analysis of the samples from the outside of the tube focused on two morphologies, referred to as the

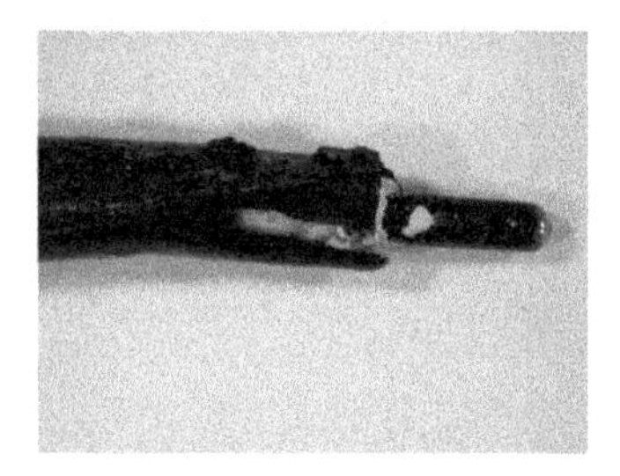

Figure 5. Photograph of a melter suction sample from the end of the test with the original feed blend. The bottom part of the ceramic tube was removed to expose the glass inside. Note patches of yellow phase (originating from the melt surface) and orange-brown material on the far right tip (originating from the melter floor). White material is the ceramic sampling tube.

"yellow" and "mottled brown/gray" areas, respectively. The "yellow" areas were found to consist of sodium and cesium molybdates (Figure 6). The "mottled brown/gray" areas consist of sodium molybdates and cesium oxides. Microscopy on the samples from the glass taken from the inside of the tube indicates that the secondary phases at the melter floor consist of calcium, barium, and sodium molybdates (Figure 7). The higher concentration of alkaline earth molybdates in the melter floor sample is consistent with previous observations from full-scale melter tests with this simulant, with the denser alkaline earth molybdates settling to the melter floor while the lighter alkali molybdates are found in higher concentration on the melt pool surface.

In summary, therefore, the DM10 melter tests with the original feed formulation reproduced the yellow phase formation that was found in full-scale melter tests and in the small-scale screening tests.

Figure 6. SEM image of yellow material from the outside of a suction tube sample. The grey areas are sodium molybdates; the bright areas are cesium molybdates.

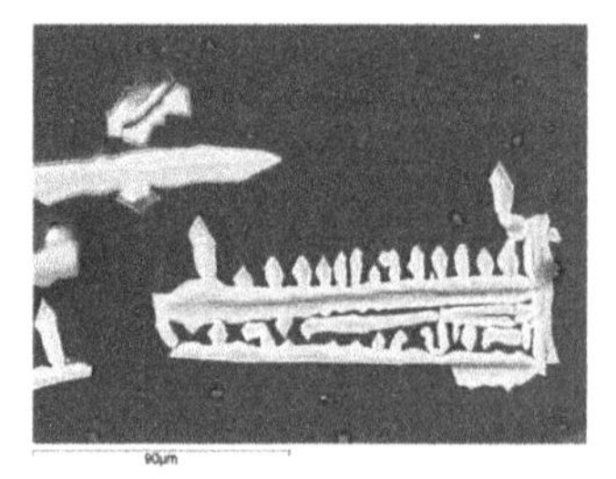

Figure 7. SEM image of calcium molybdate crystals found in a sample from the melter floor.

High Waste Loading Formulation

A sufficient quantity of the new glass frit was first prepared to support melter testing of the new glass formulation. The same melter and test conditions were used to test the high waste loading formulation. The nominal waste oxide loading of 30.4 wt% was used for two successive tests, each of 100 hours duration. Two 100-hours tests at progressively higher waste loadings of 32 wt% and 34 wt% were then conducted. The same methods of sampling of the melt surface and melter floor were conducted at the end of each of these tests as was conducted for the original formulation. In contrast to the results from the test with the original formulation, no yellow phase was evident in any of the samples from any of these tests, even at 34 wt% waste oxide loading. There were no signs of secondary phases on the outside of the dip and suction samples, indicating that there were no secondary phases on the melt surface. Furthermore, all glass inside the suction tubes was free of any visible secondary phases, indicating that molybdate phases did not form during the test and settle to the bottom of the melter. The lack of secondary phases in these samples was confirmed by analysis using SEM-EDS.

NOBLE METAL IMPACT ON YELLOW PHASE FORMATION

To investigate the effects of the type of NM (Ru, Rh, Pd) and their concentrations on yellow phase formation, VGF tests were performed on feed blends of a redistributed version of the original feed blend. The details of VGF test results and sample analysis are provided below.

Yellow Phase Formation in NM-Bearing and NM-Surrogate Feed Blends

Two Japanese HLW simulants, one with typical contents of Ru, Rh and Pd and the other with their noble metal surrogates, were tested in the VGF under the identical conditions. The noble metals surrogates were the same as those employed in previous Japanese full-scale non-radioactive tests (Fe, Co, and Ni). The feed blends were based on a redistributed version of the original feed blend (20.8 wt% waste loading) with a modified glass frit and glass former chemicals added into the HLW simulants. The top view images of the two feed blends after 30 minute VGF experiments are shown in Figure 8. Clearly, considerably more yellow phase is present on the surface of the reacted VGF sample of the NM-surrogate feed than is the case for the NM feed. This result was also confirmed in small-scale melter tests[12].

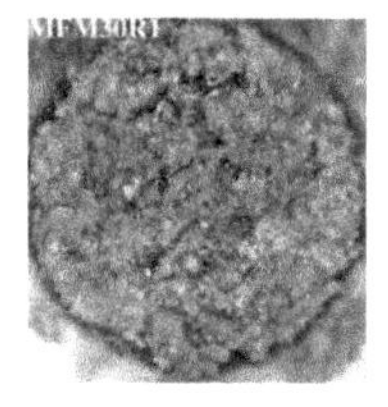

NM-Surrogate

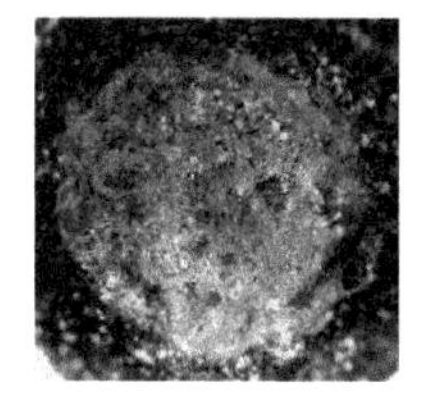

NM-Bearing

Figure 8. Top views of redistributed version of original feed blends with NM-surrogate or NM after VGF test (30 minutes).

Association of Yellow Phase Formation with Ru, Rh or Pd

To investigate the effects of type (Ru, Rh, Pd) and concentration of noble metals on yellow phase formation, additional VGF experiments were performed on the same redistributed version of the original feed blends. For each individual noble metal component (Ru or Rh or Pd), feed blends with two noble metal concentrations (nominal and maximum) were formulated for standard VGF tests. Table 1 lists the target NM values of all feed blends used in this set of feed samples.

Table 1. Concentrations of Individual NM Elements in Feed Blends for VGF Experiments.

NM Concentration	Noble Metal Oxide in Glass (wt%)		
	Ru	Rh	Pd
Maximum NM Concentration	1.28	1.28	1.28
Nominal NM Concentration	0.67	0.15	0.51

Figure 9 shows the top views of the VGF samples of single NM (nominal) feed blends. At the nominal level of each single noble metal alone, the feed overall conversion and the yellow phase formation in particular are easily distinguishable depending on the nature of the noble metal. Minor to moderate amounts of yellow phase were present on the surface of VGF samples with

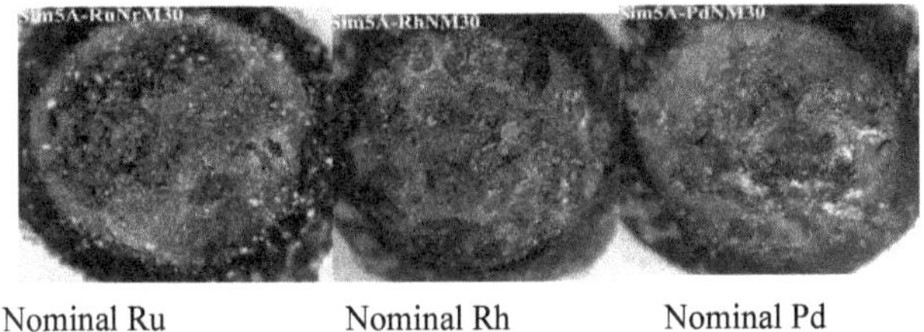

Figure 9. Top views of the redistributed version of original feed blends with Ru, Rh or Pd at nominal concentrations after VGF test (30 minutes).

nominal Ru and the reacted feed blend appeared dark and opaque, almost identical to that of the normal redistributed original feed blend (Figure 8).

In contrast to the case for Ru, visually only traces of yellow phase were evident on the surface of 30-minute VGF samples with nominal Rh or Pd (similar features were also observed from the samples with longer test times). Moreover, the feed blends with Rh- or Pd-bearing simulants had different surface features after the VGF experiments. As shown in Figure 9, considerable amounts of loose and granular materials remained on the surface of Rh-bearing VGF samples. The vitreous materials visible from areas not covered by those loose materials were dark brown, which is different from typical VGF samples for feed blends of normal NM contents. However, different from the Rh-bearing feed, VGF samples for the nominal Pd feed are characterized by dark brown scale in the central surface, the coverage of which decreased with VGF test time, exposing more underlying translucent greenish grey material.

Figure 10 shows the top views of VGF samples of single NM (maximum) feed blends. At the maximum level, considerably less yellow phase formed on the surface of the Ru-bearing feed blends in comparison to their corresponding VGF samples with nominal Ru concentrations. Nevertheless, slightly more yellow phase was present on the sample surface at longer test times. Compared to the nominal Ru VGF samples, the overall conversion for the maximum Ru VGF

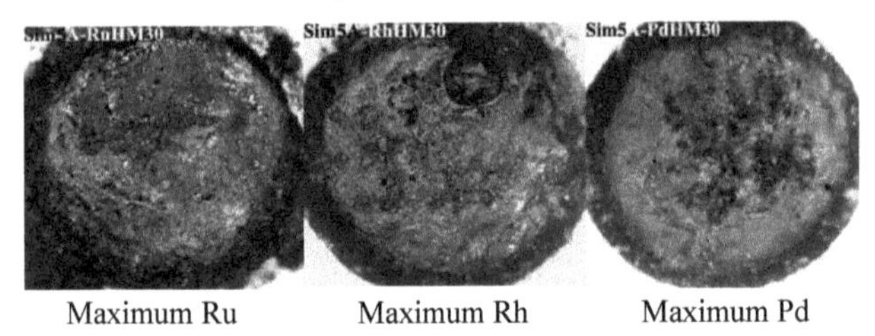

Figure 10. Top views of the redistributed version of original feed blends with Ru, Rh or Pd at maximum concentration after VGF test (30 minutes).

samples appeared slower, as suggested by less of the vitreous surface materials. The features evident in the maximum Pd feed samples are similar to those in its nominal concentration counterpart, with dark colored scales in the central portions above translucent material. In contrast, the features evident in the maximum Rh feed samples differ substantially from those in its nominal concentration counterpart, with formation of more vitreous material on the surface.

Nevertheless, the yellow phase formation appeared to be similar for the Pd and Rh VGF samples at both nominal and maximum levels.

In general agreement with surface visual inspection, analysis of the solution from surface washing of 60-minute VGF samples indicates that less salt components were present on the sample surface for the maximum Ru concentration than for that of the nominal concentration. As shown in Figure 11, the molar concentrations of major salt components dissolved in the washing solution fall close to a straight line corresponding to a 2:1 ratio of Na + Li to Mo + S. Roughly 30% less major salt components were present in the washing solutions from the VGF feed samples with the maximum Ru concentration than that of nominal Ru feed. In contrast, the changes in the major salt components are not as clear for the Rh feed. In comparison to feed at the nominal Rh level, from three repeats at the maximum Rh level, two are similar in their Mo + S content and one is substantially lower in both (Na + Li) and (Mo + S). The available data suggest that more salt components were dissolved from the surface of the maximum Pd feed than that of the nominal Pd feed. Nevertheless, the total salt contents are lowest for Pd feed samples, which agrees with the results from the visual inspection described above.

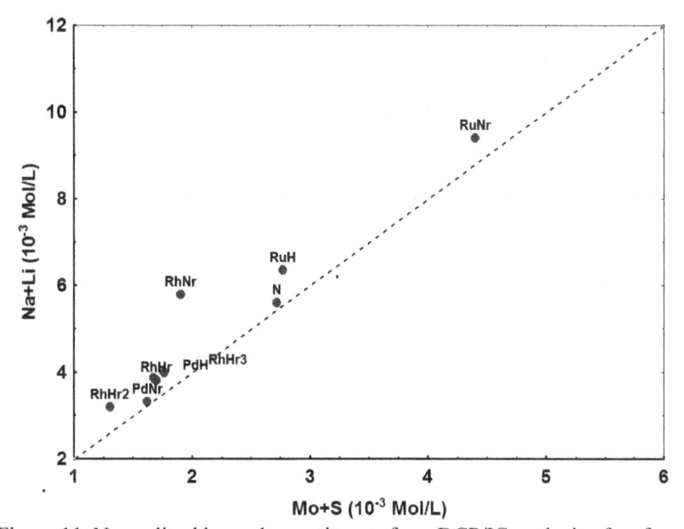

Figure 11. Normalized key salt constituents from DCP/IC analysis of surface washing solutions from 60 minute VGF samples. Data points are labeled according to the type of NM (Ru, Rh or Pd) and concentration (N: nominal; H: Maximum). The repeated VGF samples are labeled with r and its sequence number at the end of the data point label.

It is interesting to note that the maximum Ru VGF sample showed similar dissolved major salt components to that of the feed blend with all three noble metal at their normal concentration, while all other single noble metal feed samples either released more salt (nominal Ru sample) or less salt (nominal and maximum of Pd and Rh).

In general, the major salt phases present in salt patches on the six individual NM-bearing feed samples are the same as those commonly observed in the reacted VGF samples, independent of the identity and content of the noble metal element. Typically observed are sodium and cesium molybdates, sulfates, and perrhenates. No correlation was observed with regard to salt phase and the nature and concentration of noble metal element. However, the formation of a wide spread mesh of Pd mixed with various salt phases on the surface of the VGF sample (Figure 12) is noteworthy. It is likely this Pd concentrated on the sample surface contributed to the formation of the dark colored scale, as shown in Figures 9 and 10. EDS analysis indicated that Pd metal and occasionally partially oxidized Pd were the dominant Pd phases (Figure 13), as compared to Ru and Rh, which are typically present in oxide form.

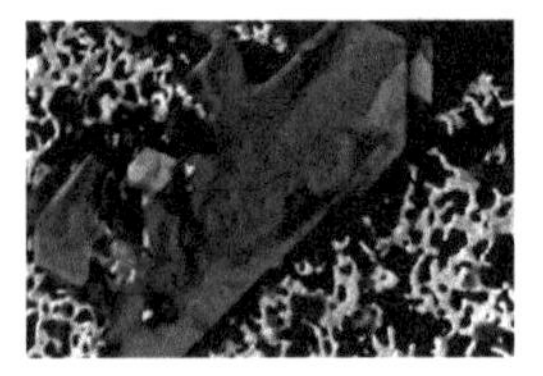

Figure 12. SEM image of Pd mesh in surface salt patches from 30 minute VGF surface sample of maximum Pd simulant. The large grey angular form in the center is alkaline/alkaline earth molybdates.

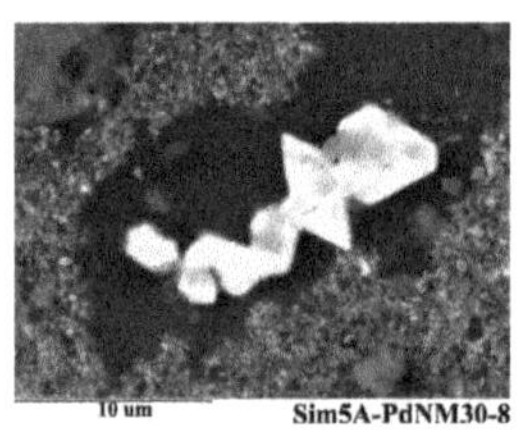

Figure 13. SEM image of Pd in surface salt patches from 30 minute VGF surface sample of nominal Pd simulant. The bright crystalline phase in the center is Pd that is partially oxidized.

CONCLUSIONS

Yellow phase development during conversion of HLW feed under melter operating conditions has been reproduced in small-scale experiments with mixed feed samples subjected to either an isothermal heat treatment or either horizontal or vertical temperature gradients. The temperature range within which an alkali molybdate dominated yellow phase forms and segregates is around 800 to 900°C, with sodium molybdate as the dominant salt. Once formed, the molybdate yellow phase tends to separate and flow to the surface of the partially molten feed materials due to its lower density, melting point, and viscosity.

Previous results showed that yellow phase formation during feed conversion can be effectively mitigated *without* modifying the composition of the final glass product by redistributing a small portion of oxides from the original glass frit to the waste simulant in order to inhibit the formation and segregation of the alkali molybdate yellow phase[5]. The results of the present work show that if the product composition is allowed to change, the redistribution approach can be used in combination with glass composition reformulation to vastly increase the robustness with respect to yellow phase suppression. This permits significant increases in the waste loading while still suppressing yellow phase formation. These results were confirmed in tests on a small-scale continuously fed joule heated ceramic melter system. These tests confirmed the formation of yellow phase with the original feed blend but showed no yellow phase formation with the modified feed blend. The results of the present work demonstrated increases in waste loading from 20.8 wt% to 34 wt%. The new glass formulation meets all of the processability and product quality requirements while preventing yellow phase formation.

The formation of yellow phase from the redistributed original feed blend varied depending on the nature and content of the noble metals present in the waste simulants. Overall, Pd and Rh were more effective in the suppression of yellow phase at both the nominal and maximum level. Ru, at its nominal level, had no visible effect on suppression of yellow phase based on the results of VGF tests. However, it became more effective at the maximum level, and was comparable to the normal HLW simulant with all three noble metal elements. The degree of conversion of the feed blends also appeared different depending on the nature of the noble metal element. In general, Ru had the least impact on the conversion of the feed blends at a given test duration while the reactions with Pd- and Rh-containing feeds appeared to be relatively sluggish, as suggested by abundant pores below the surface. The observed association between the yellow phase formation and Ru, Rh and Pd may be due to their roles as potential catalysts for redox reactions involving molybdenum, sulfate, and rhenium under melter processing conditions. Platinum group elements are well known catalysts for a variety of chemical reactions. It may be significant that at the temperatures typical for waste glass processing, Ru is likely present as RuO_2[13], Rh as oxide or metal[14], and Pd primarily as metal[15], and evidently Pd in its metallic form was most effective in yellow phase suppression in the VGF experiments reported in this work.

Ongoing work is also addressing the application of this approach to projected HLW streams with slightly different compositions, the tolerance of this strategy with respect to process variations, and larger-scale testing to support implementation at the Rokkasho facility.

ACKNOWLEDGMENTS

This work was carried out as a part of the research supported by Japan Nuclear Fuel Limited with Grant-in-Aid by the Ministry of Economy, Trade and Industry. The authors are grateful to Dr. Malabika Chaudhuri and the staff of the Vitreous State Laboratory for their assistance with this work.

REFERENCES

[1] E. Schiewer, H. Rabe and S. Weisenburger, in *Scientific Basis for Nuclear Waste Management V*, Vol. **11**, p 289, Materials Research Society (1982).

[2] R.S. Roth, J.R. Dennis, and H.F. McMurdie, in *Phase Diagrams for Ceramists*, Vol. **VI**, The American Ceramic Society, Inc. (1987).

[3] O. Pinet, J.L. Dussossoy, C. David, and C. Fillet, Glass Matrices for Immobilizing Nuclear Waste Containing Molybdenum and Phosphorus, *J. Nucl. Mat.*, **377**, 307 (2008).

[4] R.J. Short, R.J. Hand, and N.C. Hyatt, Molybdenum in Nuclear Waste Glasses - Incorporation and Redox State, in *Scientific Basis for Nuclear Waste Management XXVI*, vol. **757**, p. 141 (2002).

[5] I.L. Pegg, H. Gan, K.S. Matlack, Y. Endo, T. Fukui, A. Ohashi, I. Joseph and B.W. Bowan, Mitigation of Yellow Phase Formation at the Rokkasho HLW Vitification Facility, Paper 10107, *Proc. Waste Management Symposium*, Phoenix Arizona, USA. (2010).

[6] W. Lutze, in *Radioactive Waste Forms for the Future*, Edited by W. Lutze and R.C. Ewing, Elsevier Science Publishers B.V. (1988).

[7] W. Kot, H. Gan, and I.L. Pegg, The Process of Sulfur Incorporation in Waste Glass Melts of Various Compositions, *Ceramic Transactions*, vol. **107**, p. 441 (2000).

[8] D.A. McKeown, I.S. Muller, H. Gan, I.L. Pegg, and C.A. Kendziora, Raman Studies of Sulfur in Borosilicate Waste Glasses: Sulfate Environments, *J. Non-Crystalline Solids*, vol. **288**, p. 191-199 (2001).

[9] D.A. McKeown, I.S. Muller, H. Gan, I.L. Pegg, and W.C. Stolte, Determination of Sulfur Environments in Borosilicate Waste Glasses Using X-ray Absorption Near-Edge Spectroscopy, *J. Non-Crystalline Solids*, vol. **333**, p. 74 (2004).

[10] K.S. Matlack, S.P. Morgan, and I.L. Pegg, *VSL-00R3501-2*, Melter Tests with LAW Envelope A and C Simulants to Support Enhanced Sulfate Incorporation, Vitreous State Laboratory, The Catholic University of America, Washington, DC (2001).

[11] K.S. Matlack, I.S. Muller, W. Gong and I.L. Pegg, *VSL-07R7480-1*, Small Scale Melter Testing of LAW Salt Phase Separation, Vitreous State Laboratory, The Catholic University of America, Washington, DC (2007).

[12] K.S. Matlack, W.K. Kot, and I.L. Pegg, DM100 Melter Scale-Up Testing Using a Redistributed Formulation for Yellow Phase Mitigation, *VSL-12R2520-1*, Vitreous State Laboratory, The Catholic University of America, Washington, DC (2012)

[13] A. Bencan, M. Hovat, S. Nernik, J. Holc and M. Kosec, *J. Mater. Sci. Lett.*, **18** [19] 1563-1565 (1999).

[14] R. D. Shannon, *Solid State Commun.*, **6**, 139-143 (1968).

[15] P. Patnaik, in *Handbook of Inorganic Chemicals*, McGraw-Hill (2003).

COLD CRUCIBLE VITRIFICATION OF HANFORD HLW SURROGATES IN ALUMINUM-IRON-PHOSPHATE GLASS

S.V. Stefanovsky, S.Y. Shvetsov, V.V. Gorbunov, A.V. Lekontsev, A.V. Efimov, I.A. Knyazev, O.I. Stefanovsky, M.S. Zen'kovskaya
FSUE Radon
Moscow, Russia

J.A. Roach
Nexergy Technical, PLLC
Idaho Falls, ID, USA

ABSTRACT

Hanford site HLW surrogates were vitrified using a cold crucible inductive melting (CCIM) technology and the aluminum-iron phosphate based materials were produced under lab-scale conditions in a resistive furnace in 50 cc alumina crucibles, at small-scale unit in a ~130 mm inner diameter cold crucible, and at the bench-scale facility in a 236 mm inner diameter cold crucible. Wet slurry with a water content of ~50 wt.% was vitrified. Major process variables have been determined. At the bench-scale unit average slurry feeding, glass production and specific melt production rates, and melting ratio at vitrification of the slurry with ~50 wt.% water content with production of aluminum-iron-phosphate glass were ~1.8 kg/hr, ~0.8 kg/hr, ~4.2 kg/(dm^2·day) and ~77 kW·hr/kg of glass, respectively. The products were examined by X-ray diffraction and scanning electron microscopy with energy dispersive spectroscopy. The materials produced were found to be glass-crystalline. Magnetite- and hercinite-type spinels were present as major crystalline phases in all the samples. Orthophosphates, hematite, and zirconia-based solid solution also occurred. Ruthenium remained in oxidized form entering zirconia-based solid solution. Vitreous phase is strongly depleted with iron and enriched with sodium and aluminum.

INTRODUCTION

Vitrification is the world-wide accepted technology for immobilizing high level radioactive wastes (HLW). The current reference Joule Heated Ceramic Melter has limited throughput and waste loading capabilities for certain types of waste chemistries. Some of these include specific HLW tanks at Hanford, with potential application to the HLW calcine at Idaho, as well as some of the Low Activity Wastes at Hanford.[1,2] The development and deployment of advanced melter technologies could transform current and future HLW vitrification processes resulting in dramatic cost savings in operational and disposal costs. Moreover, the development of transformational, next generation melter concepts could pave the way for the use of vitrification for treatment of other radioactive, hazardous, or industrial wastes.

One candidate backup technology with a high chance of success is the vitrification of the Hanford HLW into a high waste loaded iron phosphate (FeP) glass using a cold crucible induction melter (CCIM).[3] Preliminary tests suggest a loading for certain Hanford HLW compositions in the CCIM FeP of as high as 60 wt%.[3,4] These FeP glasses are generally not amenable to conventional melters due to the high corrosivity of the melt at operating temperatures of 1150°C to 1200°C. However, because the CCIM does not use immersed electrodes to introduce the power into the melt, and uses a cooled outer shell that creates a protective layer of frozen glass around the perimeter (i.e., skull), the CCIM is not susceptible to these corrosion issues.[3-5]

The CCIM technology is currently used to producing borosilicate based HLW glasses in France[6] and ILW glasses in Russia.[7] It is also considered as alternative to Joule-heated ceramic melter for HLW vitrification in Russia.[7,8] Nevertheless, very little operational knowledge and design experience exists in the US for the CCIM technology[9-11] and, in general, a better understanding of the operational limits and parameters, and how to control them, must be gained prior to implementation in a radioactive environment. The purpose of this work is to evaluate suitability of the CCIM technology to producing FeP glasses with liquid-fed slurry feeding.

EXPERIMENTAL

Glass compositions and lab-scale tests

Glass formulations (Table I) were specified by Nexergy Technical PLLC. Recalculations of weight percents to molar percents shows that values of R-factor for the glasses

$$R = [Na_2O+3(Al_2O_3+Fe_2O_3)]/P_2O_5$$

is close to 3 and thus their compositions fall within orthophosphate range making the structure of glasses instable with a high tendency to devitrification.

Table I. Glass formulations.

Oxides	MS60AZ101		MS50AZ101-2		MS50AZ101-3	
	wt.%	mol.%	wt.%	mol.%	wt.%	mol.%
Al_2O_3	14.75	17.45	12.29	13.82	12.29	13.54
B_2O_3	-	-	-	-	5.00	8.07
CaO	0.84	1.81	0.70	1.43	0.70	1.40
CdO	1.30	1.22	1.08	0.96	1.08	0.94
Ce_2O_3	0.48	0.18	0.40	0.14	0.40	0.14
Cr_2O_3	0.28	0.22	0.23	0.17	0.23	0.17
Cs_2O	0.30	0.13	0.25	0.10	0.25	0.10
Fe_2O_3	22.60	17.07	18.84	13.53	18.84	13.25
La_2O_3	0.53	0.20	0.45	0.16	0.45	0.16
MnO	0.55	0.94	0.46	0.74	0.46	0.73
Na_2O	6.35	12.36	12.29	22.74	10.29	18.65
Nd_2O_3	0.39	0.14	0.33	0.11	0.33	0.11
NiO	1.00	1.62	0.83	1.27	0.83	1.25
P_2O_5	40.80	34.68	43.67	35.28	40.67	32.18
RuO_2	0.09*	0.08*	0.08*	0.07*	0.08	0.07
SiO_2	2.26	4.54	1.89	3.61	1.89	3.53
SnO_2	0.40	0.32	0.33	0.25	0.33	0.25
SO_3	0.23	0.35	0.19	0.27	0.19	0.27
ZrO_2	6.86	6.72	5.72	5.32	5.72	5.21
Total	100.00	100.00	100.00	100.00	100.00	100.00
R		3.34		2.97		3.08

* not introduced

Glasses with formulations given in Table I were prepared from reagent-grade $Al(OH)_3$, H_3BO_3, $CaCO_3$, $Cd(NO_3)_2{\cdot}4H_2O$, $Ce(NO_3)_3{\cdot}6H_2O$, $Cr(NO_3)_3{\cdot}9H_2O$, $CsNO_3$, Fe_2O_3, $La(NO_3)_3{\cdot}6H_2O$, MnO, $NaNO_3$, Nd_2O_3, $NiCl_2{\cdot}6H_2O$, H_3PO_4 (85%), RuO_2, SiO_2, $SnCl_2{\cdot}2H_2O$, Na_2SO_4, and ZrO_2. The chemicals were placed in alumina crucibles, thoroughly intermixed, and kept till complete gas release. Mixtures in crucibles were step-by-step heated in a resistive furnace to temperatures either 1500 °C (MS60AZ101) or 1350 °C (MS50AZ101-2 and MS50AZ101-3), kept at these temperatures for 1 hr and poured onto a metal plate.

The products were examined by X-ray diffraction (XRD) using a Rigaku D / Max 2200 diffractometer (Cu Kα radiation, 40 keV voltage, 20 mA current, stepwise 0.02 degrees 2θ) and scanning electron microscopy with energy dispersive spectrometry (SEM/EDS) using a JSM-5610LV+JED-2300 analytical unit.

As follows from XRD and SEM data all the melts even quenched onto a metal plate yield glass-crystalline materials. Hematite and spinel structure phases were found to be major constituents and two aluminum orthophosphate varieties were present as minor phases in both materials at 50 wt.% waste loading. At 60 wt.% waste loading aluminum phosphate also becomes a major phase whereas partly reacted zirconia and a phase with diffraction pattern close to $Na_3Fe_2(PO_4)_3$ are present as minor phases.

The sample MS50AZ101-2 with 50 wt.% waste loading is composed of major vitreous phase, hematite and spinel, and minor orthophosphate. Spinel is non-uniformly distributed within matrix glass forming aggregates of larger and smaller sized grains (Figure 1*a*).

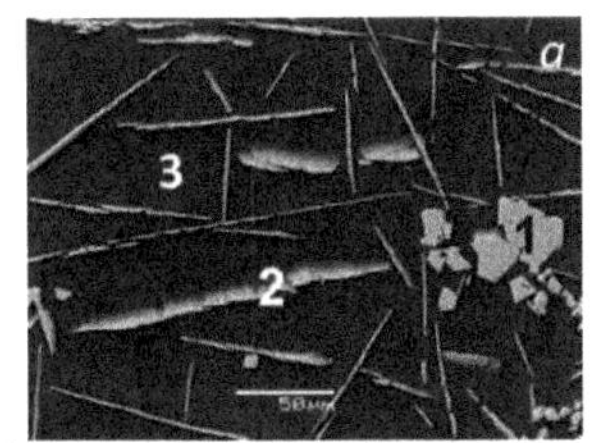

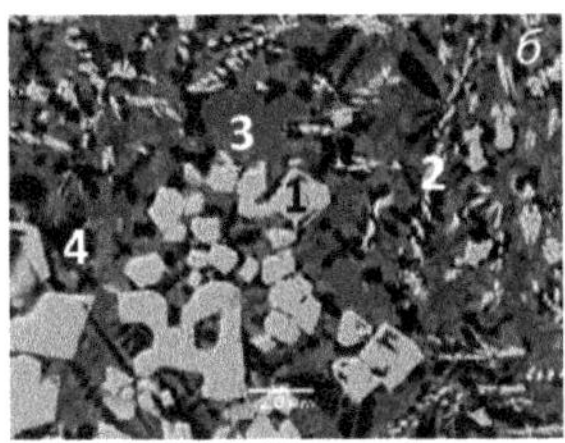

Figure 1. SEM images of MS50AZ101-2 (a) and MS60AZ101 (b) samples.
a: 1 – high-Fe spinel, 2 – hematite, 3 – glass; b: 1 – high-Fe spinel, 2 – hematite, 3 – Na-Al-Fe orthophosphate, 4 – $AlPO_4$+glass.

Chemical composition of matrix glass is some different in various areas (Table II). In the areas with low spinel content iron oxides concentration in glass is the highest and reaches up to ~25 wt.% whereas P_2O_5 concentration is the lowest among all. In the areas with high spinel content iron oxides concentration is lower (15-17 wt.%). Glass between hematite plates has intermediate iron oxides content (~18 wt.%).

Chemical composition of spinel is also variable in various zones of the sample (Table II). One variety contains higher iron oxides and lower Cr_2O_3 concentration than another one. Both them belong to magnetite/trevorite/hercinite ($FeFe_2O_4$/$NiFe_2O_4$/$FeAl_2O_4$) solid solution. One more Fe-bearing phase occurred in the material is hematite (gray narrow plates on SEM images – see Figure 1*a*). Its chemical composition is approaching to pure Fe_2O_3 (Table II).

The sample MS50AZ101-3 is composed of major glass and magnetite-type spinel and minor zirconia-based sold solution and hematite. Unlike the sample MS50AZ101-2 in the given sample magnetite-type spinel strongly prevails over hematite. Both the Fe-bearing phases are distributed in the glassy matrix. Ru remains in oxidized state entering zirconia-based solid solution containing also minor SnO_2. This phase is present as rare crystals white on SEM images.

The bulk of the sample MS60AZ101 is composed of dark matrix and aggregates of lighter (gray colored on SEM images) grains (Figure 1*b*). The matrix is inhomogeneous and consists of nanometric-sized crystals with Fe-Al-orthophosphate composition, where Fe content exceeds significantly Al content (Table II), vitreous phase, and, probably, minor hematite or magnetite. Formula of the orthophosphate phase is close to $Al_{0.4}Fe_{0.6}PO_4$. Interstitial glass has primarily aluminophosphate composition. Moreover phases with aluminum-iron phosphate and mixed sodium-aluminum-iron phosphate compositions, magnetite-type spinel, hematite and zirconia are present (Figure 1 and Table II).

Table II. Chemical Composition (wt.%) of Co-Existing Phases in the Lab-Scale Produced Materials.

Oxides	MS50AZ101-2							MS60AZ101-1							
	LSG Aver	LSG Scan	HSG	GBH	S-1	S-2	H	Fe-Al-P	Al-Fe-P	Na-Fe-Al-P-1	Na-Fe-Al-P-2	S-1	S-2	H	SS
Na_2O	17.8	19.1	18.5	17.9	-	-	-	6.9	2.3	16.3	10.5	-	-	-	-
Al_2O_3	12.5	12.4	14.4	14.7	6.4	5.9	2.3	13.5	34.1	11.4	10.4	8.2	3.8	5.5	-
SiO_2	2.6	2.4	1.4	2.8	-	-	-	3.6	2.6	5.3	4.4	-	-	-	-
P_2O_5	39.2	37.7	45.6	44.0	-	-	-	36.4	50.8	41.9	41.3	-	3.8	10.2	-
SO_3	-	0.2	0.2	0.2	-	-	-	-	-	0.3	0.3	-	-	-	-
Cl	-	-	-	-	-	-	-	-	-	0.7	0.4	-	-	-	-
CaO	0.8	0.7	0.9	1.0	-	-	-	0.9	-	1.2	1.6	-	-	-	-
Cr_2O_3	-	0.1	0.2	0.2	1.0	-	-	0.1	-	-	-	5.1	0.3	-	-
MnO	-	0.4	0.5	0.5	-	-	-	-		0.5	0.5	-	-	-	-
Fe_2O_3	25.3	22.8	15.2	18.0	76.8	72.3	94.6	36.0	8.4	20.3	26.8	75.4	79.9	77.3	4.5
NiO	1.1	1.0	0.4	0.8	15.8	21.9	-	1.4	-	-	-	10.8	9.7	3.7	-
ZrO_2	-	0.4	1.2	1.6	-	-	-	0.6	1.5	0.5	0.5	-	-	-	93.6
CdO	0.9	1.1	1.0	0.8	-	-	-	1.2	-	1.7	1.7	-	-	-	-
SnO_2	-	0.3	0.1	0.1	-	-	3.1	-	-	-	0.2	-	-	-	1.9
Cs_2O	-	0.2	0.2	0.2	-	-	-	-	-	-	0.3	-	-	-	-
La_2O_3	-	0.7	0.5	0.6	-	-	-	-	-	-	0.5	-	-	-	-
Ce_2O_3	-	0.5	0.4	0.5	-	-	-	-	-	-	0.5	-	-	-	-
Nd_2O_3		0.3	0.3	0.4	-	-	-	-	-	-	0.3	-	-	-	-
Total	100.0	100.2	101.0	100.0	100.0	100.1	100.0	100.6	99.6	99.9	100.0	99.4	97.5	96.7	100.0

LSG – low-spinel glass area, HSG – high-spinel glass area, GBH - glass between hematite plats, S-1 and S-2 – two different types of spinel, H – hematite, SS – zirconia-based cubic solid solution.

Chemical composition of aluminum-iron phosphate (black on SEM image) is close to $Al_{0.9}Fe_{0.1}PO_4$. Among the two Na-Al-Fe phosphate phases, one is some enriched with sodium and depleted with iron as compared to another one (Table II). This phase has gray color on SEM images (Figure 1*b*), its formula corresponds approximately to $Na_{2.6}Al_{1.1}Fe_{1.3}(PO_4)_3$ and it is probably responsible for the reflections assigned to sodium-iron orthophosphate $Na_3Fe_2(PO_4)_3$ on XRD pattern. Another Na-Al-Fe phosphate phase concentrates minor elements (S, Cl, Mn, Zr, Cd, Sn, Cs, La, Ce, Nd), has light-gray color on SEM images and may be attributed as glass.

Two varieties of spinel with some different iron and aluminum oxide contents and hematite were found (Table II). Their formulae may be represented as $(Fe_{0.67}Ni_{0.33})^{2+}(Fe_{1.48}Cr_{0.15}Al_{0.37})^{3+}O_4$ and $(Fe_{0.68}Ni_{0.32})^{2+}(Fe_{1.80}Cr_{0.01}Al_{0.19})^{3+}O_4$. Spinel forms relatively large (up to ~100 μm in size) individual or aggregated crystals (Figure 1*b*). Hematite is present as elongated crystals forming chains and containing trace of Al, P and Sn (Table II) probably due to capture of surrounding material. Zirconia-based solid solution (Table II) occurs as rare individual white grains.

Small-scale cold crucible test

The glass with a MS50AZ101-3 formulation was produced in a small-scale unit equipped with a 130 mm inner diameter cold crucible energized from a 60 kW generator operated at 1.76 MHz and a simplified off-gas system (Figure 2).

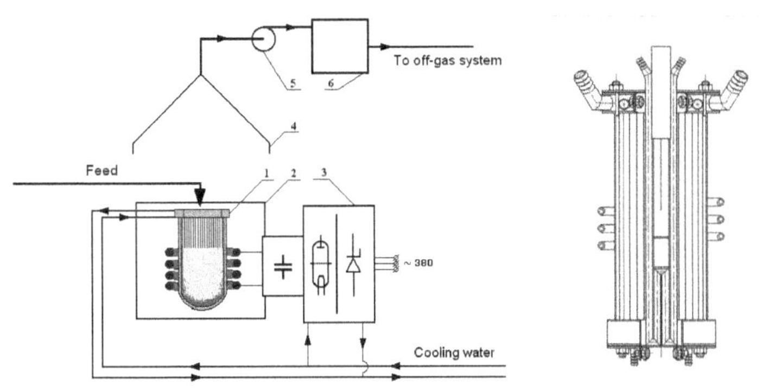

Figure 2. Flow sheet of the FSUE Radon CCIM unit (left) and the cold crucible (right).
1 – cold crucible with inductor, 2 – protective screen, 3 – high-frequency generator, 4 – off-gas unit, 5 – fan, 6 – off-gas sleeve filter.

The slurry was prepared using the same chemicals as in lab-scale tests. Stability of the slurry was assessed. It has been found that at preparation of waste surrogate followed by addition of water in amount of 50 wt.% major segregation process occurs for the first hour of keeping. No significant changes since that time occurred. After addition of glass formers foaming on the surface of solution occurred and foam remained stable. The most significant changes at intermixing of waste surrogate and glass formers took place for the first 30 minutes of keeping.

The height of inductor was 90 mm. Melt pouring level was positioned by 30 mm below the lower edge of inductor. Cooling water temperature at the cold crucible outlet was measured using a DS18 Dallas Semiconductor electronic digital thermometer and off-gas temperature at the fan outlet was measured using a chromel/alumel thermocouple. Melt surface temperature was measured using an optical fiber pyrometer with a measuring range between 350 and 1550 °C and melt temperature in the near-bottom zone was measured using a Pt/Rh thermocouple. Melt surface status was monitored online both visually and using a web-camera with a color filter.

Starting batch composition with a target formulation of (wt.%/mol.%) 17.4/32,3 Na_2O, 10.9/7,8 Fe_2O_3, 14.8/16,7 Al_2O_3, 53.4/43,2 P_2O_5 (R ≈ 2,45) was prepared from sodium orthophosphate ($Na_3PO_4 \cdot 10\ H_2O$), 85% orthophosphoric acid (H_3PO_4), ferrous oxide (Fe_2O_3) and alumina (Al_2O_3). The batch in amount of 1.6 kg was charged onto a bottom of the cold crucible and a 200 mm height/30 mm outer diameter SiC rod to initiate melting was inserted axially in the centre of the batch mixture. When starting melt was produced (at ~800 °C) the SiC rod was removed (at that the calcine arch destroyed spontaneously), melt was kept and heated to a temperature of ~1300 °C. Total duration of start-up process was about 1 hr. Input generator's power and inductor's voltage at start-up were 13.5 kW and 6.0 kV, respectively.

The MS50AZ101-3 slurry with ~50 wt.% water content was fed in portions for 3 to 5 sec followed by 12 to 25 sec breaks. The batch was fed till filling of operation volume with respect to upper coil of the inductor. The melt was kept for 15 min. At that, generator's power was built-up to maintain melt surface temperature. After keeping melt volume reduced probably due to completion of glass formation and gas release processes and melt surface temperature reduced as well (to ~950 °C). Portion of the melt was poured into a container. Melt seemed to be "shorter" than previously studied borosilicate glassmelts but had sufficient flowability and spreadability. Average slurry feed rate was 2 L/hr being determined by keeping melt surface area within the range of 950-1150 °C.

XRD patterns of the materials produced in the 130 mm inner diameter cold crucible demonstrate nearly amorphous nature of the sample produced from fresh slurry (Figure 3, *1*). Only trace of incompletely dissolved quartz was present (reflection at ~3.34 Å). Actually the specimen #1 contained significant fraction of glass produced from starting batch with simplified composition being some different from MS50AZ101-3 in particular enriched with P_2O_5 as compared to 50AZ101-3 glass and thus having lower R-factor (~2.45 rather than ~3.08 as for 50AZ101-3 glass).

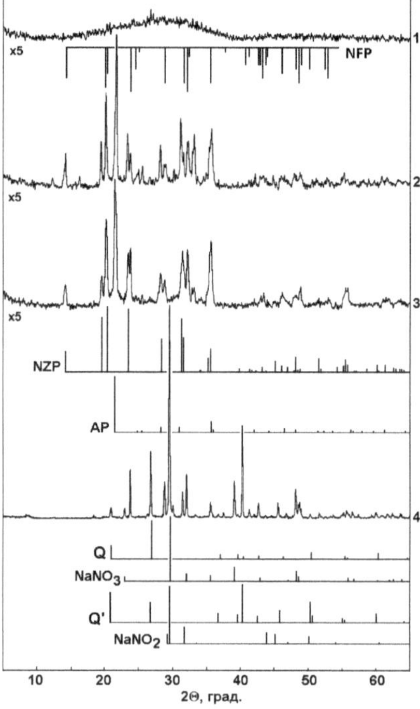

Figure 3. XRD patterns of the specimens produced in the 130 mm inner diameter cold crucible and sampled from containers after spontaneous cooling (1,2) and melt surface (3), and deposit sampled from sleeve filter (4). 1 – first pour, 2 – last pour; 3 – melt surface (calcine "cold cap"); AP – $AlPO_4$, NFP – $Na_3Fe_2(PO_4)_3$, NZP – $NaZr_2(PO_4)_3$, Q – SiO_2 (quartz).

During slurry feeding and substitution for starting glass as material composition became closer to target one product crystallization occurred (Figure 3, *2*). Aluminum orthophosphate $AlPO_4$ (berlinite), sodium-iron orthophosphate $Na_3Fe_2(PO_4)_3$ and sodium-zirconium orthophosphate $NaZr_2(PO_4)_3$ (kosnarite) were found to be major and secondary in abundance crystalline phases, respectively. Minor magnetite-type spinel was also present. Same phases are present in the upper layer (calcine, "cold cap") of the melt in the cold crucible (Figure 3, *3*). This shows that the phases are probably formed during feed heating and calcining rather than crystallized from the melt. Chemical composition of these phases is some varied. Berlinite is rather not pure $AlPO_4$ but has a composition something like $Na_{3-x}Al_{x/3}PO_4$.

Comparison of the samples with the same compositions produced in cold crucible and in alumina crucibles in resistive furnace demonstrated their phase compositional similarity taking into account difference in their thermal history. The sample MS50AZ101-2 produced at 1350 °C and poured onto a metal plate (quenched) is composed of major vitreous and iron-bearing crystalline phases hematite Fe_2O_3 and minor aluminum orthophosphate $AlPO_4$ (hexagonal variety). The sample with MS60AZ101 formulation produced at 1500 °C and quenched was found to be some higher crystallized as expected due to higher waste loading and had lower content of vitreous phase and different crystalline phases: magnetite and orthorhombic aluminum orthophosphate as major phases, and sodium-iron orthophosphate $Na_3Fe_2(PO_4)_3$ and hematite as minor phases (Figure 3).

The boron-containing samples with MS50AZ101-3 formulation both undoped and doped with RuO_2 also had significant degree of crystallinity. The Ru free specimen is composed of spinel and orthorhombic $AlPO_4$. Trace of extra phases in particular spinel also occurs.

Specimen of the carry-over deposited on sleeve filter was also examined by XRD. Sodium nitrate, $NaNO_3$, and sodium nitrite, $NaNO_2$ were found to be major phases whereas silica, SiO_2 two varieties) was present as minor phase (Figure 3, *4*). Silica is a product of mechanical carry-over of highly-dispersed sand with off-gas. One more phase is unidentified.

Bench-scale cold crucible test

The test was conducted in a bench-scale unit with a 521 mm height/236 mm inner diameter cold crucible (Figure 4) energized from a 60 kW oscillation power generator operated at a frequency of 1.76 MHz and equipped with a feeding and off-gas systems (Figure 4) previously used in tests on vitrification of high-Na/Al/Fe Savannah River Site wastes with production of borosilicate glass.[7,8] Melt is poured from a side pouring gate till a level indicated on Figure 5*a* (11). The crucible is manufactured from water-cooled 12 mm inner diameter stainless steel pipes daubed with a protective putty based on aluminum-chromium-zirconium phosphate binder and taped around with glass fabric (Figure 5*b*). The crucible is covered by a 5 mm thickness water-cooled lid. The 16 mm thickness crucible bottom, 295 mm inner diameter induction coils and off-gas pipe are water-cooled as well. The crucible is installed in a protection box (Figure 5*c*). The off-gas system consists of a coarse (sleeve) filter, heat-exchanger, condensate collector, and electric heater to prevent formation of condensate in the off-gas pipe (Figure 4). Slurry was fed using a peristaltic pump with variable productivity.

The crucible at the bottom contained residual borosilicate glass from previous tests.[7,8] Starting batch was prepared from chemicals in amount of ~10 kg. Moreover iron phosphate glass produced in 130 mm inner diameter cold crucible was used. This glass was fed on the bottom of the 236 mm inner diameter cold crucible followed by charging of the crucible with starting batch with simplified composition (Na, Al, Fe and P compounds only). SiC rod was inserted in the bulk axially.

Feed compositions were prepared in portions from the same chemicals. Chemicals except H_3PO_4 were intermixed in containers followed by adding 85% H_3PO_4 and water that caused vigorous reaction with gas release due to decomposition of nitrates, aluminum hydroxide and calcium carbonate. Four containers of feed - two with MS50AZ101-3 composition (Ru free and Ru-bearing) and two with MS60AZ101 composition (also Ru free and Ru-bearing) were prepared. In total 16272 kg of slurry (83.441 kg with MS50AZ101-3 and 79.279 kg with MS60AZ101 compositions). Each of the slurries had water content of ~50 wt.%.

The test was conducted in four steps: 1) process start-up and starting melt production; 2) processing of slurry with MS50AZ101-3 composition; 3) processing of slurry with MS60AZ101 composition; 4) process shut-down. Summary of process variables is given in Table III.

Starting melt was produced at a value of oscillating power within the range of 37.7-59.4 kW for 7 hrs 05 min. In total 10 kgs of batch have been fed in the cold crucible and glass in amount of 6 kg has been obtained.

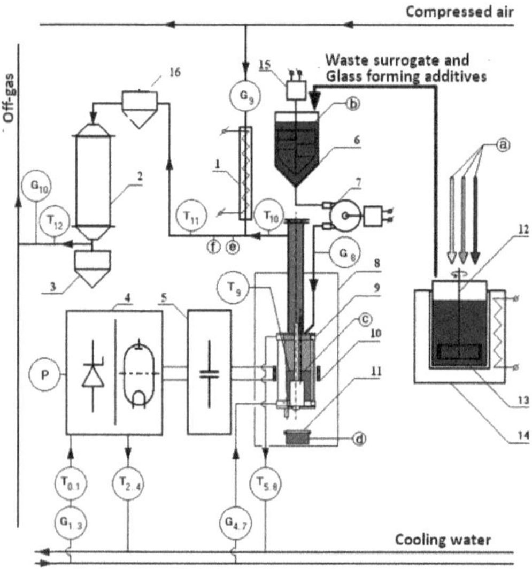

Figure 4. Flow Sheet of the CCIM Based Bench-Scale Plant and Sampling Points.
Equipment: 1- electric heater of compressed air, 2- heat-exchanger, 3- condensate collector, 4- high-frequency energizer, 5- battery of capacitors, 6- batch vessel, 7- feeder, 8- process box, 9- cold crucible, 10- inductor, 11- container for glassmelt, 12- stirrer, 13- slurry batch preparation mixer, 14- electric furnace, 15 – stirrer, 16 - filter. Sampling points: a- raw materials (source reagents), b- slurry, c- "yellow phase", d- glass, e- aerosols, f- off-gas. The parameters to be controlled: $T_{0...8}$ – cooling water temperature, T_9 – melt temperature, T_{10} – off-gas temperature at crucible outlet, $T_{11...13}$ – off-gas temperature at heat-exchanger inlet and outlet, $G_{1...7}$ – cooling water flow rate in the induction system, G_8 – slurry feeding rate, G_9 – compressed air flow rate, G_{10} – off-gas flow rate.

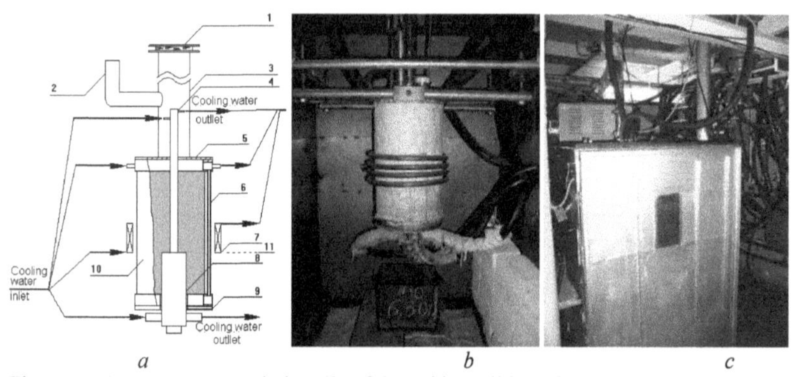

Figure 5. The scheme (*a*) and view (*b*) of the cold crucible and protection box (*c*).
a: 1 – window for visual monitoring, 2 – off-gas sampling pipe, 3 – off-gas pipe, 4 – pouring gate holder, 5 – crucible lid, 6 – crucible wall, 7 – inductor, 8 – pouring gate, 9 – crucible bottom, 11 – pouring level

At the step 2 the MS50AZ101-3 slurry was fed in portions in amount of 3 to 5 kg followed by keeping till complete melt homogenization and pouring into canisters. Every next portion was fed after melting of ~50% of previous portion of feed. At first the feed not containing RuO_2 was processed then the RuO_2-bearing slurry was fed. Total duration of the step 2 was 49 hr 05 min (49.1 hr). Average slurry feeding and glass production rates were ~1.7 kg/hr and ~0.7 kg/hr, respectively. Specific melt production rate and melting ratio were ~3.7 kg/(dm^2·day) and ~85 kW·hr/kg of glass, respectively.

Table III. Major CCIM process variables.

Major process variables	Start	MS50AZ 101-3	MS50AZ 101-3Ru	MS60AZ 101	MS60AZ 101Ru
Vitrification duration, hr:min	7:05	22:25	26:40	21:35	19:35
Average oscillating power, kW	47.7	57.2	59.3	60.3	60.1
Mass of vitrified slurry (dry batch at start), kg	(10)	41.72	41.72	39.64	39.64
Mass of produced glass, kg	6	16.60	16.60	17.50	17.50
Melt surface temperature, °C	1020-1070				
Melt temperature at a depth of 50 mm, °C	1450-1550				
Off-gas temperature (T10, Fig. 4), °C	200-350				
Average slurry feeding rate, kg/hr	-	1.86	1.57	1.84	2.02
Average glass production rate, kg/hr	0.86	0.74	0.62	0.81	0.89
Average specific energy expenses, kW·hr/kg	-	30.8	37.8	32.8	29.8
Average melting ratio, kW·hr/kg	55.5	77.3	95.6	74.4	67.5
Specific melt production rate, kg/(dm^2·day)	4.72	4.06	3.41	4.45	4.89

At Step 3 similarly to Step 2 Ru free slurry was fed first followed by feeding of the Ru-bearing slurry. Total duration of the step 3 at feeding of slurry MS60AZ101 was 41 hr 10 min (41.2 hr). Average slurry feeding and glass production rates were ~1.9 kg/hr and 0.9 kg/hr, respectively. Specific melt production rate and melting ratio were ~4.7 kg/(dm^2·day) and ~70 kW·hr/kg of glass, respectively.

In total average slurry feeding, glass production and specific melt production rates and melting ratio at vitrification of the slurry with ~50 wt.% water content with production of FeP glass were ~1.8 kg/hr, ~0.8 kg/hr, ~4.2 kg/(dm^2·day) and ~77 kW·hr/kg of glass, respectively.

Heat balance was obtained taking into account heat losses for cooling of generator (Q_e), cooling of cold crucible (Q_c), with off-gas including slurry water evaporation and heating of vapor (Q_g), and heat power for melting after completion of water evaporation (glass formation, melt homogenization and melt heating - Q_m). Effective power is determined as

$$Q_{eff} = Q_m + Q_g$$

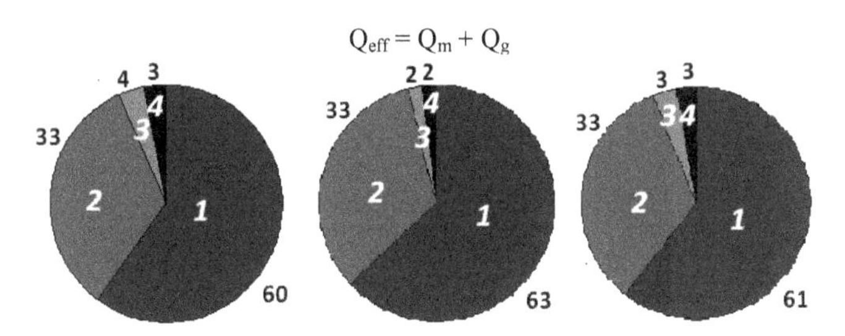

Figure 6. Heat power distribution (%) at various stages of vitrification process: slurry feeding (left), melt homogenization (middle) and average values (right).
1 – crucible cooling, *2* – generator cooling, *3* – off-gas, *4* – melt heating.

Electric losses arose from electric energy transformation in the generator were not taken into account. Passport value of efficiency of the generator established by its manufacturer is 73%. Figure 6 demonstrates low efficiency (4 to 7%) due to very high heat losses for crucible and generator cooling. These losses also include heat expenses for water evaporation.

XRD patterns of the materials sampled from canisters after cooling in air are shown on Figure 7. The specimen sampled from canisters #0 and #1 contains significant fraction of starting glass including glasses remained in the cold crucible after previous tests and placed on the bottom of cold crucible. Its composition (wt.%): 26.0 Na_2O, 5.3 Al_2O_3, 7.1. Fe_2O_3, 4.9 B_2O_3, 17.0 SiO_2, 39.4 P_2O_5 is markedly different from the target one (Table IV and Figure 7*a*). It is composed of vitreous and several crystalline phases. Sodium-aluminum orthophosphate $Na_{3-x}Al_{x/3}PO_4$ and undecomposed hexagonal structure sodium sulfate Na_2SO_4 were found to be major phases. A number of reflections due to minor source and partly reacted compounds also occur.

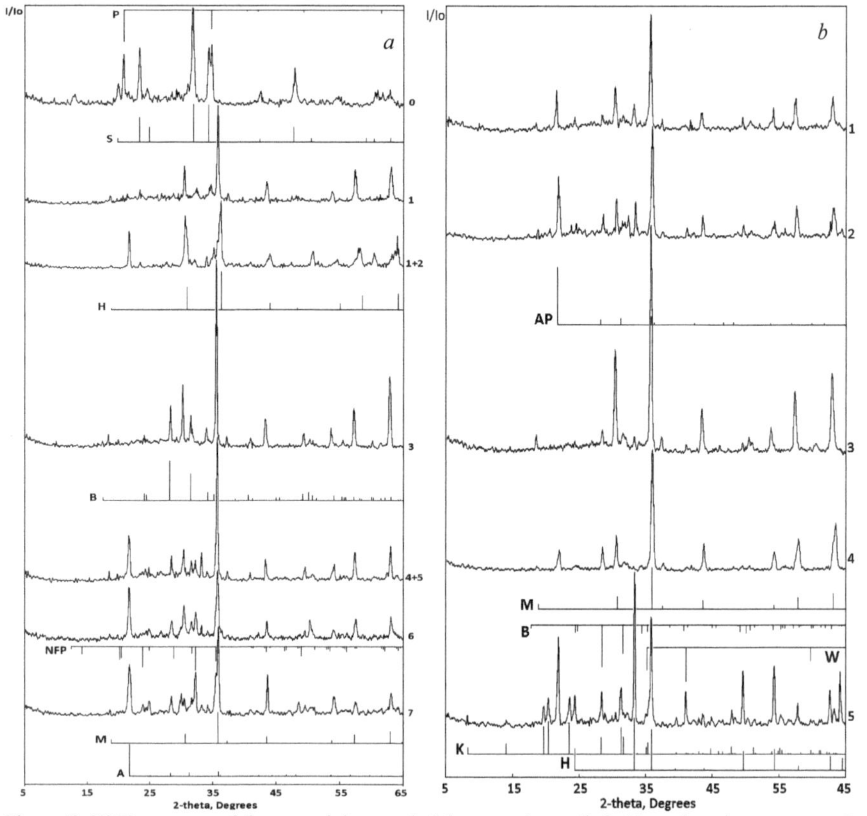

Figure 7. XRD patterns of the materials sampled from canisters (left, Nos of canisters are on the right) and during melt pouring (right) in alumina crucibles (1-3) and from residual ("dead") volume (4) and "skull" in the cold crucible (5). A – $AlPO_4$ (berlinite), AP – $AlPO_4$ (orthorhombic), B – monoclinic ZrO_2 (baddeleyite), H – $FeAl_2O_4$ (hercinite), K – $NaZr_2(PO_4)_3$; NFP – $Na_3Fe_2(PO_4)_3$, M – Fe_3O_4, P – $Na_{1-x}Al_{x/3}PO_4$, S – Na_2SO_4, W – FeO (wuestite).

The specimen sampled from canister #2 is predominantly crystalline. It is composed of two spinel structure phases different in chemical composition (Table V) and lattice parameters, aluminum orthophosphate-based, sodium aluminosilicate-based, and zirconia-based (tetragonal ZrO_2-based solid solution – $Zr_{0.78}Sn_{0.16}Fe_{0.06}O_{1.97}$) phases distributed in a matrix vitreous phase with very complex chemical composition (Tables IV and V). Larger spinel crystals are some enriched with alumina and depleted with iron oxides: $(Mn_{0.03}Fe_{0.27}Ni_{0.40}Zn_{0.30})^{2+}(Al_{0.86}Cr_{0.02}Fe_{1.12})^{3+}O_4$ whereas smaller crystals are enriched with iron oxides – $(Mn_{0.05}Fe_{0.35}Ni_{0.25}Zn_{0.35})^{2+}(Al_{0.72}Cr_{0.01}Fe_{1.27})^{2+}O_4$. Both them relate to hercinite/trevorite/magnetite solid solution. Matrix is composed of high-phosphate and high-silicate constituents. High-phosphate phase concentrates minor components (Table IV).

The specimen sampled from canister #3 is composed of major magnetite-type spinel – $(Fe_{0.74}Ni_{0.26})^{2+}(Al_{0.49}Cr_{0.12}Fe_{1.39})^{3+}O_4$, secondary in abundance partly reacted baddeleyite (monoclinic zirconia) and minor glass (Tables 4 and 5 and Figures 7*a* and 8). Moreover, rare grains of mixed Fe/Zr oxide were revealed. The specimens sampled from canisters ##4, #5 (4+5) and #6 (6) have similar phase composition but extra aluminum orthophosphate-based phase is also present. The latter is dark-gray bulk supposedly formed by glass with sodium-aluminum-iron phosphate composition containing other minor components: SiO_2, CaO, MnO, CdO, La_2O_3, Ce_2O_3. Chemical composition of the spinel phase is based on magnetite/trevorite solid solution with dominance of magnetite (Table V). White inclusions are zirconia-based solid solution containing Ru (Table IV). Black constituent is aluminum orthophosphate (Table IV).

The material sampled from Canister #6 consists of dark-gray matrix with black inclusions, gray individual or aggregated grains of spinel phase and needle-like white crystals (Figure 8). Dark-gray matrix is supposed to be phosphate-based glass. Black is aluminum orthophosphate (Table IV). Chemical composition of spinel is close to magnetite – $(Ni_{0.22}Fe_{0.78})^{2+}(Al_{0.43}Cr_{0.09}Fe_{1.48})^{3+}O_4$. White crystals are zirconia-based solid solution (Table V).

Table IV. Chemical Composition (wt.%) of Phosphate Phases in the Materials Sampled from Canisters and at Melt Pouring.

Oxides	Canister 2		Canister 3			Canister 6		Canister 7				Pour in Can 2	Pour in Can 7
	Lighter	Darker	Lighter	Darker	Actual	Lighter	Darker	Lighter	Darker	Black	Actual		
Na_2O	25,74	25,76	10,94	5.55	12.48	13.73	2.38	14.01	19.07	1.61	7.00	20.61	15.16
Al_2O_3	20,26	30,46	16,71	32.28	14.76	13.71	35.80	8.70	15.58	38.11	14.07	19.40	13.03
SiO_2	21,48	33,11	6,21	5.11	9.02	5.25	3.06	4.37	6.17	2.15	3.43	12.57	5.19
P_2O_5	10,20	6,80	44,91	48.06	17.89	41.64	54.85	44.29	47.24	54.88	29.40	30.41	42.15
SO_3	0.21	=-	-	-	-	0.19	-	0.22	0.22	-*	-	-	-
CaO	4,80	1,13	1,32	0.66	1.03	1.25	-	2.50	0.98	-	0.77	1.76	1.58
Cr_2O_3	0.08	-	-	-	0.12	-	-	-	-	-	0.22	-	-
MnO	0,35	-	0,65	-	0.65	0.54	-	1.10	0.32	-	0.48	0.67	0.40
Fe_2O_3	4,15	2,66	16,37	8.61	34.94	19.59	3,01	27.40	16.80	3.24	35.53	9.06	16.35
NiO	0.09	-	-	-	1.95	-	-	-	-	-	1.83	-	-
ZnO	3.02	-	-	-	-	-	-	-	-	-	-	1.11	-
ZrO_2	3,02	-	-	-	5.85	-	-	-	-	-	6.01	-	3.96
CdO	1.19	-	1,42	-	-	1.13	-	2.71	0.17	-	-	-	1.55
SnO_2	0.19	-	-	-	-	-	-	-	-	-	-	-	-
Cs_2O	0.22	-	0,23	-	-	-	-	0.73	1.03	-	-	-	0.24
La_2O_3	0.57	-	0,45	-	-	0.64	-	0.68	1.27	-	-	0.83	0.46
Ce_2O_3	0.39	-	0,43	-	-	0.39	-	0.54	1.62	-	-	0.59	0.42
Nd_2O_3	0,24	-	0,35	-	-	0.39	-	0.42	0.54	-	-	0.78	-
Total	96,20	99,92	99,97	100.27	98.69	98.44	100,0	106.71	110.47	100.00	98.74	97.78	100.47
B_2O_3*	5.00	-	-	-	-	-	-	-	-	-	-		

Table V. Chemical Composition (wt.%) of Oxide Phases in the Materials Sampled from Canisters and at Melt Pouring.

Oxide	Canister 2			Canister 3			Canister 6		Canister 7		Pour in Can 2		Pour in Can 7	
	S-1	S-2	O	S	O-1	O-2	S	O	S	O	S	O	S	O
Al_2O_3	20.58	16.93	-	11.13	-	-	9.58	-	9.25	-	7,59	-	8.32	-
Cr_2O_3	0.81	1.03	-	4.15	-	-	2.96	-	1.93	-	0,30	-	0.49	-
MnO	1.02	1.29	-	-	-	-	-	-		-	0,64	-	1.52	-
Fe_2O_3	52.37	59.35	4.5	75.51	6.03	49.81	79.38	4.43	85.84	8.52	81,48	4,06	82.83	4.94
NiO	14.16	8.52	-	8.72	-	-	7.18		6.59	-	5,35	-	7.00	-
ZnO	11.41	13.01	-	-	-	-	-	-	-	-	2,93	-	-	-
ZrO_2	-	-	93.6	-	87.78	30.34	-	89.46	-	65.55	-	84,58	-	87.15
SnO_2	-	-	1.9	-	4.63	4.23	-	6.11	-	10.24	-	4,59	-	5.39
RuO_2	-	-	-	-	2.18	5.33	-	-	-	5.77	-	-	-	2.94
Total	100.35	100.13	100.0	99.51	100.61	89.71	99.10	100.00	103.61	90.08	98,29	93,23	100.15	100.42

The Ru-bearing specimen from canister #7 is composed of two orthophosphate phases (sodium-iron and sodium-aluminum based) different in color on SEM images (lighter and darker, respectively), aluminum orthophosphate (black on SEM image), magnetite-type spinel (gray grains) and zirconia-based solid solution (white on SEM images) – see Figures 7 and 8 and Tables IV and V. No separate Ru-bearing phase has been found. Ru is present as isomorphic dopant in the zirconia-based phase $(Zr,Sn,Ru)O_2$ (also containing trace of Fe).

Chemical compositions of both the sodium-aluminum-iron phosphate phases have R ≈ 3.2, i.e. correspond to orthophosphate compositional range and are suggested to be crystalline phases. They probably form the finest aggregates $Na_3Fe_2(PO_4)_3$ and $Na_3Al_2(PO_4)_3$ grains with simplified formula $Na_2(Fe,Al)_2(PO_4)_3$. $AlPO_4$ is present as a separate phase. This is in a good agreement with XRD data.

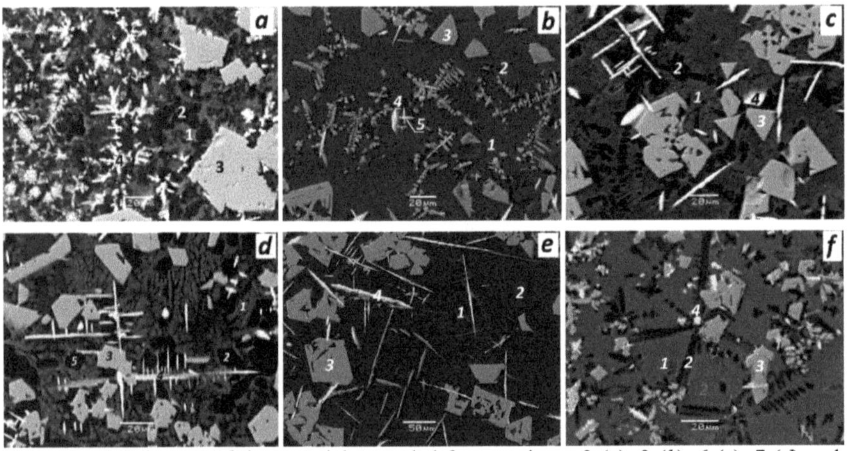

Figure 8. SEM images of the materials sampled from canisters 2 (*a*), 3 (*b*), 6 (*c*), 7 (*d*) and at melt pouring in canisters #2 (*e*) and #7 (*f*). *1* and *2* – lighter and darker phosphate phases (see Table IV), *3* – spinel, *4* – oxide phase, *5* – $AlPO_4$ (Table IV).

Specimens were also sampled at melt pouring in 50 cc alumina crucibles and cooled in air. Higher cooling rate as compared to the samples from canisters resulted in much lower content of phosphate phases segregated at late stages of crystallization than in the samples from canisters. As a result all the materials sampled at melt pouring are majorly composed of glass and spinel (magnetite or/and hercinite). Zr or Al phosphates and hematite are present as secondary in abundance or minor phases. Minor partly reacted monoclinic or tetragonal zirconia containing other quarterly charged ions (Sn^{4+}, Ru^{4+}) occurred as well (Figure 7*b*).

The specimen sampled from canister #2 is composed of vitreous and crystalline phases, mainly spinel. Difference in phase composition of the materials sampled at melt pouring is as follows. The material 101/1 sampled at pouring of Ru free MS50AZ101-3 glassmelt contains a phase close to sodium-aluminum orthophosphate (Figure 7*b 1,2*) whereas the Ru-bearing material 101/2 (Figure 7*b*, *2*) sampled at pouring of MS60AZ101 glassmelt contains aluminum orthophosphate based phase ($AlPO_4$) and higher baddeleyite and spinel content. Moreover, position of major reflections due to spinel phase on XRD pattern of the second material corresponds to hercinite rather than magnetite as for the first one. Same material poured onto a metal plate experienced the fastest cooling rate contains only spinel (magnetite) and trace of partly-reacted baddeleyite because no time to crystallize phosphate-based phases. SEM data show that the sample 101/1 is composed of dark-gray bulk with Na-Al-Fe phosphate composition, light-gray spinel grains and needle-like white crystals of zirconia-based phase (Figure 7*b* and Tables IV and V). Chemical composition of the bulk is some variable but remains within the orthophosphate compositional range. Spinel and zirconia-based solid solution have rather constant chemical composition (Table V). The material 101/2 sampled at pouring of Ru-bearing MS60AS101 glass (Figure 7*b* and Tables IV and V) is similar in phase composition to previous sample 101/1. It is composed of phosphate matrix, spinel, and zirconia-based solid solution containing RuO_2 looking gray, light-gray and white on SEM images, respectively. The darkest on SEM images elongated crystals are aluminum orthophosphate containing minor Na, Si and Fe oxides ($Al_{0.79}Na_{0.14}Fe_{0.10}P_{0.91}Si_{0.06}O_4$).

Chemical composition of the material located below pouring level in the cold crucible ("dead volume") is some different from the material produced from fresh feed because contains some fraction of the batch with different composition used for production of starting melt. Nevertheless, similarly to other materials obtained it is composed of magnetite-type spinel, baddeleyite and some glass (Figure 7*b*, *4*). As it is seen at higher magnification trace of phosphate phases is also present. As expected "skull" has the most complex phase composition because contains a number of unreacted source, partly-reacted and newly-formed phases such as iron oxides: hematite Fe_2O_3, magnetite Fe_3O_4 and wuestite FeO, kosnarite $NaZr_2(PO_4)_3$, baddeleyite ZrO_2 (monoclinic structure solid solution), berlinite $AlPO_4$ (Figure 7, *5*). Actual chemical composition of all these phases is different from nominal one and, as a result, position of reflections on XRD pattern connected to lattice parameters differs from tabular values for standards.

Occurrence of crystalline phases in the materials will impact on their performance. A study of the effect of phase composition on materials properties is in progress.

As followed from off-gas analytical data the most volatile constituents were Na, Cl and S. Carry-over took place in gaseous, aerosol and mechanical forms. The latter form results in losses of iron oxides, silica, and some phosphates.

CONCLUSION

Lab-scale studies demonstrate that both quenched and slowly cooled melts yield glass-crystalline materials composed of vitreous phase, spinel and orthophosphate phases as well trace of unreacted or partly-reacted baddeleyite. Vitrification test in a small-scale plant equipped with a 130 mm inner diameter cold crucible was conducted. Average slurry feed rate was up to 2 L/hr being determined by keeping melt surface area within the range of 950-1150 °C. The products were mostly glass-crystalline and contained orthophosphates distributed in the vitreous matrix. The four-days test was conducted in a bench-scale unit with a 521 mm height/236 mm inner diameter cold crucible energized from a 60 kW oscillation power generator operated at a frequency of 1.76 MHz and equipped with a feeding and off-gas systems. Mixture of waste surrogate and glass formers were fed into the cold crucible as a slurry with a water content of ~50 wt.%. Major process variables at vitrification of the slurry with ~50 wt.% water content with production of iron-phosphate glass were as follows: average slurry feeding rate ~1.8 kg/hr, glass production rate ~0.8 kg/hr, specific melt production rate ~4.2 kg/(dm^2·day), and melting ratio ~77 kW·hr/kg of glass. The process had low efficiency – 4-7% due to high expenses for water evaporation and cooling of cold crucible and HF generator. All the samples of final product had high degree of crystallinity and contained Fe-bearing phases (spinel, hematite) and orthophosphates. Ru was not reduced to metallic form but entered zirconia-based solid solution.

REFERENCES

[1] E.C. Smith, T.A. Butler, B. Ciorneiu, B.W. Bowan, II, K.S. Matlack, and I.L. Pegg, Advanced Joule Heated Melter Design to Reduce Hanford Waste Treatment Plant Operating Costs, *Waste Management 2011 Conf.* Feb. 27 – March 3, 2011, Phoenix, AZ, 2011, CD-ROM 11131.
[2] D.K. Peeler, F.C. Johnson, and T.B. Edwards, Technology Development to Reduce Mission Life, Life Cycle Costs, and Glass Volumes for US High Level Waste Vitrification Facilities, *Waste Management 2011 Conf.* Feb. 27 – March 3, 2011, Phoenix, AZ 2011, CD-ROM. 11461.
[3] A.S. Aloy, R.A. Soshnikov, A.V. Trofimenko, D. Gombert, D. Day, C.-W. Kim, Iron-Phosphate Glass (IPG) Waste Forms Produced Using Induction Melter with Cold Crucible, *Mater. Re. Soc. Symp. Proc*. **807**, 187-192 (2004).
[4] C.W. Kim, C.S. Ray, D. Zhu, D.E. Day, D. Gombert, A. Aloy, A. Moguš-Milanković, and M. Karabulut, Chemically Durable Iron Phosphate Glasses for Vitrifying Sodium-Based Waste (SBW) Using Conventional and Cold Crucible Induction Melting (CCIM) Techniques, *J. Nucl. Mater.* **322**, 152-164 (2003).
[5] D.E. Day, R.K. Brow, C.S. Ray, C.W. Kim, S.T. Reis, J.D. Vienna, D. Peeler, F.C. Johnson, E.R. Hansen, G. Sevigny, N. Soelberg, I.L. Pegg, and H. Gan, Iron Phosphate Glass for Vitrifying Hanford AZ102 LAW in Joule Heated and Cold Crucible Melters, *Waste Management 2012 Conf.* Feb. 26 – March 1, 2012, Phoenix, AZ, 2012, CD-ROM 12240.
[6] S. Naline, F. Gouyaud, V. Robineau, C. Girold, and B. Carpenier, Vitrification 2010 - A Challenging Waste Project to Retrofit a Cold Crucible Inductive Melter at the La Hague Plant French Vitrification, *Waste Management 2011 Conf.* March 7 –11, 2010, Phoenix, AZ, 2011, CD-ROM 10382.
[7] V.V. Lebedev, D.Y. Suntsov, S.V. Shvetsov, S.V. Stefanovsky, A.P. Kobelev, F.A. Lifanov, and S.A. Dmitriev, CCIM Technology for Treatment of LILW and HLW, *Waste Management 2010 Conf.* March 7-11, 2010, Phoenix, AZ, 2010, CD-ROM 10209.
[8] V.V. Lebedev, S.V. Shvetsov, S.V. Stefanovsky, A.P. Kobelev, CCIM Technology For HLW Treatment. Nowadays Conditions And Opportunities, *2011 ISRSM Proc. Int. Symp. on Radiation Safety Management*, November 2-4, 2011, Gyeongju, Rep. of Korea, 2011, 260-267.
[9] D. Gombert, J. Richardson, A. Aloy, D. Day, Cold Crucible Design Parameters for Next Generation HLW Melters, *Waste Management 2002 Conf.* February 24-28, 2002, Tucson, AZ (2002) CD-ROM.
[10] J.A. Roach and J.G. Richardson, Technical Development of New Concepts for Operation and Control of Cold Crucible Induction Melters for Vitrification of Radioactive Waste, *Waste Management 2006 Conf.* February 26 – March 2, 2006, Tucson, AZ (2006) CD-ROM.
[11] M. Delauney, A. Ledoux, J.-L. Dussossoy, P. Boussier, J. Lacombe, C. Girold, C. Veyer, and E. Tchemitcheff, Vitrification of Representative Simulant of DWPF SB4-Type Waste in a CCIM Industrial Scale Demonstration on the CEA Marcoule Platform, *Waste Management 2010 Conf.* March 7-11, 2010, Phoenix, AZ (2010) CD-ROM 10063.

HAFNIUM AND SAMARIUM SPECIATION IN VITRIFIED RADIOACTIVE INCINERATOR SLAG

G.A. Malinina, S.V. Stefanovsky
FSUE Radon
Moscow, Russia

A.A. Shiryaev,
Institute of Physical Chemistry and Electrochemistry RAS
Moscow, Russia

Y.V. Zubavichus
NRC Kurchatov Institute
Moscow, Russia

ABSTRACT

Incineration of solid radioactive waste yields slag which has to be processed into solid monolithic form with high chemical durability, radiation resistance, and strong mechanical integrity. On the whole the incinerator slags are comprised by the pseudo-ternary system (Na,K,Ca)(Al,Fe)(Si,Al)O_4 – (Ca,Mg,Fe)SiO_3 – $Ca_3(PO_4)_2$. Hf and Sm were used as tetravalent and trivalent actinide surrogates, respectively. Slag surrogate with a target chemical composition (wt.%) 6 Na_2O, 9 K_2O, 15 CaO, 15 Al_2O_3, 10 Fe_2O_3, 30 SiO_2, 10 P_2O_5, 5 (HfO_2 or Sm_2O_3) was produced from reagent-grade Na, K and Ca carbonates, Al, Fe and Hf or Sm oxides, $Ca_3(PO_4)_2$ and silica by heating at 900 °C for 6 hrs. The slag surrogate and its mixtures with glass forming additives ($Na_2Si_2O_5$ or $Na_2B_4O_7$) were melted depending on composition at temperatures of 1000 to 1500 °C (unfluxed slag). As follows from XRD and SEM data the products at slag loading of ~50 wt.% and lower were predominantly amorphous whereas the higher loaded products were glass-crystalline. Sm is incorporated in glass at slag loading of ~50 wt.% and lower. At higher waste loadings Sm is partitioned between vitreous and britholite-type phases. Hf was found to be partitioned among vitreous phase and hafnia-based solid solution. As follows from EXAFS/XANES data Sm is present in a trivalent state as Sm^{3+} ions with the coordination number by oxygen from 6 to ~9. Higher values of coordination number are found to be characteristic of specimens produced with the use of the flux in which a significant fraction of Sm enters into the britholite-type crystalline phase. Hafnium entering predominantly vitreous phase is tetravalent and its coordination number in the first shell is 5.8±0.5 and R—O distance is 2.065±0.005 Å. The second coordination shell is weakly appeared.

INTRODUCTION

Slags generated at high=temperature treatment of solid organic or mixed organic/ inorganic radioactive wastes are normally non-uniform, dusting and low chemically durable materials which must be consolidated and transformed into solid, monolithic forms with low leachability of radioactive and hazardous elements and high mechanical integrity such as glass and glass-ceramic materials. Analysis of numerous slag batches from Radon incinerator and reference data showed that majot slag components are Si, Al, Fe, Ca, Mg, Na, K oxides. Some slag batches may contain phosphorus compounds. On the whole chemical composition of incinerator slags ia compsised by a pseudo-ternary system (Na,K,Ca)(Al,Fe)(Si,Al)O_4 – (Ca,Mg,Fe)SiO_3 – $Ca_3(PO_4)_2$. Slags also concentrate heavy metals and actinides (U, Np, Pu, Am). Depending on solid waste treated actinide content may be widely varied and sometimes reach fractions of percent.[1]

In our previous works we proposed sodium silicates and tetraborate as glass forming additives.[1-5] Phase composition and the structure of slag-bearing aluminosilicate and borosilicate based materials were determined by X-ray diffraction (XRD) and scanning electron microscopy (SEM) and described in details earlier.[2-5] Hf and Sm were used as tetravalent and trivalent actinide surrogates, respectively.[3-5] The goal of this study is to determine local Hf and Sm environment in the structure of vitreous slag waste forms using X-ray absorption fine structure (XAFS) spectroscopy at near-edge (XANES) and extended (EXAFS) ranges.

EXPERIMENTAL

Baseline composition of slag surrogate was (in wt.%_: 6 Na_2O, 9 K_2O, 15 CaO, 15 Al_2O_3, 10 FeO, 30 SiO_2, 10 P_2O_5, 5 HfO_2, method of its preparation and vitrification conditions were described in details in our previou works.[4,5] Chemical composition of the materials and melting temperatures are given in Table I.

Table I. Chemical Composition of the Slag-Bearing Materials.

Oxides	$Na_2Si_2O_5$ flux			$Na_2B_4O_7$ flux			No flux	$Na_2Si_2O_5$ flux			$Na_2B_4O_7$ flux			No flux
	Slag surrogate to flux mass ratio													
	50:50	75:25	85:15	50:50	75:25	85:15	100:0	50:50	75:25	85:15	50:50	75:25	85:15	100:0
	Hf 50Si	Hf 75Si	Hf 85Si	Hf 50B	Hf 75B	Hf 85B	HfMS	Sm 50Si	Sm 75Si	Sm 85Si	Sm 50B	Sm 75B	Sm 85B	Sm MS
	Oxide concentration, mol.%													
Na_2O	22.2	15.5	12.5	21.5	14.9	12.1	7.7	21.7	14.9	12.0	21.0	14.3	11.6	7.3
K_2O	3.3	5.3	6.1	3.5	5.4	6.3	7.6	3.2	5.1	5. 9	3.4	5.2	6.1	7.2
CaO	9.2	14.8	17.2	9.7	15.2	17.5	21.2	9.0	14.2	16.5	9.5	14.6	16.8	20.0
Al_2O_3	5.1	8.1	9.5	5.3	8.4	9.6	11.6	4.9	7.8	9.1	5.2	8.1	9.2	11.1
Fe_2O_3	2.2	3.5	4.0	2.3	3.6	4.1	5.0	4.7	7.4	8.6	4.9	7.6	8.7	10.5
SiO_2	54.9	47.7	44.6	18.1	28.4	32.7	39.5	53.6	46.0	42.7	17.7	27. 3	31.3	37.5
P_2O_5	2.4	3.9	4.5	2.6	4.0	4.6	5.6	2.4	3.8	4.3	2. 0	3.9	4.4	5.3
B_2O_3	-	-	-	36.1	18.9	11.5	0.0	-	-	-	35.3	18.2	11.0	-
HfO_2	0.8	1.3	1.5	0.9	1.3	1.6	1.9	-	-	-	-	-	-	-
Sm_2O_3	-	-	-	-	-	-	-	0.5	0.8	0.9	0.5	0.8	0.9	1.1
T, °C	1300	1350	1400	1050	1250	1300	1500	1300	1350	1400	1050	1250	1300	1500

XAFS spectra were recorded at the Structural Materials Science (STM) beamline of the synchrotron source at NRC Kurchatov Institute.[6] The samples were measured at room temperature either as dispersed powder or as pellets pressed from powder mixed with sucrose in transmission mode using a Si(220) channel-cut monochromator and two air-filled ionization chambers. Fluorescence spectra were also acquired. Powders of reagent-grade monoclinic oxides HfO_2 and Sm_2O_3 were used as standards and measured under identical conditions. Experimental XAFS spectra were fitted in R-space using an IFEFFIT package;[7] FEFF8[8] was used for calculation of phase shift from crystal structures of corresponding oxides and silicates.

RESULTS AND DISCUSSION

Figure 1 demonstrates similarity of X=ray absorption spectra of all the Hf-bearing samples. Absorption maximum is located at 9564 eV in spectra of both HfO_2 and Hf-bearing materials. The constant value of the absorption maximum points to the same Hf valence state (IV) and similar coordination environment.

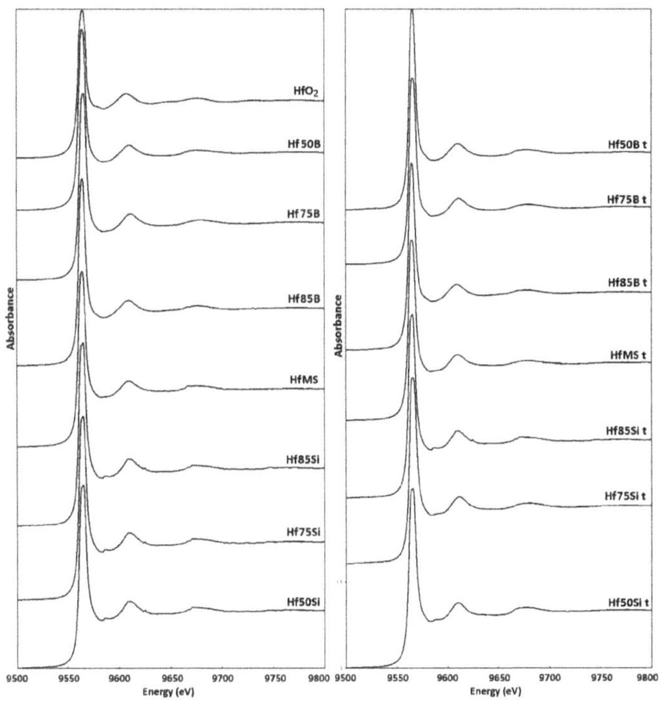

Figure 1. XANES of Hf L_3 edge in the slag-bearing quenched (left) and annealed (t – right) materials.

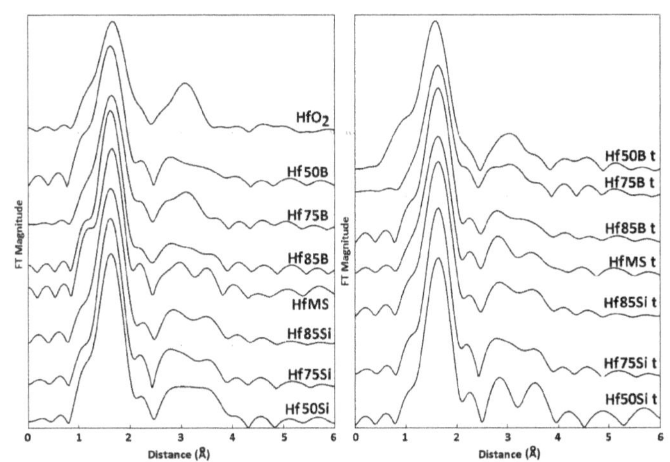

Figure 2. FT EXAFS spectra of Hf L_3 edge in the slag-bearing quenched (left) and annealed (t – right) materials (no phase shift correction).

Table II. Fitting Results for FT of Hf L_3 edge EXAFS Spectra.

Sample	CN	σ	R—O_{av}	R—M	Sample	CN1	σ	R—O_{av}	R—M
Hf50Si	6.3	0.006	2.07	3.30/3.84	Hf50Si t	6.3	0.006	2.07	3.25/3.90
Hf75Si	5.3	0.004	2.06	3.22/3.88	Hf75Si t	5.3	0.004	2.06	3.20/3.90
Hf85Si	5.8	0 005	2.06	3.25/3.89	Hf85Si t	5.8	0.005	2.06	3.26/3.90
HfMS	6.1	0.007	2.07	3.35/3.90	HfMS t	5.7	0.007	2.07	3.20/3.88
Hf50B	5.8	0.006	2.06	3.22/3.70	Hf50B t	5.8	0.006	2.06	3.43/3.85
Hf76B	5.7	0.007	2.06	3.17/3.50	Hf75B t	5.7	0.007	2.06	3.15/3.50
Hf85B	6.1	0.004	2.07	3.20/3.54	Hf85B t	6.1	0.004	2.07	3.20/3.60
HfO_2	7.0	0.001	2.15	3.53					

Fourier transforms (FT) of Hf EXAFS spectra (Figure 2) also indicate similar Hf environment in all the materials (Table II). Typical average Hf—O distance in the first oxygen coordination shell is 2.06 – 2.07 Å and coordination number is 5.8±0.5. These data are similar to those obtained for sodium trisilicate (NS3)[9] and lanthanide borosilicate (LaBS)[10,11] glasses. At that no appreciable difference between the quenched and annealed samples was found (Table II).

Ionic radius of seven-coordinated Hf^{4+} ion (0.76 Å) is close to that of Zr^{4+} ion (0.78 Å)[12] and monoclinic hafnia has the same structure as monoclinic zirconia (baddeleyite). In the first coordination shell Hf^{4+} ions have seven neighboring oxygens – three at a distance of ~2.07 Å and four at a distance of ~2.21 Å (average Hf—O is 2.15 Å).[13] Slightly longer Hf—O distance 2.06-2.07 Å) as compared to that in NS3 glass (2.04-2.05 Å)[9] and variation of Hf CN in the first coordination shell as well as occurrence of weak peaks due to the second and subsequent coordination shells on FT EXAFS spectra point to some minor contribution of crystalline hafnia to the spectra of slag-bearing materials.

As follows from XRD and SEM data the quenched slag-bearing samples at 50 wt.% slag content were predominantly amorphous whereas at higher slag content they contained minor silicophosphate and Fe oxide phases formed crystals with a size of nanometers to first microns. No Hf-bearing phases were found. The annealed samples had higher degree of crystallinity and contained silicate or phosphate (at 50 wt.% slag content) or aluminosilicate (nepheline), silicophosphate (nagelschmidtite) and Fe-bearing oxide (hematite and spinel) phases (at 75 wt.% slag content and higher), and individual Hf-enriched grains. So, in the quenched samples Hf enters either vitreous phase or nanometric-sized grains of crystalline phases. In the annealed samples Hf is also partititoned among vitreous and crystalline phases including hafnia or hafnia-based solid solution.[5] Thus XAFS spectra of the materials are superpositions of the spectra due to Hf in several phases. However similarity of the spectra for all the samples indicates its preferable incorporation in only one of the phases and as it follows from positions of absorption maximum in XANES and values of Hf—O distances and Hf CN on FT EXAFS spectra this phase is believed to be glass.

The second Hf coordination shell is split into two subshells (Table II) with shorter and longer Hf—M distances. This indicates variability of Hf environment: the second neighbour may be not only Hf but different element, for example Si or B.

XANES spectra of Sm L_3 edge in the quenched and annealed slag-bearing materials are similar. The spectra of the annealed materials shown on Figure 3 demonstrate that absorption maximum is constant (~6718 eV) in the spectra of the materials 50Si t, 75Si t, 85Si t, 50B t and 75B t but increases by 1-1.5 eV in the spectra of the materials 85B t and MS t approaching to the value being characteristic of Sm_2O_3 (~6720 eV) Shift towards higher energies may be due to variation of average Sm CN. Small difference in absorption maxima (≤2 eV) points to minor effect of factors responsible for this shift. XANES spectra for all the samples are typical of

Sm(III) corresponding to $2p \rightarrow 5d$ transition in Sm^{3+} ions. The response due to Sm(II) is positioned in the spectra of reduced borate or irradiated borosilicate glasses at 6713 eV.[14,15] and is absent in spectra of our materials.

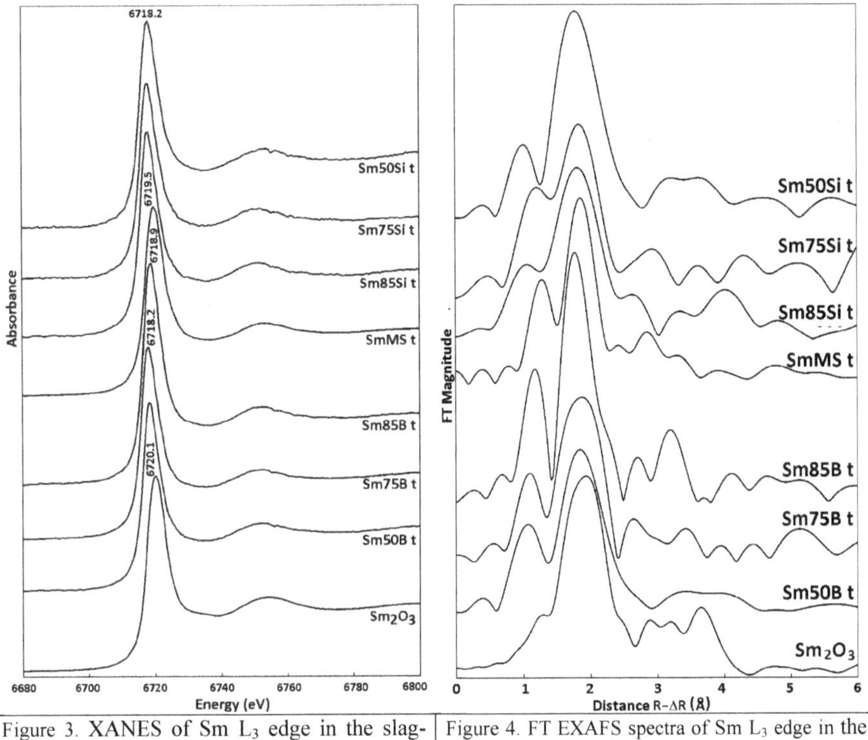

Figure 3. XANES of Sm L_3 edge in the slag-bearing annealed materials.

Figure 4. FT EXAFS spectra of Sm L_3 edge in the slag-bearing annealed materials (no phase shift correction).

FT EXAFS spectra of Sm L_3 edge in the annealed slag-bearing materials show well-determined first coordination shell whereas the second coordination shell is not clearly appeared (Figure 4). Minor contribution due to the second shell is supposedly present only in the spectra of the materials with high slag content. FT for Sm_2O_3 is very similar to that presented in ref.[14]

In the melted slag and materials produced with sodium disilicate flux the Sm—O distance and Sm CN in the first coordination shell ranges between 2.29 Å and 2.38 Å and between 6 and 9, respectively (Table III). In the materials produced with sodium tetraborate flux in the first shell average Sm—O distance was found to be 2.4 Å at CN=6-7 except the sample Sm85B t where CN=8.5. These values are in a good agreement with reference data.[14-18] At that no clear dependence of both Sm—O distance and Sm CN from composition (slag loading) was found. All the peaks due to Sm—O distance in the first shell are either asymmetric or broadened exhibiting Sm partitioning between at least two co-existing phases or/and variability of its local environment in one or more phases. Longer-range order in Sm distribution may be suggested only in the structure of materials with the highest slag crystalline constituent contents (Figure 4).

As followed from XRD and SEM data the slag-bearing samples doped with Sm after pouring onto a metal plate (quenched) are composed of major amorphous and minor crystalline phases. The crystal size ranges between tens of nanometers and first microns. Fraction of crystalline constituent grows with slag loading. After annealing the samples had higher degree of crystallinity and crystals in them had larger size (up to tens of microns) and were more regular. Major phases were found to be britholite, nepheline and magnetite-type spinel.[3,4] It has been supposed that in the quenched samples only britholite might be a host phase for Sm among crystalline phases.

Table III. Fitting Results for FT of Sm L_3 edge EXAFS Spectra.

Sample	R, Å	CN	σ	Образец	R, Å	CN	σ
Sm50Si t	2.38	8.9±2.3	0.1	Sm50B t	2.40	6.3±1.2	0.05
Sm75Si t	2.26	9.0±1.4	0.09	Sm75B t	2.40	7.0±2.0	0.12
Sm85Si t	2.36	6.2±1.5	0.016	Sm85B t	2.30	8.5±1.6	0.11
SmMS t	2.29	7.5±1.1	0.09	Sm_2O_3	2.45	7.0	0.05

Britholite $(Na,Ca,REE)_{10}(Si,P)_6O_{24}(OH,F)_2$ is isostructural with apatite. In the absence of volatile anions its formula may be represented as $(Na,Ca,REE)_{10}(Si,P)_6O_{25}$ (Si>P).[19] In the structure of apatite Ca^{2+} ions form single or triple columns linked by $[PO_4]$ tetrahedra and are located in two different sites seven- and nine-coordinated by oxygen. At coupled isomorphic substitution $Ca^{2+} \rightarrow RE^{3+}$ and $PO_4^{3-} \rightarrow SiO_4^{4-}$ RE^{3+} ions replace Ca^{2+} ions in their sites. At that symmetry is lowered from hexagonal with space group $P6_3/m$ for apatite to $P6_3$. As a result average CN for RE^{3+} becomes about 8 and average RE—O distance of 2.533 Å to 2.637 Å.[19]

In the quenched samples Sm enters predominantly vitreous phase whereas in the annealed samples Sm is partitioned between vitreous and britholite-type phases. Nevertheless in our samples although the britholite-type phase is revealed by XRD and SEM its effect on Sm partitioning remains rather uncertain. In the first shell the highest Sm CN values take place in the annealed samples Sm50Si t and Sm75Si t containing minor britholite and in the sample Sm85B where content of this phase is markedly higher. At the same time in the samples Sm75B and SmMS where britholite are major phase or one of the major phases Sm CN values are 7 to 7.5, i.e. within the values typical of glasses. This demonstrates that Sm partitioning both between glass and crystalline phase (britholite), and within crystalline phase itself over the sample bulk is rather inhomogeneous and depends on a number of factors such as synthesis conditions including heating and cooling rates, sampling point from the ingot, elemental concentration in the sample, etc.

Suggestions on formation of the second Sm coordination shell in the highly slag-loaded samples point to tendency to ordering of Sm distribution due to its entering crystalline phase probably britholite. Relatively low CN values for the first shell may be due to either low content of this phase or preferable Sm accommodation in the vitreous phase.

CONCLUSION

In the slag-bearing glass-crystalline materials produced with sodium disilicate or sodium tetraborate flux Hf exists in a tetravalent form as Hf^{4+} ions partitioned among vitreous and crystalline phases mainly hafnia-based solid solution. Average Hf—O distance and Hf CN in the first coordination shell are 2.06-2.07 Å and 5.8, respectively, being some different from those of monoclinic hafnia (2.15 Å and 7). No appreciable difference between Hf environment in the quenched and annealed materials was found. The second Hf coordination shell in the materials was not clearly manifested.

In both the quenched and annealed samples Sm was found to be trivalent. In the quenched materials and annealed materials at low slag loadings Sm^{3+} ions enter predominantly vitreous phase whereas in the annealed materials with the highest slag loadings (75-100 wt.%) Sm^{3+} ions are partitioned between the vitreous and britholite-type phases in favor of the first one. Distribuion of Sm^{3+} ions over the bulk of the materials is rather inhomogemenous and depends on a number of factors primarily synthesis conditions.

REFERENCES

[1] S. Stefanovsky, F. Lifanov, and I. Ivanov, Glass forms for Incinerator Ash Immobilization, Proc. *XVI International Congress on Glass, Oct. 4-9, 1992, Madrid, Spain, Bol. Soc. Esp. Ceram. Vid.* **31 C** [3] 209-214 (1992).
[2] T.N. Lashtchenova, F.A. Lifanov, and S.V. Stefanovsky, Incorporation of Radon Incinerator Ash in Glass and Glass Crystalline Materials, *Proc. Waste Management 1997 Conf., March 2-6, 1997, Tucson, AZ CD-ROM*, 13-17 (1997).
[3] G.A. Malinina, O.I. Stefanovsky, and S.V Stefanovsky, Glass Ceramics for Incinerator Ash Immobilization, J. Nucl. Mater., **416**, 230-235 (2011).
[4] G. A. Malinina, S. V. Stefanovsky, and O. I. Stefanovskaya, Phase Composition and Structure of Boron_Free and Boron_Containing Sodium Aluminum Iron Silicate Glass Materials for Solid Radioactive Waste Immobilization, Glass Physics and Chemistry, **38**, [3], 280–289 (2012).
[5] G.A. Malinina, S.V Stefanovsky, and O.I. Stefanovsky, Phase Composition and Structure of Glass-Crystalline Materials Doped with Hafnium Oxide as a Matrices for Immobilization of Radioactive Slags, Phys. Chem. Mater. Treat. (Russ), [2] 75-82 (2012).
[6] A.A. Chernyshov, A.A. Veligzhanin, and Y.V. Zubavichus, Structural materials science end-station at the Kurchatov synchrotron radiation source: Recent instrumentation upgrades and experimental results, *Nucl.Instrum.Meth.Phys.Res.***A603**, 95-98 (2009).
[7] B. Ravel and M.Newville, ATHENA, ARTEMIS, HEPHAESTUS: Data analysis for X-ray absorption spectroscopy using IFEFFIT. *J. Synchrotron Rad.*, **12**, 537-541 (2005).
[8] A.L.Ankudinov and J.J. Rehr Relativistic spin-dependent X-ray absorption theory. Phys.Rev. **B56**, 1712-1716 (1997).
[9] M. Wilke, J. Dubrail, and S. Simon, XAFS of Hf in Silicate Glasses and Minerals, *Helmholtzzentrum Potsdam – Deutsches GeoForschungsZentrum GFZ, Annual Report,* Potsdam, Germany (2010), http://photon-science.desy.de/annual_report/files/2010/20101092.pdf.
[10] A.A. Shiryaev, Ya.V. Zubabichus, S.V. Stefanovsky, A.G. Ptashkin, and J.C. Marra. XAFS of Pu and Hf LIII Edge in Lanthanide-Borosilicate Glass, *Mat. Res. Soc. Symp. Proc*. **1193**, 259-265 (2009).
[11] S.V. Stefanovsky, A.A. Shiryaev, I.V. Zubavichus, and J.C. Marra. Hafnium Environment in LaBS and ABS Glasses for Plutonium Immobilization, *XXII International Congress on Glass*, Sept. 20-25, Bahia, Brazil, 2010. Book of Abstracts. 2010. P. 114.
[12] R.D. Shannon, Revised Effective Ionic Radii and Systematic Studies of Interatomic Distances in Halides and Chalcohenides, *Acta Cryst*. **A32**. 751-767 (1976).
[13] F.P. Wells, Structural Inorganic Chemistry, Clarendon Press, Oxford, UK (1986).
[14] Y. Shimizugawa, N. Sawaguchi, K. Kawamura, and K.Hirao, X-ray Absorption Fine Structure of Samarium-Doped Borate Glasses. *J. Appl. Phys*. **81**, 6657-6661 (1997).
[15] E. Malchukova, B. Boizot, G. Petite, and D. Ghaleb, Optical Properties and Valence State of Sm Ions in Aluminoborosilicate Glass Under β-Irradiation, *J. Non-Cryst. Solids*. **353**, 2397-2402 (2007).
[16] D.T. Bowron, R.J. Newport, B.D. Rainford, G.A. Saunders, and H.B. Senin, EXAFS and X-Ray Structural Studies of $(Tb_2O_3)_{0.26}(P_2O_5)_{0.74}$, Metaphosphate Glass. *Phys. Rev*. **B51**, 5739-5745 (1995).

[17] D.T. Bowron, G.A. Saunders, R.J. Newport, B.D. Rainford, and H.B. Senin, EXAFS Studies of Rare-Earth Metaphosphate Glasses, *Phys. Rev.* **B53**, 5268–5275 (1996).
[18] R. Anderson, T. Brennan, G. Mountjoy, R.J. Newport, and G.A. Saunders, An EXAFS Study of Rare-Earth Phosphate Glasses in the Vicinity of Metaphosphate Composition, *J. Non-Cryst. Solids*. **232-234**, 286-292 (1998)..
[19] R. Oberti and L. Ottolini, Della Ventura G., Parodi G.C. On the Symmetry and Crystal Chemistry of Britholite: New Structural and Microanalytical Data, *Amer. Mineral.* 86, 1066–1075 (2001).

Author Index